IMAGE 2.0

IMAGE 2.0:
INTEGRATED MODELING OF GLOBAL
CLIMATE CHANGE

Edited by

JOSEPH ALCAMO

with papers by

The IMAGE Project

National Institute of Public Health
and Environmental Protection (RIVM), the Netherlands

Dutch National Research Programme
on Global Air Pollution and Climate Change (NOP)

Reprinted from *Water, Air, and Soil Pollution,*
Volume 76, Nos. 1–2, 1994

SPRINGER SCIENCE+BUSINESS MEDIA, B.V.

Library of Congress Cataloging-in-Publication Data

ISBN 978-0-7923-2860-5   ISBN 978-94-011-1200-0 (eBook)
DOI 10.1007/978-94-011-1200-0

Front Cover: Simulation of land cover by IMAGE 2 model for Conventional Wisdom scenario,
year 2025. See Alcamo *et al.*, (part 2), and Zuidema *et al.*, this volume.

Front Cover Design: Martin Middelburg

Book Design and Layout: Martin Middelburg

Printed on acid-free paper

TABLE OF CONTENTS

FOREWORD

During the UN Conference on Environment and Development (UNCED) in 1992 much attention was given to global environmental problems. The Framework Convention on Climate Change and other international agreements, such as the Montreal Protocol for the protection of the ozone layer, are ample demonstration that governments take the issues of global change seriously.

Many of the international political agreements could not have been developed had it not been for the underpinning provided by scientific reserch and assessment. There are three major programmes established to reduce the scientific uncertainties related to global environmental change: the World Climate Research Programme (WCRP since 1980), the International Geosphere-Biosphere Progrqamme : A Study of Global Change (IGBP, since 1986) and the Human Dimensions of Global Environmental Change (HDP, since 1990).

Results from these and other scientific programmes can be made more useful for the policy process if they are evaluated and synthesized. At the international level, this is carried out by the Intergovernmental Panel on Climate Change (IPCC) for issues about climate change. It is hoped that the HDP/IGBP/WCRP initiative System for Analysis, Research and Training (START) can perform a similar function at the regional level for global change issues..

Simulation models are crucial as a means to formalize the synthesis of data from different disciplines. Models can also have predictive capabilities and be used to develop and analyze scenarios of global environmental change. Only few laboratories in the world have taken the bold step to attempt the integration of sub-models of the climate system, the global biogeochemical cycles and the human/societal components. This volume reports on such a major undertaking and is an important step towards an integrated approach to global change science.

Models attempting to link the physical, biogeochemical and societal subsystems at the global scale are, by necessity, simplifications. Their major role is to show the interdependence of the three subsystems, provide a formal structure for synthesis, and identify major weaknesses in our understanding. Any attempt to do so is brave, as scientists have a tendency to analyze and criticize the model subcomponent they are most familiar with while losing the overall objective of the integrated model development. The IMAGE 2 model is important in demonstrating our current ability to model the complex global system. It will stimulate further refinement and development within RIVM and will also lead to the development of similar models in other laboratories and institutions. The authors should be congratulated for making available a model description that will stimulate an essential debate and further model development. This will lead to improved scientific assessments on which policy decisions must be based.

Professor Thomas Rosswall
Executive Director, International Geosphere-Biosphere Programme (IGBP)

PREFACE

The main purpose of this publication is to document the development and testing of the IMAGE 2.0 model, together with a selection of its applications. One of the main objectives of IMAGE 2.0 is to link science with policy, but in this publication we emphasize the scientific rather than policy aspects of the model, because a strong scientific foundation is necessary before a model can be useful for policy analysis.

IMAGE 2.0 is a type of earth systems model, a new category of simulation tool made possible by two recent developments. The first is rapid progress in understanding the workings of the global system based on new data that is rapidly becoming available. These data have come from comprehensive measurement programs such as *TOMS* (Total Ozone Mapping Spectrometer), *ALE/GAGE* (Atmospheric Lifetime Experiment/Global Atmospheric Gases Experiment), and *ERBE* (Earth Radiation Budget Experiment), as well as impressive efforts to compile global data bases through *IGAC* (International Global Atmospheric Chemistry Programme), *IGBP-DIS* (International Geosphere Biosphere Programme - Data and Information System), and other programs. The second development making earth systems models possible is the increase in power and utility of computer hardware and software which has allowed more and more institutes and researchers to handle the simulations of large geographic and dynamic systems.

Apart from being made possible by the above advances, the IMAGE 2.0 modeling approach evolved from two directions. First, it stems from the earlier, global-average version of IMAGE (now referred to as "IMAGE 1" which was developed at the National Institute of Public Health and Environmental Protection of the Netherlands (RIVM) under the leadership of Jan Rotmans. Rotmans' determination and skill led to one of the first integrated models of climate change (Rotmans, 1991; Rotmans *et al.*, 1991), which coupled calculations of energy, emissions, climatic consequences and sea level rise within a single framework. Second, IMAGE 2.0 evolved from developments in global change modeling that took place during the 1980s at the International Institute for Applied Analysis, Laxenburg, Austria. In particular, the BIOME model (Prentice *et al.*, 1993) and the RAINS model (Alcamo, et al. 1990) contributed ideas about rule-based simulations, process-based models applied on a geographic scale, and spatial mapping, which led to the geographically-explicit calculations of IMAGE 2.0. The crucial advantage of geographically-explicit modeling is that it increases the opportunity to test global models against data from comprehensive inventories and/or field campaigns. Hence, model development can keep pace with the state of understanding of global systems as new data become available.

As with other new developments in science, the field of earth systems modeling has already taken many different directions depending on the type of questions researchers wish to address. It is likely that these varied efforts will lead to rapid and exciting advances in the coming years in understanding and simulating the global system.

Joseph Alcamo
Leader, Project on Modeling Global Climate Change, RIVM

References

Alcamo, J., R. Shaw, and L. Hordijk (eds): 1990, *The RAINS Model of Acidification: Science and Strategies in Europe,* Kluwer Academic Publishers, Dordrecht, Boston, 402 pp.

Prentice, I.C., W. Cramer, S.P. Harrison, R. Leemans, R.A. Monserud, and A.M. Solomon: 1992, A global biome model based on plant physiology and dominance, soil properties and climate, *J. Biogeogr.*, 19:117-134.

Rotmans, J.: 1990, *IMAGE: An Integrated Model to Assess the Greenhouse Effect,* Kluwer Academic Publishers, Dordrecht, Boston, 289 pp.

Rotmans, J., H. de Boois, R. Swart: 1990, An integrated model for the assessment of the greenhouse effect, *Climatic Change,* **16**: 331-356

ACKNOWLEDGEMENTS

The development of the IMAGE 2.0 model has been funded by MAP Project Number 481507 of the Dutch Ministry of Housing, Physical Planning and Environment; and NOP Project Numbers 851037, 851040, 851042, 851044, and 851045 of the Dutch National Research Programme on Global Air Pollution and Climate Change (NOP). The terrestrial environment research of the IMAGE Project is an Activity of the Core Project "Global Change and Terrestrial Ecosystems" (GCTE), of the International Geosphere-Biosphere Programme (IGBP).

The work presented herein has benefited from the constructive criticism of an international review panel of the IMAGE Project organized by the NOP in 1993. This panel consisted of L. Hordijk (Convener), J. Edmonds, J.Goudriaan, J. Grasman, M. Hulme, T. Johansson, P. Love, B. Metz., A. Rahman, A. Solomon, and A. van Ulden.

The IMAGE Project is especially indebted to F. Langeweg and R. Swart of the National Institute of Public Health and Environmental Protection, the Netherlands for their continued support of the development and application of IMAGE 2. The production of this special issue was greatly aided by many reviewers who handled these papers quickly and on short notice, by the efforts of the Journal of *Water, Air and Soil Pollution* Special Issues Editor, J. Wisniewski, and the Senior Editor, B. McCormac, and by the skills of M. Middelburg who was responsible for its design and layout.

Joseph Alcamo

MODELING THE GLOBAL SOCIETY-BIOSPHERE-CLIMATE SYSTEM: PART 1: MODEL DESCRIPTION AND TESTING

J. ALCAMO, G.J.J. KREILEMAN, M.S. KROL, G. ZUIDEMA

National Institute of Public Health and Environmental Protection (RIVM)
P.O. Box 1, 3720 BA, Bilthoven, the Netherlands

Abstract. This paper describes the IMAGE 2.0 model, a multi-disciplinary, integrated model designed to simulate the dynamics of the global society-biosphere-climate system. The objectives of the model are to investigate linkages and feedbacks in the system, and to evaluate consequences of climate policies. Dynamic calculations are performed to year 2100, with a spatial scale ranging from grid (0.5° x 0.5° latitude-longitude) to world regional level, depending on the sub-model. The model consists of three fully linked sub-systems: Energy-Industry, Terrestrial Environment, and Atmosphere-Ocean. The *Energy-Industry* models compute the emissions of greenhouse gases in 13 world regions as a function of energy consumption and industrial production. End use energy consumption is computed from various economic/demographic driving forces. The *Terrestrial Environment* models simulate the changes in global land cover on a grid-scale based on climatic and economic factors, and the flux of CO_2 and other greenhouse gases from the biosphere to the atmosphere. The *Atmosphere-Ocean* models compute the buildup of greenhouse gases in the atmosphere and the resulting zonal-average temperature and precipitation patterns. The fully linked model has been tested against data from 1970 to 1990, and after calibration can reproduce the following observed trends: regional energy consumption and energy-related emissions, terrestrial flux of CO_2 and emissions of greenhouse gases, concentrations of greenhouse gases in the atmosphere, and transformation of land cover. The model can also simulate long term zonal average surface and vertical temperatures.

Keywords: integrated modeling, integrated assessment, greenhouse gas emissions, global change, climate change, land cover change, C cycle.

1. Introduction

Scientific and policy questions about the global system of society, biosphere and climate are by nature multi-disciplinary, and have local, regional and global aspects. Nevertheless, most global change research focuses on either a single aspect or spatial scale of the system. The objective of the IMAGE 2.0 model described in this paper is to fill in some multi-disciplinary gaps in global change research by providing a disciplinary and geographic overview of the society-biosphere-climate system. It is our belief that this approach can provide new scientific information about the relative importance of linkages/feedbacks in the society-biosphere-climate system, and new policy information linking human activity with its consequences on the global biosphere and climate. The purpose of this paper is to summarize the development and testing of the model, emphasizing its scientific foundation; a companion paper (Alcamo *et al.*, 1994) presents some preliminary applications of the model.

Earlier versions of IMAGE (*Integrated Model to Assess the Greenhouse Effect*) are described in Rotmans (1990) and Rotmans *et al.* (1990), and are part of the ESCAPE framework presented in CEC (1992). The IMAGE 1.0 model proposed a global-average integrated structure for climate change issues by combining (1) an energy-model for greenhouse gas emissions, (2) a global C cycle model and (3) highly parameterized mathematical expressions for global radiative forcing, atmospheric temperature response, and sea level rise (Rotmans, 1990). The global-average calculations of IMAGE 1.0 were useful for evaluating policies at both the Dutch national level and international level (e.g. IPCC, 1990). Following this work, the developers of IMAGE 1.0 contributed to the ESCAPE framework, which combined parameterized global-average climate calculations with grid-based impact calculations for Europe (CEC, 1992). As part of the ESCAPE framework, an innovative approach was taken to estimate emissions from energy (CEC, 1992) and land use (Bouwman *et al.*, 1992) for world regions. The IMAGE 2.0 model contains elements of these two submodels together with several other new submodels.

In comparison to previous integrated models, IMAGE 2.0 covers not only the entire globe, but also performs many calculations on a global grid (0.5° x 0.5° latitude-longitude); this spatial resolution increases model testability against measurements, allows an improved representation of feedbacks, and provides more detailed information for climate impact analysis (discussed further in Sec. 1.2). Moreover, the submodels of IMAGE 2.0 are in general more process-oriented and contain fewer global parameterizations than previous models, which enhances the scientific credibility of calculations (NRP, 1993). Of course, these developments also add greatly to the computational and data handling tasks of the model.

1.1   OBJECTIVES OF THE MODEL

The scientific goals of the IMAGE 2.0 model are:
- To provide insight into the relative importance of different linkages in the society-biosphere-climate system;
- To investigate the relative strengths of different feedbacks in this system;
- To estimate the most important sources of uncertainty in such a linked system and
- To help identify gaps in knowledge about the system in order to help set the agenda for climate change research.

The policy-related goals of the model are:
- To link important scientific and policy aspects of global climate change in a geographically-explicit manner in order to assist decision making;
- To provide a dynamic and long-term (50 to 100 years) perspective about the consequences of climate change;
- To provide insight into the cross-linkages in the system and the side effects of various policy measures;
- To investigate the influence of economic trends and technological development on climate change and its impacts and
- To provide a quantitative basis for analyzing the costs and benefits of various measures (including preventative and adaptive measures) to address climate change.

These objectives steer the design and development of the model. The accomplishment of some of these goals are illustrated by model applications in Alcamo *et al.* (1994).

1.2 MODEL SUB-SYSTEMS

The model consists of three fully linked sub-systems of models (Figure 1):
- The Energy-Industry System;
- The Terrestrial Environment System and
- The Atmosphere-Ocean System.

The *Energy-Industry* models compute the emissions of greenhouse gases in 13 world regions as a function of energy consumption and industrial production. End use energy consumption is computed from various economic driving forces. The following submodels are included: Energy Economy, Energy Emissions, Industrial Production, and Industrial Emissions.

IMAGE 2.0
Framework of Models and Linkages

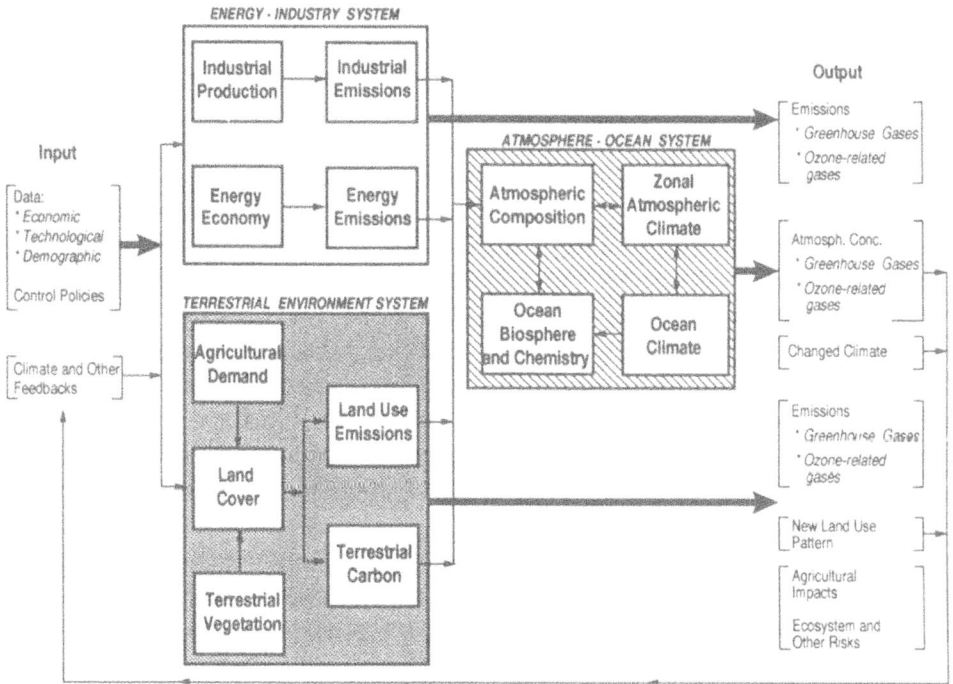

Figure 1: Schematic diagram of IMAGE 2.0 showing its framework of models and linkages.

The *Terrestrial Environment* models simulate the changes in global land cover on a grid-scale based on climatic and economic factors. The role of land cover and other factors are then taken into account to compute the flux of CO_2 and other greenhouse gases from the biosphere to the atmosphere. This sub-system includes the following submodels: Agricultural Demand, Terrestrial Vegetation, Land Cover, Terrestrial Carbon, and Land Use Emissions.

The *Atmosphere-Ocean* models compute the buildup of greenhouse gases in the atmosphere and the resulting zonal-average temperature and precipitation patterns. The following sub-models are included: Atmospheric Composition, Zonal Atmospheric Climate, Oceanic Climate, and Oceanic Biosphere/Chemistry.

One of the main scientific contributions of IMAGE 2 is its representation of many of the important feedbacks and linkages between models in these sub-systems, and between sub-systems. The sub-systems are described in Sections 2 to 4 of this paper.

1.3   TEMPORAL AND SPATIAL SCALES

In order to provide a long-term perspective about the consequences of climate change, the model's time horizon extends to year 2100. The time steps of different submodels vary, depending on their mathematical and computational requirements, but typically vary from one day to five years.

Another goal of the model is to provide as much information as possible on a grid of 0.5° latitude by 0.5° longitude. This is because nearly all potential impacts of climate change (e.g., impacts on ecosystems, agriculture, and coastal flooding) have a strong spatial variability. Moreover, land use-related greenhouse gas emissions (e.g., N_2O from soils or NH_4 from agricultural activities) greatly depend on "local" environmental conditions and human activity. In addition, climate feedbacks, such as the effect of temperature on soil respiration or the effect of changing CO_2 levels on plant productivity also vary substantially from location to location. There are two additional reasons for computing grid-scale information. First, policy makers are interested in regional/national policies to address climate change. Indeed most climate policies are location-specific (e.g. sequestering carbon in forest plantations, or reducing N_2O by modifying agricultural practices). Second, grid-scale information makes model calculations more testable against observations as compared to more aggregated models.

Nevertheless, we are unable at this time to provide grid-scale calculations for all components of climate change. In particular, this is infeasible for economic calculations because of the difficulty in specifying economic/demographic factors (e.g. trade relationships, technological development, and similar data) on a country- or sub-country scale for the entire world over the long time horizon of the model. As an intermediate step, economic calculations are performed for 13 world regions (Figure 2), which follows common practice in global economic studies. The criterium for grouping countries together in a particular region is mainly economic similarity, and our grouping conforms somewhat to that used by the IPCC, OECD, U.N., and other international organizations. A list of countries comprising the world regions is given in Appendix 1. (We should note that

World regions in IMAGE 2

1	Canada	6	Eastern Europe	10	China + C.P. countries
2	USA	7	CIS	11	East Asia
3	Latin America	8	Middle East	12	Oceania
4	Africa	9	India + S.Asia	13	Japan
5	OECD Europe				

Figure 2: Aggregation of countries into world regions in IMAGE 2.0.

these organizations themselves do not have a common method for aggregating countries.) However, the IMAGE 2.0 model has the additional requirement that countries within a region be adjacent or nearby because of the model's approach to global land cover simulation (see below).

1.4 A NOTE ON MODELING APPROACH

Since the IMAGE 2.0 model is based on large global data sets and poorly-understood global processes, it is unavoidable that many parameters are ill-defined and have large degrees of freedom. With this in mind, our basic approach is to propose submodels that have comparable level of process detail, and to adjust a limited number of parameters with greatest degree of freedom to obtain model calculations in reasonable agreement with 1970-90 data. We selected the period 1970-90 because of data availability, although we intend to test the model against data from a longer historical period. We first tested individual submodels and adjusted their parameters, then linked and tested the three sub-systems of models, and finally the fully linked model. In the following paragraphs we identify the key parameters that can be adjusted in each of the submodels. Results of the fully linked model are presented in Section 5.

Of course, this procedure does not ensure that adjusted parameters and other inputs will be correct for scenario analysis under changed economic and environmental conditions; nevertheless, it does indicate the adequacy of the model in explaining global changes that occurred during the 1970-90 period, such as the increase in energy-related emissions, estimated changes in deforestation rate and terrestrial carbon fluxes, and the build-up of various greenhouse gases in the atmosphere.

2. Energy-Industry System of Models

The objective of the *Energy-Industry* system of models is to compute the emissions of greenhouse gases from world regions as a function of energy consumption and industrial production. The models are designed especially for investigating the effectiveness of improved energy efficiency and technological development on future emissions in each region, and can be used to assess the consequences of different policies and socio-economic trends on future emissions.

2.1   ENERGY ECONOMY MODEL

The *Energy Economy* model divides the energy economy of each world region (Figure 2) into five energy sectors (Table 1) and computes the demand for end use "heat" and "electricity" in each of these sectors (six energy carriers are included, Table 1). Such an end use approach makes it easier to assess the potential of energy conservation measures in reducing overall energy consumption and, in turn, in reducing greenhouse emissions. Details of the models are given in de Vries *et al.* (1994).

The end use heat and electricity is computed from elasticity functions that relate

TABLE 1

End Use Energy Carriers and Sectors in Energy Economy Model

Energy Carriers	Energy Sector	Corresponding Activity Level
Coal	Industry	Value-added industrial production
Gas		
Oil	Transportation	Number of vehicles
Fuelwood	Residential	Personal expenditures
Other Biomass	Commercial	Value-added commercial services
Electricity	Other	GNP

"activity levels" of each sector (Table 1) with end use energy consumption. Elasticity coefficients are derived for each world region with data from 1970 to 1990. The computation of end use energy also includes an energy conservation function which relates energy prices from 1970 to 1990 with the motivation for energy conservation during that period. The main adjustable variables in the model are the parameters of this energy conservation function, and the above elasticity coefficients, the main driving forces in the model are regional changes in population and GNP.

The computed end use electricity in each sector is converted into a required power plant capacity by taking into account an average power conversion rate for each region. The final step of calculations is to compute primary consumption of energy by specifying a region-specific fuel mix to deliver the end use heat in each sector, and to fulfill the required power plant capacity.

2.2 ENERGY EMISSIONS MODEL

The *Energy Emissions* model applies emission coefficients to the energy consumed in each energy sector to compute the amount of CO_2, CH_4, N_2O, and other greenhouse gases released from each region (Table 2). These sector-specific emission coefficients are obtained from the literature, but adjusted within their known range until good agreement is obtained between model calculations and data for global and regional emissions in 1990. In addition, the model takes into account emissions of CH_4 that are related to fuel transportation/transformation such as CH_4 leakage from natural gas pipelines. Because the Energy Emissions model is linked to the Energy Economy model, the model user can investigate emission control strategies related to the energy system; e.g. the feasibility of reducing emissions by altering the fuel mix in a sector, or by improving the efficiency of different technologies used to provide energy. In addition, control strategies of non-CO_2 gases can be investigated by prescribing reductions of gases due to abatement technologies.

2.3 INDUSTRIAL PRODUCTION AND INDUSTRIAL EMISSIONS MODELS

The *Industrial Production* and *Emissions* models are used to compute emissions of greenhouse gases or their precursors that are not directly associated with energy combustion. (Emissions coming from the combustion of fuel by industry are taken into account in the "Industry" sector of the Energy Emissions model described above.) Examples of these emissions are: halocarbon emissions from refrigerators and industrial products, CO_2 from the cement industry, and VOCs from chemical manufacturing. Because emissions are related to the level of industrial activity, future activity is computed by the Industrial Production model which uses a simple indexing method to compute future industrial output in different regions.

2.4 LINKAGES OF ENERGY/INDUSTRY WITH THE REST OF IMAGE 2.0

In the Atmospheric Composition model (see below), emissions from the Energy-Industry

TABLE 2

Main sources of greenhouse gases accounted for in the IMAGE 2.0 model.

Source	Greenhouse Gas	IMAGE 2.0 Submodel**
Fuel combustion in end use sectors (Table 1).	CH_4, CO, CO_2, NO_x, N_2O, VOC	EE
Energy conversion, transformation and transportation.	CH_4, CO, CO_2, NO_x, N_2O, VOC	EE
Industrial processes	N_2O, VOC, halocarbons	IE
Cement manufacturing	CO_2	IE
Wetland rice fields	CH_4	LUE
Natural wetlands	CH_4^*	LUE
Landfills	CH_4	LUE
Animal enteric fermentation	CH_4	LUE
Animal waste	CH_4, N_2O	LUE
Natural soils	N_2O^*	LUE
Fertilized soils	N_2O^*	LUE
Aquatic sources	CH_4, N_2O	LUE
Biomass burning (Deforestation, agricultural waste burning, savanna burning)	CH_4^*, CO^*, CO_2^*, NO_x^*, N_2O^*, VOC^*	LUE, TC
Soil Respiration	CO_2^*	TC

*   Emissions calculated on grid-scale. Regional totals calculated for other sources.
**   EE = Energy Emissions model,   IE = Industrial Emissions model,
     LUE = Land Use Emissions model,   TC = Terrestrial Carbon model

sub-system of IMAGE 2.0 are added to emissions of CO_2 and other greenhouse gases coming from the terrestrial biosphere; the model then computes the resulting build-up of greenhouse gases in the atmosphere. The Energy-Industry sub-system is also linked to other parts of IMAGE 2.0 via the demand for fuelwood; this demand is computed in the Energy Economy model and then used by the Land Cover model (described below) to compute new deforested areas. (However, only part of the fuelwood demand is assumed to lead to deforestation). The Terrestrial Carbon model (also described below) computes the change in terrestrial carbon flux owing to fuelwood extraction. Biofuel demands computed by the Energy Economy model are also taken into account in the Land Cover model as an additional demand for land competing with the demand for land to satisfy food demand. This linkage allows the model to comprehensively evaluate the effect of using biofuels on greenhouse gas emissions. No feedbacks of global change on energy use are taken into account.

3. Terrestrial Environment System of Models

The aim of the *Terrestrial Environment* system of models is to simulate global land cover on a grid-basis, and its effect on greenhouse gas emissions and carbon flux from the biosphere to the atmosphere. This set of models can be used to evaluate the effectiveness of land use policies for controlling the build-up of greenhouse gases including changed agricultural practices or sequestration of carbon in forest plantations. They can also be used to evaluate the impact of climate change on global ecosystems and agriculture.

3.1   AGRICULTURAL DEMAND MODEL

Transformations of global land cover are strongly related to changing land use for agricultural products, which require croplands, pasture land, rangelands, and managed forests. The purpose of the Agricultural Demand model is to estimate the societal demands for agricultural products that lead to significant land use demand. These calculations are performed for the same 13 world regions as used in Energy-Industry system (Figure 2) and using the same main driving factors (population, GNP), and are detailed in Zuidema *et al.* (1994).

The Agricultural Demand model begins by calculating the per capita human consumption for different crops and meat products based on an assumed elasticity relationship between consumption and per capita income. These elasticity coefficients are estimated from 1970-90 data of the U.N. Food and Agriculture Organization, and are the main adjustable parameters in the model. Total human (as opposed to animal) demands for these products are then computed for a given income and population scenario. The same procedure is followed to compute the total demand for meat in a region.

The demand for meat is converted into a required number of livestock, and feed requirements of animals are divided into "concentrates" from crops and crop residues, and "roughage" from rangeland and pasture land. The demand for animal feed concentrates adds to the demand for crops computed above for the human population. These two factors together lead to the total regional demand for crops and other non-meat agricultural products. This demand is converted into actual cropland demand in each region by taking into account export and import from each region. The demand for animal feed roughage is converted into a demand for grassland by taking into account the average amount of rangeland required per animal.

World timber production and trade is currently not included in the model, but demand for fuelwood is computed in the Energy Economy model (described above). Similarly, world food trade is presently prescribed rather than computed by the model, although it is planned to include a simple trade model in the next version.

3.2   TERRESTRIAL VEGETATION MODEL

Global land cover transformations depend not only on demand for land, but also on its potential. The purpose of this model is to compute "potential land cover", i.e. the potential vegetation and potential crop productivity in a global grid for a given climate and soil. This model is described in Leemans and van den Born (1994). Calculations are performed on a spatial grid of 0.5° latitude by 0.5° longitude.

Two different models make up the Terrestrial Vegetation model:

A modified version of the BIOME model (Prentice *et al.*, 1992) is used to compute potential vegetation. The basic idea of this model is that global vegetation patterns can be derived by conglomerations of *plant functional types*, each having definable environmental constraints. If environmental conditions can be described at a particular site (e.g., mean coldest-month temperature, and effective temperature sums), then the types of plants which occur can also be predicted. Different plant types are then aggregated into the land cover categories of IMAGE 2.0 (described below under Land Cover model.) This aggregation procedure is the main way to calibrate the model to current vegetation. In the present version of the model the use of the BIOME model is limited. Land cover is initialized according to Olson *et al.* (1985). The BIOME potential vegetation is allocated to a cell only when it is abandoned agricultural land. Full use of the BIOME model will be made in the next version of the IMAGE 2 model.

The FAO Crop Suitability model (FAO, 1978) computes the potential productivity of eight crop classes on a global grid. These are the "constraint-free rain-fed yields" based exclusively on local climatic conditions. First, the potential growing period of a particular crop is estimated from local temperature and moisture conditions. If the growing period is found to be sufficient for a particular crop, then its potential productivity is estimated from a simple relationship between the crop's rate of photosynthesis and local climatic variables.

The climate-related yields computed by the FAO model are adjusted for grid-specific soil conditions by a "soil factor" (ranging from 0.1 to 1.0) which takes into account three soil quality indicators: (i) nutrient retention and availability, (ii) level of salinity, alkalinization, and toxicity, (iii) and rooting conditions for plants (related to soil texture), all three assumed to be constant in time.

3.3   LAND COVER MODEL

This model simulates land cover transformations on a global grid by reconciling the regional demand for land use with the "local" potential for land. The model is described in detail in Zuidema *et al.* (1994). The two main inputs to the model are (1) regional demands for cropland and rangeland (from the Agricultural Demand model) and fuelwood demand (from the Energy Economy model), and (2) "local" potential for land (from the Terrestrial Vegetation model). Table 3 gives an overview of the factors taken into account to compute global land cover.

Global land cover is divided into 17 categories (see Figure 5a) on a global grid of 0.5° latitude by 0.5° longitude; these categories are aggregated from the 51 categories in Olson's (1985) data base. We use Olson's data to initialize the model in 1970. Land use demands for different types of cropland are satisfied by a single land cover category --

TABLE 3

Factors Included in Calculation of Land Use Demand and Land Cover Potential in Terrestrial Environment System of IMAGE 2.0.

Factors For Land Use Demand	Factors for Land Cover Potential
Demand for 8 classes of crops	Climate-limited potential productivity of 8 classes of crops
Demand for 4 classes of livestock	Reduction to crop productivity due to local soil conditions
Demand for fuelwood	Land requirement per unit livestock
Land for biofuels and plantations	Climate-limited potential vegetation types (other than crops)

"agricultural land", while the demand for rangeland can be satisfied by several types of grassland, depending on what exists "locally" -- warm grassland, cool grassland and tropical dry savanna. We assume that in Africa, India plus S. Asia and East Asia, urban residential fuelwood demand leads to forest clearing and subsequently to another land cover type (Zuidema *et al.*, 1994).

The basic idea of the Land Cover model is to change gridded land cover within a "world region" until the total demands for the region are satisfied. (For this reason, countries within a region must be geographically close to one another.) In a sense the model produces pictures of future landscapes logically consistent with demand and potential. Of course, satisfying the demands of a world region by increasing agricultural production anywhere within its territory is an unsatisfactory assumption. Yet the alternative -- taking into account country-scale or grid-scale land use demands -- is infeasible at this time because of the data and analysis necessary for estimating these demands for the entire world and over a long projection period. The model takes into account several types of land conversion including many crucial to global change such as tropical deforestation arising from demand for cropland and rangeland or demand for fuelwood, or reforestation of abandoned agricultural land.

In order to satisfy the regional agricultural demand correctly, we would need to fully comprehend the basic driving forces of land cover transformations. Since these forces are in general poorly understood, our approach is to prescribe transparent, logical "rules" to match land demand with potential. These land use rules are tested to see if they can explain known transformations of land cover from 1970 to 1990. If they can explain historical changes, they can also provide some basis for simulating future land use changes. These rules, together with a "management factor" (which takes into account the effect of technology on yield per hectare),. are the main adjustable variables in the model. As an example of their straightforward nature -- there are two weighted rules that guide the assignment of new agricultural areas: (1) new areas should occur adjacent, if possible, to existing agricultural land (because of availability of transport, infrastructure, and population), and (2) new areas should occur in locations of highest potential productivity. Similar sets of rules are specified to satisfy fuelwood and rangeland demand. The model is calibrated by adjusting the management factor until the computed coverage of agricultural land in each region agrees with FAO estimates (FAO, 1992).

3.4   TERRESTRIAL CARBON MODEL

Because CO_2 is the most important contributor to global radiative forcing, it is of particular importance to accurately estimate its global sources and sinks. Elsewhere in IMAGE 2.0, we compute the source of CO_2 from fossil fuel combustion (Energy Emissions model), and its net geochemical sink in the ocean (Ocean Biosphere/Chemistry model). In the Terrestrial Carbon model we estimate the sources, sinks, and reservoirs of C in the terrestrial biosphere resulting from natural processes and human disturbances. The model is described in detail in Klein Goldewijk *et al.* (1994). This is a grid-scale adaption of the terrestrial part of the C cycle model proposed by Goudriaan and Ketner (1984).

The model describes the CO_2 flux of the biosphere by computing Net Ecosystem Productivity (NEP), i.e. the difference between CO_2 assimilated by Net Primary Productivity (NPP) and CO_2 released by organic decay processes. NPP is pre-defined by land cover category, but scaled to temperature and moisture availability in order to take into account a possible change in productivity if local climate changes. The initial NPP is the main adjustable parameter of the model.

To properly take into account organic decay processes, the reservoirs and fluxes of C within each land cover type are estimated. First, the C in living plant material is partitioned into different biomass components: leaves, branches, stems and roots. Next, rates of decay to non-living matter (litter, humus and charcoal) are assigned based on turnover times of the living plant materials. Last, the decay of non-living matter leads to release of CO_2.

It is often assumed that C in natural ecosystems is in equilibrium with the atmosphere and thus have a NEP of zero, implying no net release of CO_2 to the atmosphere. However, when land is converted from one type to another, this balance is disturbed. To compute CO_2 flux due to human disturbances, we make a consistent set of assumptions regarding the fate of biomass under each type of conversion (e.g. forest to grassland, or grassland to agriculture). For instance, when a tropical forest grid cell is converted to either grassland or agricultural land, it is assumed that certain fractions of its biomass components are burned on-site, immediately releasing CO_2. In addition, the remaining biomass is allocated to non-living C reservoirs which may accelerate decay processes, and the new land cover type often has a lower NPP; the sum of these changes may result in an additional long term low flux of CO_2 from this location. Other land conversions are treated in a similar way, with assumptions being made about the fraction of biomass burned and the allocation of remaining biomass.

The model also takes into account various feedbacks to the biosphere including the effect of moisture availability and temperature on soil respiration, and the effect of temperature and CO_2 on NPP ("CO_2 fertilization"). The CO_2 fertilization effect, in turn, takes into account local limitations to enhanced C uptake due to temperature, moisture availability, nutrient availability and elevation. Both global warming and the rising concentration of CO_2 constantly change the equilibrium state of C in the natural ecosystem to which the actual ecosystem is approaching.

3.5 Land Use Emission Model

In this model we compute emissions of CO, CH_4, NO_x, N_2O, and VOC stemming from different types of land use/cover (Table 2). (The flux of CO_2 from the terrestrial environment is computed by the Terrestrial Carbon model described above). The Land Use Emissions model is described in Kreileman and Bouwman (1994). Land cover patterns and transformations computed by the Land Cover model are used for calculations in this submodel, together with emission coefficients and additional data. The emission coefficients are the main adjustable parameters of the model. Some of the more important processes covered by the model are:

Biomass Burning (Forest, Grassland, Savanna, Agricultural waste). The clearing and burning of vegetation releases not only CO_2 but other greenhouse gases as well. The emissions of CO, CH_4, N_2O, NO_x ,and VOCs are estimated by using a fixed ratio between these compounds and emitted CO_2 (as calculated by the Terrestrial Carbon model). The C release by seasonal burning of savanna and agricultural waste is assumed to be compensated for by subsequent vegetation. For these sources only the net emissions of non CO_2 gases are taken into account.

Animals, Livestock. Enteric fermentation of animals is a significant source of atmospheric CH_4. To obtain total regional emissions, the total number of animals in a region are computed by the Agricultural Demand model and this number is multiplied by a fixed emission factor per animal.

Natural Wetlands. Anaerobic biological processes in wetlands are estimated to be a large source of CH_4. The area of wetlands is included in the land cover data base, and are multiplied by a fixed emission factor to obtain CH_4 emissions.

Irrigated Rice Fields. Anaerobic decomposition in rice paddies is an important source of atmospheric CH_4. The total rice-growing area is estimated by the Land Cover model described above based on regional rice demands, and this area is multiplied by a fixed emission factor to compute CH_4 emissions.

Other Sources of Greenhouse Gases. Additional sources of greenhouse gases included in the model are landfills, animal waste, agricultural residue burning, and natural soils (Table 2).

3.6 Linkages within the Terrestrial Environment System

The models within the Terrestrial Environment system are linked in a straightforward way. Nevertheless, some of these linkages deserve attention since they represent important feedback processes that are not usually considered.

In the grid-based part of the computations climate indices play an important role. These indices determine not only the potential vegetation at a specific location, but also the potential productivity of crops at that location, some of the main quantities in the C cycle simulation (eg. NPP) and emissions of non CO_2 gases (eg. CH_4 from rice paddies). The climate indices are, of course, shared by the models within the Terrestrial Environment system. In this way the changes in land cover, the changes in the terrestrial C cycle, and the changes in land cover related emissions of non CO_2 gases are inherently consistent.

Another obvious link concerns the effect of changing land cover on emissions. Increasing agricultural demand can lead to expanding agricultural area and shrinking area of natural vegetation. In this way anthropogenic influences also affect the natural emissions of greenhouse gases, a feedback process that is usually not represented in emissions scenarios (IPCC, 1990; IPCC, 1992).

3.7   LINKAGES OF THE TERRESTRIAL ENVIRONMENT SYSTEM WITH THE REST OF IMAGE 2.0

Emissions from the terrestrial environment are combined with emissions from the global energy system (calculated by the Energy/Industry models) and are input to the Atmosphere-Ocean models in order to compute the build-up of greenhouse gas concentrations and the resulting radiative forcing and global temperature and moisture patterns. The computed change in climate and CO_2 are then fed back to the Terrestrial Environment System to calculate changes in potential vegetation, potential crop productivity, NPP, and so on.

Changes in land cover are used by the Zonal Atmospheric Climate model to re-compute latitudinal-average albedo. Changes in albedo may affect the zonal energy balance.

We have already referred to the link between the Industry-Energy and Terrestrial Environment subsystems via the demand for fuelwood and biomass. The additional demand for land owing to these fuel demands are taken into account in the Land Cover model.

4. Atmosphere-Ocean System of Models

The purpose of this system of models is to compute the build-up of greenhouse gases in the atmosphere and the resulting change in global temperature and precipitation patterns. A detailed overview of this system is given in de Haan *et al.* (1994). The basic idea of these models is to compute transient changes in climate resulting from changes in greenhouse gas emissions, and to do it in a way that is more computationally economic than a general circulation model (GCM). This makes it feasible to dynamically link the Atmospheric-Ocean models with the Energy-Industry and Terrestrial Environment sets of models described above, and makes it possible to use the entire IMAGE 2.0 model iteratively for policy analysis.

Unlike most GCMs used up to now, the model includes not only CO_2, but all major greenhouse gases explicitly in its radiative scheme. Consequently the model can be used to analyze such problems as the effect of a shift in energy supply from oil to biofuels. Such a shift will reduce CO_2 emissions but enhance non-CO_2 emissions.

The tradeoff of using the present set of atmosphere-ocean models is a lower degree of scientific realism as compared to a GCM, in that IMAGE 2.0 parameterizes many atmospheric/oceanic processes that are explicitly described in a GCM. As an example, zonal heat circulation in IMAGE 2.0 is parameterized as a diffusion process by a few simple expressions based on observations, whereas it is usually explicitly calculated in GCMs. Furthermore, IMAGE 2.0 requires GCM results to scale down zonal average temperature and precipitation patterns to the grid level (see discussion of linkages below).

4.1   ATMOSPHERIC COMPOSITION MODEL

As a starting point, CO_2 and other emissions as well as C sequestration computed by the Terrestrial Environment System are combined with emissions from the Industry-Energy System and are input to the Atmospheric Composition model. This model is described in Krol and van der Woerd (1994). The temporal trends of the average tropospheric concentration of the following gases are computed: CO_2, CH_4, N_2O, CFCs and other halogenated species, O_3, OH radicals, and CO. The influence of emissions of N_2O and NMHCs on atmospheric composition are accounted for. Although simple, the model does explicitly describe the non-linear relationship between chemical species most relevant to radiative forcing.

The calculation of CO_2 takes into account the fluxes between the atmosphere, ocean and biosphere. Computation of CH_4 includes anthropogenic and natural sources, and losses by deposition onto the biosphere, dispersion from the troposphere to stratosphere, and oxidation by OH radicals. Because of the importance of the OH sink, the level of OH is explicitly computed, taking into account loss through CO oxidation. Tropospheric levels of N_2O and CFCs are computed using fixed atmospheric lifetimes.

To compute tropospheric O_3, the model takes into account its precursors, NO_x, CH_4 and CO. For the lower stratosphere, the level of O_3 is parameterized by the availability of free Cl and in the upper stratosphere by the levels of Cl and active N and ambient temperature. The parameterizations used for O_3 and other calculations in the Atmospheric Composition model are derived from "multi one-dimensional" atmospheric chemistry models (Thompson et al., 1989).

The level of sulphate aerosol is computed in a very provisional way by scaling the current spatial distribution of sulphate aerosol to the trend of global sulphur emissions. We distinguish between the sulphate aerosol originating from anthropogenic and biogenic sources. Hence, for scenario analysis, we can in a very preliminary fashion, take into account the different trends in anthropogenic versus biogenic sulphur emissions.

4.2   ZONAL ATMOSPHERIC CLIMATE MODEL

In the second part of the computations, the Zonal Atmospheric Climate model uses the computed atmospheric levels of greenhouse gases and sulphate aerosol together with other data, to compute the earth's energy balance (de Haan et al., 1994). First, the radiative forcing due to greenhouse gases (H_2O, CO_2, CH_4, N_2O, O_3, CFC-11 and CFC-12) is computed using shortwave and longwave parameterizations from MacKay and Khalil (1991). Also included is the cooling effect of backscattering of solar radiation on sulphate aerosol. For radiative forcing calculations, the atmosphere is divided into 18 layers.

Next, the earth's heat balance is computed in eight vertical layers and along 10° latitudinal bands using a set of heat balance equations for the atmosphere and surface heat exchanges with land and ocean (Peng *et al.*, 1982). Temperature is the only prognostic variable in the model in that all processes are parameterized relative to a temperature field which depends on latitude and altitude. Horizontal heat and moisture transport is treated in a simple way by assigning constant diffusion coefficients, and vertical transport by an altitude-dependent diffusion coefficient. As noted previously, surface albedo includes the effect of changing land cover computed by the Land Cover model. Both sea-ice coverage of the oceans and snow coverage of land are parameterized to surface temperature, and cloud cover is represented as a single effective layer. The height and optical thickness of clouds are the main calibration variables in the model.

The main output of the model is annual average surface air temperatures in 10° latitudinal bands covering the entire globe, both land and ocean. The model also makes a crude estimate of average latitudinal precipitation rate by taking the difference between computed rates of surface evaporation and divergence of H_2O.

4.3   Oceanic Climate Model

In order to accurately calculate the energy balance and CO_2 level of the atmosphere, it is necessary to account for the ocean's role as an enormous sink of heat and CO_2. This is done in the IMAGE 2.0 ocean models (de Haan *et al.*, 1994). In the Oceanic Climate model, the world's oceans are divided into vertical layers of 400m thickness, and horizontal segments of 10° latitude, in order to link with the latitudinal bands of the Zonal Atmospheric Climate model.

Advective circulation in the world's oceans is prescribed in the model based on the "conveyor belt" concept of Broeker and Peng (1982), and includes both horizontal flow, as well as upwelling and downwelling estimates. Dispersion coefficients are adjusted so that the model agrees with observed profiles of ^{14}C and ^3H. By representing the ocean's circulation patterns, the model can estimate the dispersal of heat via horizontal and vertical transport.

4.4   Ocean Biosphere/Chemistry Model

To correctly compute CO_2 levels in the atmosphere it is also important to take into account the ocean's physical/chemical uptake of CO_2, as well as the flux of C from oceanic phytoplankton. These processes are covered by the Ocean Biosphere/Chemistry model (de Haan *et al.*, 1994; Klepper *et al.*, 1993). The physical-chemical uptake of CO_2 at the ocean-atmosphere interface is treated in the conventional way by taking into account temperature-dependent equilibria between atmospheric CO_2 and aquatic carbonate compounds. The role of oceanic phytoplankton in releasing and assimilating CO_2 is represented by non-linear equations relating nutrients, light, and temperature to phytoplankton growth. Also included is the process by which non-living organic matter sinks to lower oceanic layers, serving as a "carbon pump" of atmospheric CO_2. While this carbon pump is not thought to be important now, it may play a role in future build-up of atmospheric CO_2 if oceanic circulation changes. The sensitivity of the model to such changes is investigated by changing the ocean circulation in a prescribed way, see for instance de Haan *et al.* (1994). The Ocean Biosphere/Chemistry model and the Ocean Climate model are defined on the same grid and share the description of dispersion.

The model incorporates a variety of feedbacks from the atmosphere to the oceanic environment including the effect of temperature on CO_2 aqueous solubility, the effect of temperature and CO_2 on phytoplankton productivity, and the effect of temperature on the recycle rate of non-living matter to phytoplankton. Many feedback processes influencing the oceanic C cycle have been identified. Feedbacks vary both in their effect on the C cycle and in their uncertainty (Klepper *et al.*, 1993). Here only the fairly well established processes are included.

4.5 LINKAGES WITHIN THE ATMOSPHERE-OCEAN SYSTEM

Within the Atmosphere Ocean system many climate related feedbacks exist. Obviously changes at the interface of atmosphere and ocean are treated consistently in both models.

Warming of the atmosphere affects atmospheric composition. Increase of temperature in the troposphere leads to higher H_2O mixing ratios. This has a positive effect on the formation of OH radicals, thus enhancing the atmospheric sink for CH_4 and other gases.

Increases of sea surface temperature affect not only the equilibrium concentration of dissolved C, but also the sea ice coverage. This changes the area of the open sea where gas exchange is possible. This effect is important, especially since the CO_2 solubility is largest in colder waters, ie. at high latitudes, where the sea ice is located.

4.6 LINKAGES BETWEEN THE ATMOSPHERE-OCEAN SYSTEM AND THE REST OF IMAGE 2.0

At each model time step, new computed patterns of surface air temperature and precipitation are input to the Terrestrial Environment System to calculate changes in potential vegetation, which leads to a new land cover pattern, land use emissions, and so on. Computed temperature and precipitation are also fed back to other parts of the Terrestrial Environment System, for example, to modify the rates of primary productivity in the Terrestrial Carbon model, and to modify the rate of N_2O emissions computed by the Land Use Emissions model. The coupling of land cover and surface albedo has already been mentioned. At this point there is no linkage between the moisture/heat flux to the atmosphere assumed in the Zonal Atmospheric Climate model and global land cover computed by the Land Cover model, although this linkage is a high priority in the next version of IMAGE 2.

Since output data from the Atmosphere-Ocean models are in the form of latitudinal averages, these data must be transformed to the global grid of IMAGE 2.0 calculations. This transformation is performed with data from the current climate data base of Leemans and Cramer (1991) and with interpolated results from a general circulation model of

Manabe and Wetherald (1987).

The transformation of zonal average climate changes to grid-scale changes is, of course, a source of model uncertainty. To investigate this uncertainty we compare downscaling results using four different GCMs. Model runs were performed with identical, Business as Usual type forcing functions, while using results of climate change due to CO_2 doubling from four different GCMs (GFDL, GISS, OSU, UKMO). On the time horizon of the model (1970 to 2100) differences in global quantities were small. Differences in CO_2 concentrations never exceeded 15 ppmv, or 2%, differences in CH_4 concentrations were limited to 0.02 ppmv (less than 1 %), global temperatures were only 0.04°C apart. Larger differences appeared in the areas of agricultural land (3% global, but up to 20% in specific regions) and the geographic distribution of agricultural land. Especially at higher latitudes, where the climate change signal is strongest, significant differences were found in, for instance, the total forested area and the northward shift of agricultural area.

5. Model Testing

5.1   COMPARISON TO 1970-90 DATA

The individual submodels and fully linked version of IMAGE 2.0 have been tested against measurements and other data from 1970-90. This period was selected because of the availability of data, although we plan to use data from a longer historical period for additional validation tests. In this section we briefly review results for the entire linked model. The results of individual submodels are examined in more detail in the papers quoted in sections 2,3 and 4 of this paper.

For these tests it was necessary to assume or estimate the initial states of all models in 1970. Submodels with highly dynamic behaviour were initialized in 1900 in order to establish their non-equilibrium states in the year 1970. For the initialization of the Terrestrial Carbon model, it was assumed that the various reservoirs of terrestrial C were in steady state in 1900; using this steady state as an initial condition, the model was then run to 1970, using as input the historical changes in CO_2 and climate. To initialize the Zonal Atmospheric Climate, Ocean Climate, and Ocean Biosphere/Chemistry models, these models were linked and run to equilibrium by setting the atmospheric concentrations of greenhouse gases constant at their estimated 1900 levels. Starting from this equilibrium state, the linked models were then run in transient mode from 1900 to 1970, using historical concentrations of CO_2 and other greenhouse gases.

As to the results -- The Energy Economy model calculates the consumption of energy in five sectors, and these figures are quite close to official data from 1970-90 for all regions. Sample results are presented for OECD Europe and East Asia in Figure 3. The computed energy consumption differs somewhat from official data, but the model captures the slump in energy consumption in OECD Europe after 1980, and the steady rise in energy consumption in East Asia during this period. Results for other regions (de Vries *et*

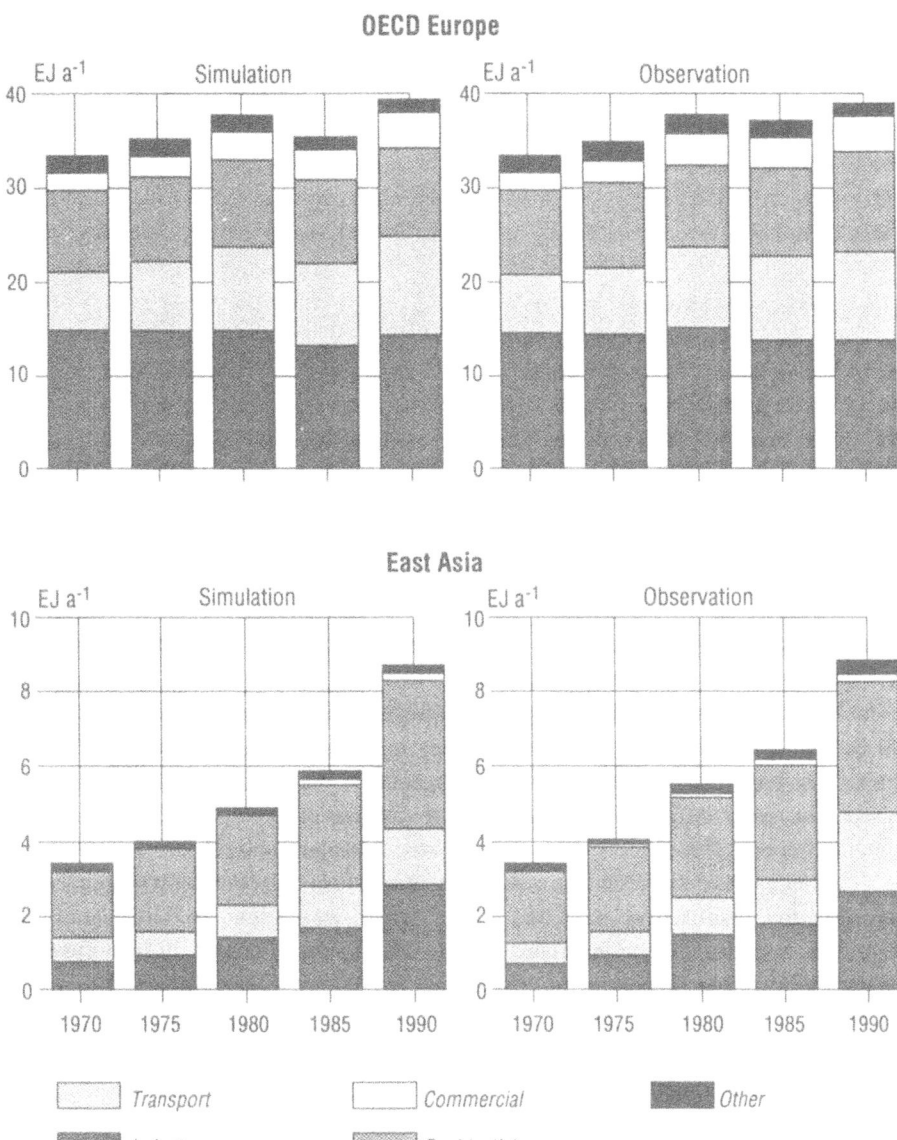

Figure 3: Comparison of end use energy consumption by sector as calculated by IMAGE 2.0 and data from International Energy Agency, 1970-90 in OECD Europe and East Asia.

al., 1994) are similarly good. It seems that the simple economic-energy relationships contained in the Energy Economy model are adequate for reproducing recent trends.

Based on the computed sectoral energy consumption and other data, the Energy-Emissions model computes energy-related emissions of greenhouse gases. Estimates from this model are within ±25% of 1990 data from the four IPCC world regions (Figure 4).

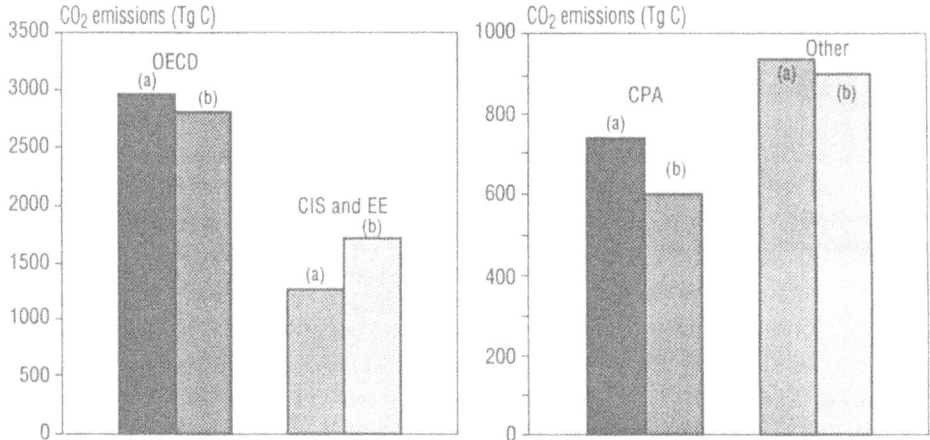

Figure 4: Comparison of calculated CO_2 emissions with IPCC estimates for four world regions. (a) IMAGE 2.0 calculations, (b) Data from IPCC (1990) (CIS and EE = Commonwealth of Independent States and Eastern Europe; CPA = Centrally Planned Asia).

This mainly shows that the emission categories and coefficients in the model have been correctly selected and calibrated, at least for 1990.

The Land Cover model computes global changes in land cover in the period 1970 to 1990 by taking into account actual demand for agricultural commodities together with estimated potential crop productivity and vegetation. As noted earlier in this paper, estimates of potential crop productivity and vegetation come from the Terrestrial Vegetation model. This model is shown elsewhere (Leemans and van den Born, 1994) to give a reasonable representation of areas of potential wheat and other crop productivity.

To illustrate results from the Land Cover model, we present the initial land cover (1970) of Europe and Africa in Figure 5a, and the computed change in land cover (i.e. disappeared land cover) between 1970 and 1990 in Figure 5b. The computed changes in these two decades are very different for the two continents. In Europe, decreasing demand for agricultural land due to increased crop yields led to reversion of agricultural land (indicated by red grid cells in Figure 5b) to other types of land cover (mostly deciduous forest and grassland). By contrast, increasing yields could not keep pace with the crop requirements of Africa's increasing population (even after taking into account food imports during these decades). Added to crop demands were increased demands for fuelwood and rangeland, also stemming from increased population. Taking these demands for land together, the Land Cover model computes an expansion of agricultural land at the expense of tropical forest in Zaire, grasslands and woodlands in Ethiopia, and forest and grasslands in West Africa (Figure 5b). As noted previously, the amount of agricultural land in different regions is calibrated to agree with FAO estimates for 1990. However, the rate of deforestation is indirectly computed by the model and therefore serves as a check of the model's ability to simulate actual land cover changes. Computed regional total deforestation agrees with FAO estimates (Table 4). Moreover, Zuidema et al. (1994) show

5A

Agricultural land
Ice
Cool (semi)desert
Hot desert
Tundra
Cool grass/shrub
Warm grass/shrub
Xerophytic woods/shrub
Taiga
Cool conifer forest
Cool mixed forest
Temp deciduous forest
Br.leav./warm mixed forest
Tropical dry/savanna
Tropical seasonal forest
Tropical rain forest
Wetland

Figure 5(a): Initial (1970) land cover assumed in IMAGE 2.0 for Europe and Africa. Source: Aggregation of land cover types in Olson, *et al* (1985).

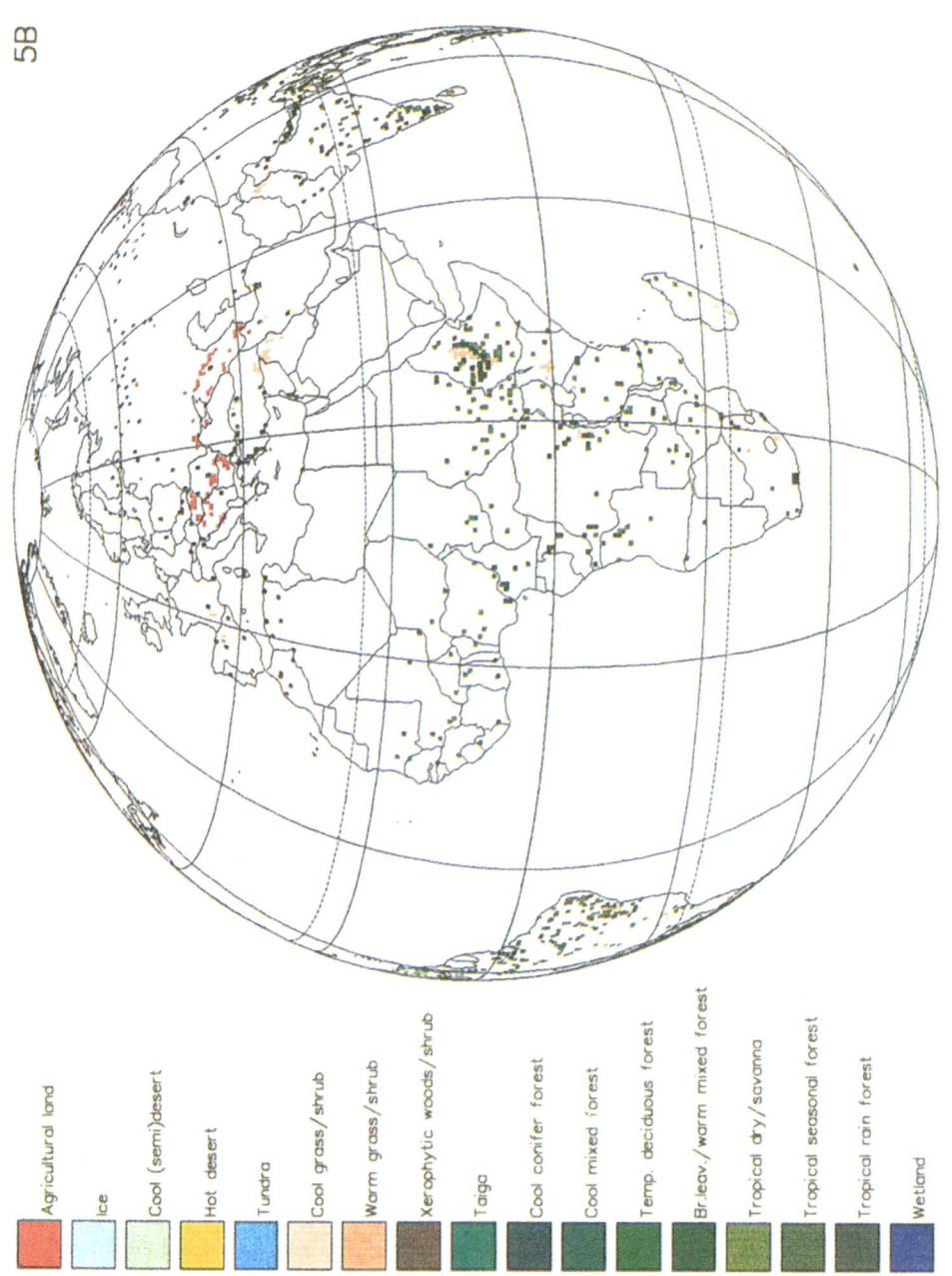

Figure 5 (b): Computed changes in land cover, 1970-90, in Europe and Africa.

TABLE 4

Deforestation rates in various world regions (Mha a^{-1}).

		Latin America	Africa	Asia	Other regions	World
Period 1970 - 1980						
Myers (1980)	All countries [1]	4.2	2.3	3.9	-	10.4
Houghton *et al.* (1983)	All countries [1]	2.9	2.9	3.4	0.1	9.2
FAO (1982)	Tropical countries	5.6	3.7	2.0	-	11.3
IMAGE 2.0	Tropical countries	5.6	3.2	5.1	-	13.8
IMAGE 2.0	All countries	6.9	3.5	7.7	5.3	23.4
Period 1980 - 1990						
WRI (1992)	Tropical countries	8.3	5.0	3.6	-	16.9
FAO (1993)	Tropical countries	7.4	4.1	3.9	-	15.4
IMAGE 2.0	Tropical countries	5.9	3.1	4.8	-	13.7
IMAGE 2.0	All countries	6.7	3.4	6.4	8.7	25.2

[1] Assuming a limited number of conversions of land cover

that computed deforestation rates for the 1980s are within a factor of two of FAO estimates for countries with the highest rate of deforestation. Nevertheless, we expect the spatial accuracy of calculations to improve in the next version of IMAGE 2 when factors such as irrigation and soil degradation are included, and locations of roads and rivers are taken into account.

After land is converted, the Terrestrial Carbon model computes the resulting change in CO_2 flux from the terrestrial biosphere due to human disturbance. As noted previously, this model also takes into account NPP and soil respiration. For 1980, we estimate that land cover changes resulted in a net flux of about 1.0 Gt C a^{-1} to the atmosphere, while uptake to the terrestrial biosphere was approximately 1.9 Gt C a^{-1} (Table 5). The Ocean Climate and Ocean Biosphere/Chemistry models compute an oceanic uptake of 1.4 Gt C

TABLE 5

Global Carbon Budget for 1980 (Gt C or, equivalently, Pg C).

Author	Land cover changes	Terrestrial biosphere uptake	Ocean uptake	Fossil fuel emission	Net accumulation in the atmosphere
Houghton *et al.* (1983)	-0.7 - 2.1	-	1.5 - 2.5	4.5 - 5.9	-
Detwiler & Hall (1988)	0.3 - 1.7	-	1.8 - 2.5	4.8 - 5.8	
Tans *et al.* (1990)	0.4 - 2.6	-	0.6-2.4	5.3	-
Goudriaan (1992)	1.0	1.2	2.0	5.0	2.8
IPCC (1992)	0.6 - 2.5	-	1.2 - 2.8	5.3	-
Sedjo (1992)	1.0	0.7	1.8 - 2.5	4.8 - 5.8	2.9
IMAGE 2.0	1.0	1.9	1.4	5.1	2.8

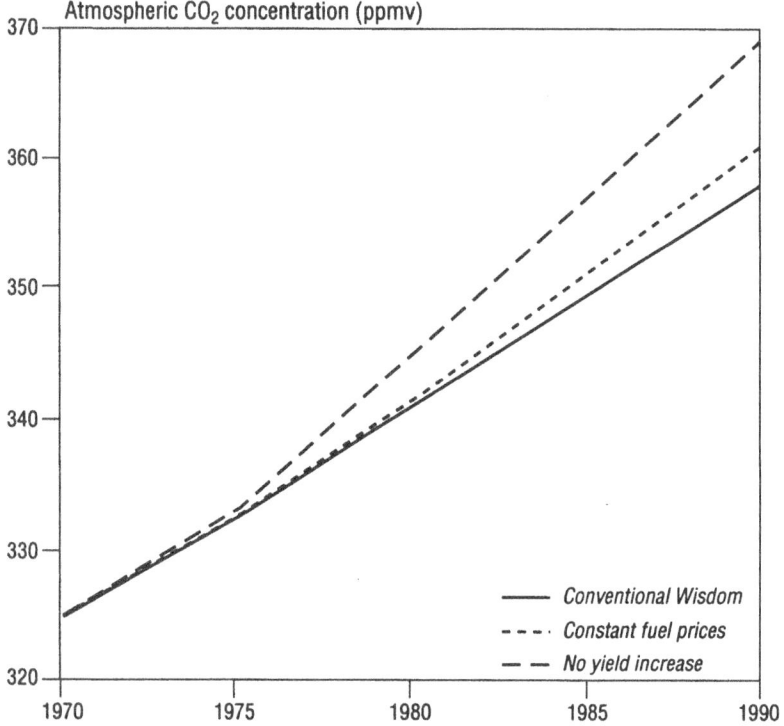

Figure 6: Computed atmospheric concentration of CO_2, 1970-90. Comparison of base case with constant fuel prices and constant yield per hectare (Sec. 5.2). Note scale begins at 320 ppm.

a^{-1}, and the Energy Emissions model estimates fossil fuel emissions to be 5.1 Gt C a^{-1}. This implies a net accumulation of 2.8 Gt C a^{-1} in the atmosphere. The computed accumulation of C in the atmosphere leads the Atmospheric Composition model to slightly overestimate atmospheric CO_2 concentration in 1990 (Figure 6).

The spatial distribution of CO_2 sources and sinks for 1980 (Figure 7) shows that the northern boreal forests act as extensive sinks. Red grid cells scattered in Figure 7 indicate relatively high fluxes of CO_2 to the atmosphere. These fluxes arise from biomass burning associated with recently cleared forest and other land. Unfortunately, there are no comprehensive data sets for comparing our spatial CO_2 flux calculations. However, the calculated total regional CO_2 fluxes due to anthropogenic disturbances are consistent with other estimates (see Klein Goldewijk *et al.*, 1994).

The Land Use Emissions model also updates its estimates of CH_4, N_2O, and other emissions (Table 2) when new land cover is computed. For instance, the computed distribution of N_2O emissions from natural soils in Latin America (Figure 8) reflects the clearing of forests and the changed soil and environmental conditions.

As for the validation of the Atmospheric-Ocean set of models in IMAGE 2.0, the Atmospheric Composition model is able to reproduce the observed trends in CH_4, N_2O,

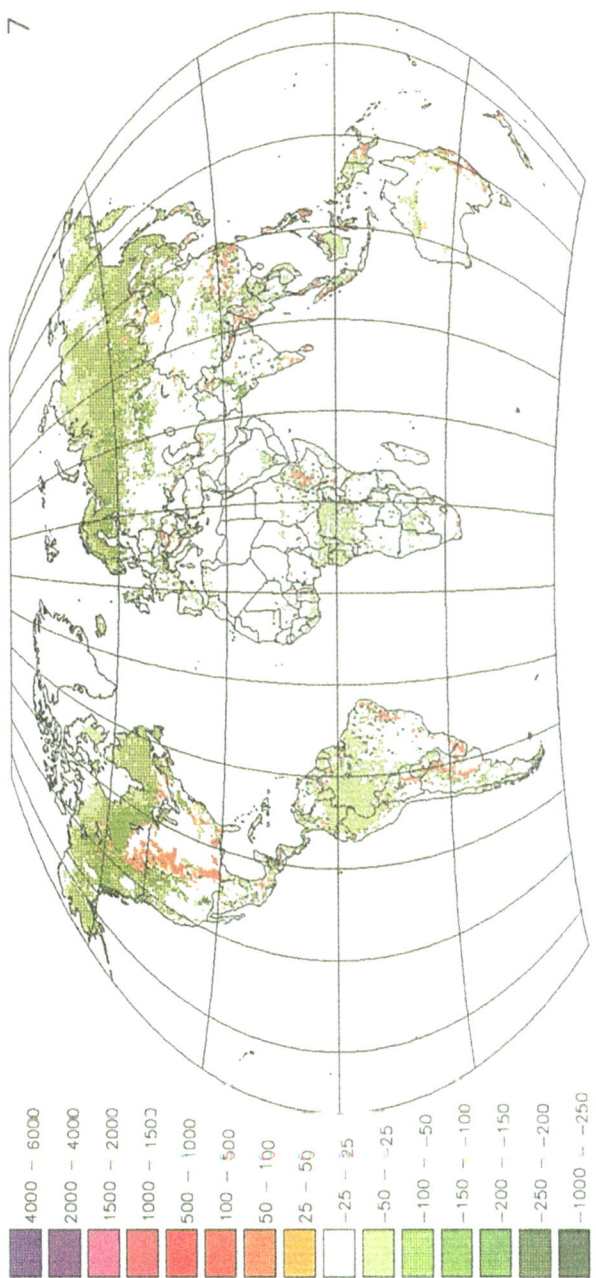

Figure 7: Computed spatial distribution of terrestrial CO_2 fluxes, 1990 (t C km^{-2} a^{-1}).

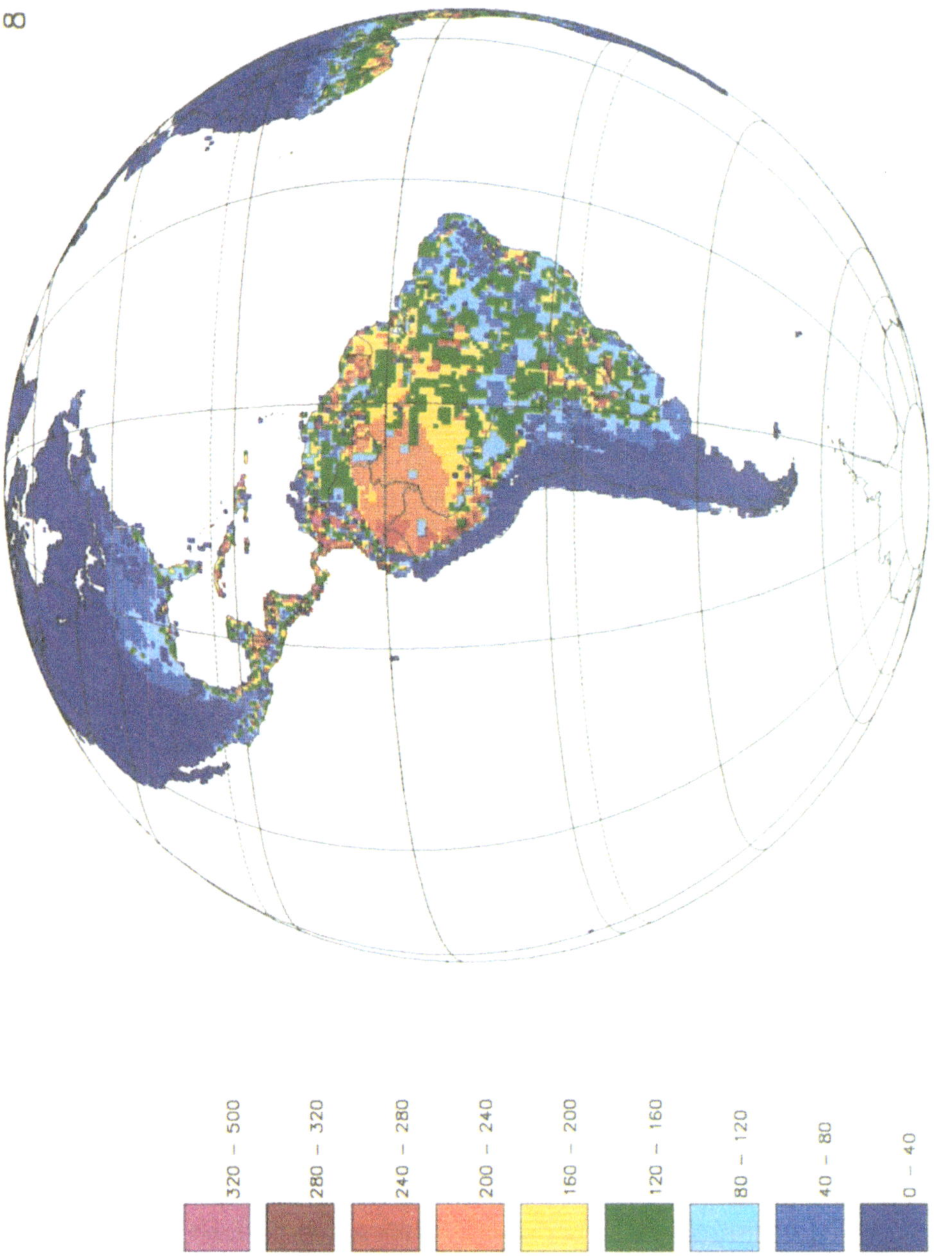

Figure 8: Computed N_2O emissions from natural soils in Latin America, 1990 (kg N_2O-N km^{-2} a^{-1}).

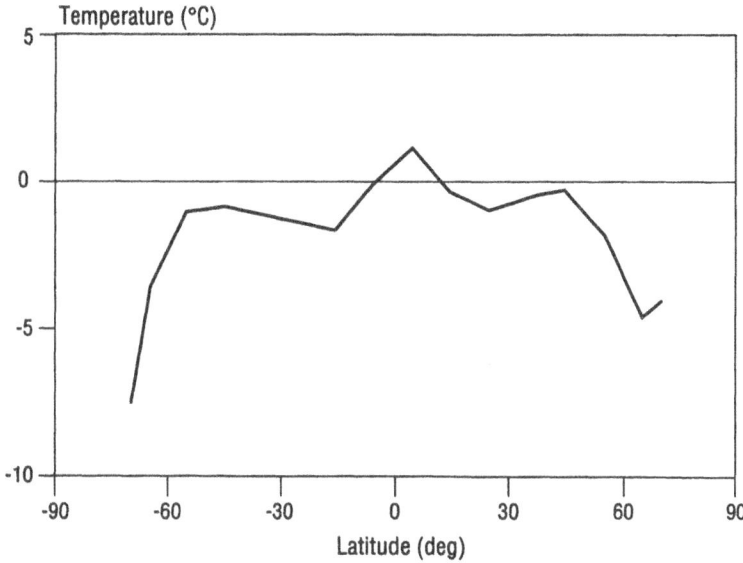

Figure 9: Deviation of computed surface temperatures (1970) from climate average data of Oort (1983).

O_3, and other greenhouse gases. As noted above, the build-up of CO_2 is slightly overestimated. Using the current composition of the atmosphere, the Atmosphere-Ocean models can reproduce fairly well the current latitudinal profile of surface temperatures (Figure 9). Furthermore, both the climate sensitivity of the climate model and the uptake of C by the ocean are within their accepted uncertainty ranges (IPCC, 1992). More details of the validation of these models are given in de Haan *et al.* (1994).

Apart from comparisons with data, the IMAGE 2.0 model is also being tested by mathematical uncertainty analysis; i.e., by assigning uncertainties to model inputs, and then propagating these uncertainties through model equations to estimate the uncertainty of model output. Preliminary results are available for the Energy Economy, Energy Emissions and Atmospheric Composition model. For the Atmospheric Composition model, an uncertainty analysis of input parameters (for a reference emissions scenario) produced an output uncertainty of CH_4 of about 25% (coefficient of variation) in year 2100, with one of the main sources of uncertainty being the parameterization of sensitivity to NO_x in the model (Krol, 1994).

Summing up, we noted in section 1.4 that each submodel was first tested separately, and a limited number of parameters were adjusted in each submodel so that good agreement was obtained with data. Above, we reported results of the fully linked model, which indicate similarly good agreement with data. This is encouraging because of the non-linearities that arise when the model is fully linked. In another paper (Alcamo *et al.*, 1994) we also find that the model is stable for reference runs up to year 2100. Nevertheless, this first comparison with data is inadequate for identifying the model's strengths and weaknesses. Since the model will be used for prognosis up to year 2100, it

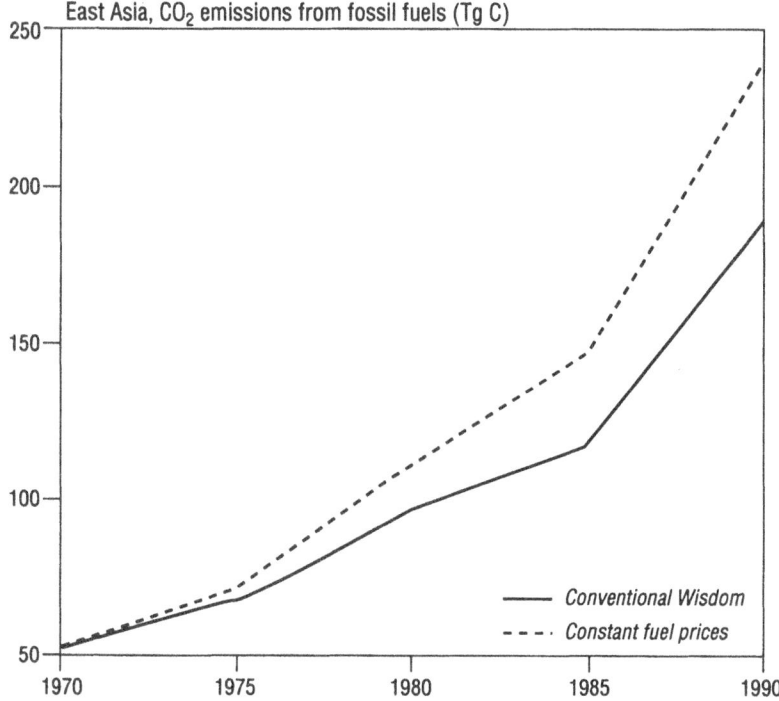

Figure 10: CO_2 emissions from fossil fuels, East Asia, 1970-90. Comparison of base case with constant fuel prices.

should be tested against a longer historical period, which we intend to do. Moreover, to understand the relative importance of different model uncertainties, a comprehensive sensitivity and uncertainty analysis is required. This effort is currently underway as noted above.

5.2 PRELIMINARY SENSITIVITY CALCULATIONS

The complete coupled model can provide insight into linkages in the society-biosphere-system for the period 1970-90. As an example, if fuel prices are set constant between 1970-90 in the Energy Economy model, we compute the energy consumption that would have occurred if higher prices had not spurred conservation during these decades. The consequences on energy-related CO_2 emissions are depicted on Figures 10 and 11. For East Asia, emissions in 1990 would have been 241 Tg C as compared to 190 in the base case; while the world total would have been 6634 Tg C as compared to 5887 Tg C in the base case. As for the atmospheric concentration of CO_2 in 1990, constant prices result in 361.1 ppm as compared to 358.3 ppm in the base case (Figure 6).

As another example, we investigated the affect on the system of improved crop yield (due to technology rather than climate) which occurred in every region during the 1970s

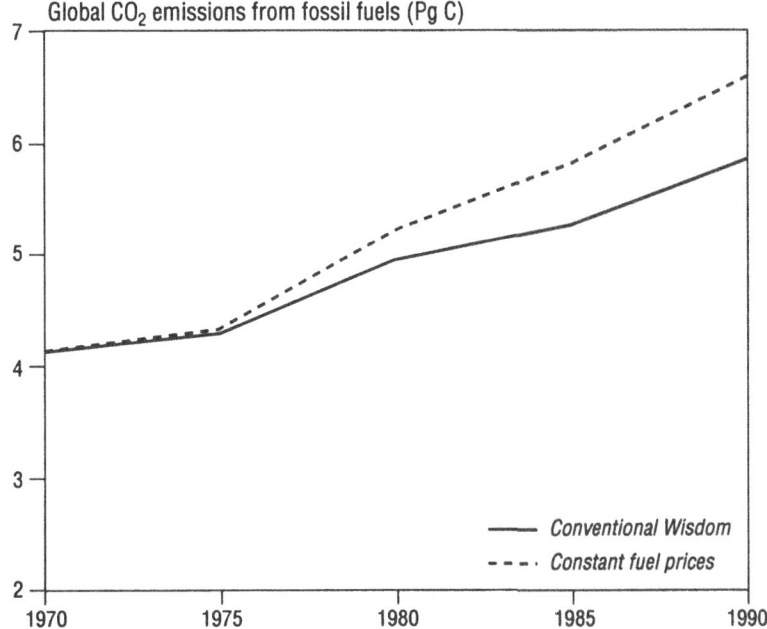

Figure 11. Global CO_2 emissions from fossil fuels, 1970-90. Comparison of base case with constant fuel prices.

and 80s. This was done by running the model with a constant yield per hectare for this period, and comparing results to a base run which takes into account actual yield increases. If yield per hectare had not increased, substantially more agricultural land would have been required to satisfy the demand for crops and livestock, and this would have led to higher deforestation rates (Figure 12). For Latin America, a constant yield per hectare results in 898 Mha of remaining forest in 1990, as compared to 957 Mha in the base case. For Africa these figures were 1188 Mha, compared to 1227 Mha; and for the India plus S.Asia region, the model computes that nearly all of its forest area would have been consumed if yield per hectare had not increased (85 Mha *vs.* 1 Mha). Associated with increased deforestation, is increased biomass burning, and consequently, greater CO_2 fluxes from the terrestrial biosphere (Table 6). This leads to an atmospheric concentration of CO_2 in 1990 of 369.3 ppm as compared to 358.3 ppm in the base case (Figure 6).

6. Discussion and Conclusions

As noted above, sensitivity and uncertainty analyses will provide insight into the greatest opportunities for model improvement. In the meantime, Table 7 presents a partial list of some planned model improvements based on recommendations of an International Review Panel of the IMAGE 2.0 model (NRP, 1993), and other information.

With regards to model applications, the purpose of this paper was to present the

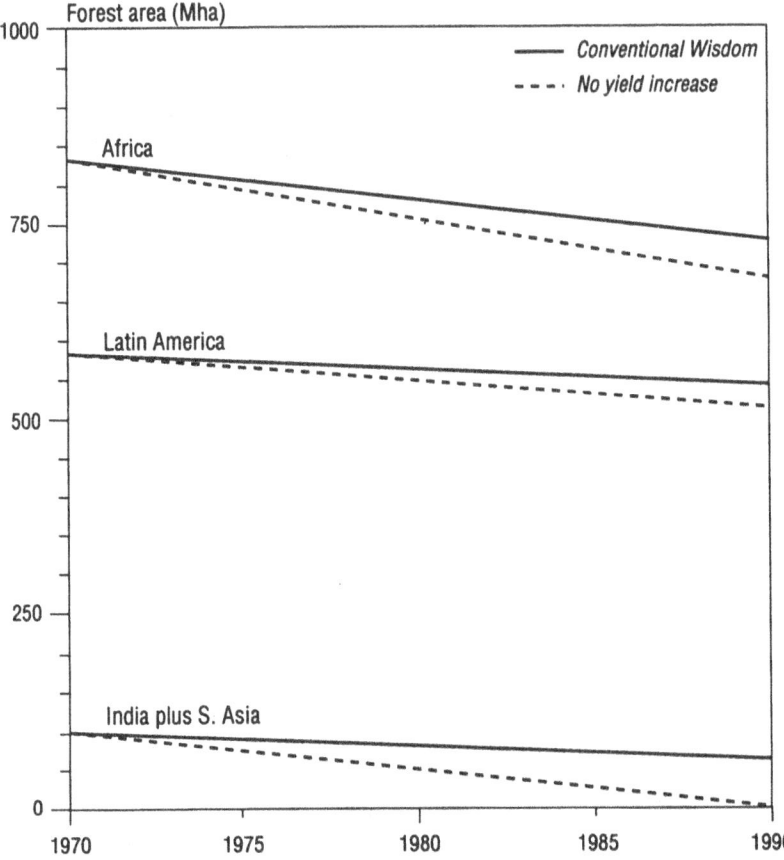

Figure 12: Change in forest area in Africa, Latin America and India plus South Asia. Comparison of base case with constant yield per hectare.

scientific development and testing of the IMAGE 2.0 model rather than how it can be used. Preliminary applications of the model are presented in Part 2 of this paper (Alcamo *et al.*, 1994).

In summing up, the main achievement of the model is its presentation of a geographically-detailed, global, and dynamic overview of the linked society-biosphere-climate system. The model represents in some detail the relation between society's economic and demographic trends and the generation of greenhouse gas emissions. (Of course, our representation of society is quite primitive compared to various global macro-economic models.) It is a first attempt to simulate in geographic detail the transformation of land cover as it is affected by climatic, demographic, and economic factors. It links explicitly and geographically the changes in land cover with the flux of CO_2 and other greenhouse gases between the biosphere and atmosphere, and conversely, takes into account the effect of climate in changing productivity of the terrestrial and oceanic biospheres. At the same time it dynamically couples emissions from society and the

TABLE 6

CO_2 fluxes from the terrestrial biosphere, 1990 (Pg C). Comparison of base case with constant yield per hectare.

	Base case	Constant yield per hectare
Canada	-0.36	-0.40
USA	-0.07	0.13
Latin America	0.14	0.29
Africa	-0.10	-0.05
OECD Europe	-0.14	-0.13
Eastern Europe	-0.02	0.02
CIS	-0.82	-0.92
Middle East	-0.03	-0.05
India + S. Asia	0.09	0.37
China + C.P. countries	-0.19	0.06
East Asia	0.31	0.89
Oceania	-0.02	-0.01
Japan	-0.01	-0.01
Global	-1.23	0.19

TABLE 7

Partial list of some model improvements planned for IMAGE 2.0

Energy-Industry Sub-system:

• Addition of energy supply constraints on energy prices and end use consumption

• Addition of price-driven mechanism for selecting fuel mix

• Experimental linkage of IMAGE 2.0 with global macro-economic model

Terrestrial Environment Sub-system

• Addition of soil degradation and irrigation processes

• Addition of information on population density, roads, and rivers

• Implementation of trade model for agricultural and wood products

• Improvement of land use rules

• Implementation of dynamic vegetation calculations

Atmosphere-Ocean Sub-system

• Implementation of description of cryosphere

• Implementation of sea level rise calculations

• Improved spatial definition of greenhouse gases and their precursors, where necessary (particularly O_3 and its precursors)

• Improved treatment of atmosphere/terrestrial fluxes

biosphere with chemical and physical processes in the atmosphere and ocean, and feeds climate changes back to the terrestrial and oceanic biospheres.

Linking the entire system together, the model was able to reproduce 1970-90 trends in energy consumption, emissions related to the energy/industrial system, terrestrial fluxes of CO_2 and other greenhouse gases to the atmosphere, the build-up of greenhouse gases in the atmosphere, and long-term current climate patterns. Geographic detail is available for many of these calculations. This was accomplished by adjusting a limited number of parameters as described in the text.

Because of its components and spatial resolution, the model is particularly well-suited to investigate the scientific and policy-oriented goals laid out in the beginning of this paper. In particular it has the potential to provide new insight into the linkages and feedbacks of the global society-biosphere-climate system.

Acknowledgements

The authors are indebted to their colleagues who contributed to the development of IMAGE 2.0: G.-J. van den Born, A.F. Bouwman, B.J. de Haan, M. Jonas, K. Klein Goldewijk, O. Klepper, J. Krabec, R. Leemans, J.G. van Minnen, K. Olendrzyński, J.A. Olivier, J. Rotmans, R. Swart, M. Vloedbeld, H.J.M. de Vries, H.J. van der Woerd, R. van den Wijngaart. This research was supported by the Dutch Ministry of Housing, Physical Planning and the Environment (MAP numbers 482505-482509), as well as the Dutch National Research Program on Global Air Pollution and Climate Change, Grant Numbers 851037, 851040, 851042, 851044 and 851045.

Appendix 1

Countries comprising the 13 world regions in the IMAGE 2.0 framework of models.

1. CANADA
Canada

2. USA
United States of America

3. LATIN AMERICA

Argentina	Guyana
Belize	Haiti
Bolivia	Honduras
Brazil	Jamaica
Chile	Mexico
Colombia	Nicaragua
Costa Rica	Panama
Cuba	Paraguay
Dominican Rep.	Peru
Ecuador	Puerto Rico
El Salvador	Suriname
Falkland Is.	Trinidad and Tobago
French Guiana	Uruguay
Guatemala	Venezuela

4. AFRICA

Algeria	Madagascar
Angola	Malawi
Benin	Mali
Botswana	Mauritania
Burkina Faso	Morocco
Burundi	Mozambique
Cameroon	Namibia
Centr. African Rep.	Niger
Chad	Nigeria
Congo	Rwanda
Cote d'Ivoire	Senegal
Djibouti	Sierra Leone
Egypt	Somalia
Equatorial Guinea	South Africa
Ethiopia	Sudan
Gabon	Swaziland
Gambia	Tanzania
Ghana	Togo
Guinea	Tunisia
Guinea-Bissau	Uganda
Kenya	Western Sahara
Lesotho	Zaire
Liberia	Zambia
Libya	Zimbabwe

5. OECD EUROPE

Austria	Ireland
Belgium	Italy
Denmark	Netherlands
Finland	Norway
France	Portugal
Germany	Spain
Greece	Sweden
Greenland	Switzerland
Iceland	United Kingdom

6. EASTERN EUROPE

Albania	Poland
Bulgaria	Romania
former Czechoslovakia	former Yugoslavia
Hungary	

7. CIS
former USSR

8. MIDDLE EAST

Afghanistan	Neutral Zone
Cyprus	Oman
Iran	Qatar
Iraq	Saudi Arabia
Israel	Syria
Jordan	Turkey
Kuwait	United Arab Emirates
Lebanon	Yemen

9. INDIA plus SOUTH ASIA

Bangladesh	Nepal
Bhutan	Pakistan
India	Sri Lanka
Myanmar	

10. CHINA plus CENTRALLY PLANNED COUNTRIES

China	Laos
Kampuchea	Mongolia
Korea, Dem.People's Rep.	Vietnam

11. EAST ASIA

Brunei Darussalam	Malaysia
East Timor	Papua New Guinea
Indonesia	Philippines
Korea, Rep. of	Thailand

12. OCEANIA

Australia	New Caledonia
Fiji	New Zealand

13. JAPAN
Japan

References

Alcamo, J., G.J. van den Born, A.F. Bouwman, B.J. de Haan, K. Klein Goldewijk, O. Klepper, J. Krabec, R. Leemans, J.G.J. Olivier, A.M.C. Toet, H.J.M. de Vries and H.J. van der Woerd: 1994, Modeling the global society-biosphere-climate system, Part 2: Computed scenarios, *Wat. Air Soil Pollut.*, **76** (this volume).

Broecker, W.S. and T.H. Peng: 1982, *Tracers in the sea*, Eldigo Press.

Bouwman, A.F., L. van Staalduinen and R.J. Swart: 1992, *The IMAGE land use model to analyze trends in land use related emissions*, Report 222901009, RIVM, Bilthoven, the Netherlands.

CEC (Commission of the European Communities), Directorate General for Environment, Nuclear Safety, and Civil Protection: 1992, *Development of a framework for the evaluation of policy options to deal with the greenhouse effect.*

Detwiler, R.P. and C.A.S. Hall: 1988, Tropical forests and the global carbon cycle, *Science*, **239**: 42-47.

FAO (World Food and Agriculture Organization): 1978, *Report on the agro-ecological zones project*. FAO: Rome.

FAO: 1982, *Tropical forest resources.*

FAO: 1992, *AGROSTAT-PC, Computerized information series: User manual, Population, Land use, Production, Trade, Food balance sheets, Forest products.* Edition October 1992, FAO, Rome.

FAO: 1993, *Summary of the Final Report of the Forest Resources Assessment 1990 for the Tropical World*, FAO, Rome, Italy.

Goudriaan, J.: 1992, Biosphere structure, carbon sequestering potential and the atmospheric ^{14}C carbon record, *J. Exp. Bot.*, **43**: 1111-1119.

Goudriaan, J., and P. Ketner: 1984, A simulation study for the global carbon cycle including man's impact on the biosphere, *Clim. Change*, **6**: 167-192.

de Haan, B.J., M. Jonas, O. Klepper, J. Krabec, M.S. Krol and K. Olendrzyński, K.: 1994, An atmosphere-ocean model for integrated assessment of global change, *Wat. Air Soil Pollut.*, **76** (this volume).

Houghton, R.A., J.E. Hobbie, J.M. Melillo, B. Moore, B.J. Petterson, G.R. Shaver and G.M. Woodwell: 1983, Changes in the carbon content of terrestrial biota and soils between 1860 and 1980: a net release of CO_2 to the atmosphere, *Ecol. Monogr.*, **53**: 235-262.

IPCC: 1990, J.T. Houghton, G.J. Jenkins and J.J. Ephraums (eds), *Climate Change. The IPCC Scientific Assessment*, Cambridge Univ. Press.

IPCC: 1992, J.T. Houghton, B.A. Callender and S.K. Varney (eds), *Climate Change 1992. The Supplementary Report to the IPCC Scientific Assessment*, Cambridge Univ. Press.

Klein Goldewijk, K., J.G. van Minnen, G.J.J. Kreileman, M. Vloedbeld and R. Leemans: 1994, Simulating the carbon flux between the terrestrial environment and the atmosphere, *Wat. Air Soil Pollut.*, **76** (this volume).

Klepper, O., B.J. de Haan, P. Saager and M.S. Krol: 1993, *Oceanic uptake of anthropogenic CO_2: mechanisms and modeling*, Report 481507004, RIVM, Bilthoven, the Netherlands.

Kreileman, G.J.J. and A.F. Bouwman: 1994, Computing land use emissions of greenhouse gases, *Wat. Air Soil Pollut.* **76** (this volume).

Krol, M.S.: 1994, Uncertainty analysis for the computation of greenhouse gas concentrations in IMAGE, in: J. Grasman and G. van Straten (eds), *Predictability and Nonlinear Modeling in Natural Sciences and Economics*, Kluwer.

Krol, M.S. and H.J. van der Woerd: 1994, Atmospheric composition calculations for evaluation of climate scenarios, *Wat. Air Soil Pollut.*, **76** (this volume).

Leemans, R. and G.J. van den Born: 1994, Determining the potential global distribution of natural vegetation, crops, and agricultural productivity, *Wat. Air Soil Pollut.*, **76** (this volume).

Leemans, R. and W. Cramer: 1991, *The IIASA database for mean monthly values of temperature, precipitation and cloudiness on a global terrestrial grid*, Research Report RR-91-18, International Institute of Applied Systems Analyses, Laxenburg, pp. 61.

MacKay, R.M. and M.A.K. Khahil: 1991, Theory and development of a one dimensional time dependent radiative convective climate model, *Chemosphere*, **22**: 383-417.

Manabe, S. and R.T. Wetherald: 1987, Large scale changes of soil wetness induced by an increase in atmospheric carbon dioxide, *J. Atm. Sci.*, **44**: 1211-1235.

Myers, N.: 1980, *Report of survey of conversion rates in tropical moist forests*, National Research Council, Washington D.C., USA.

NRP (Dutch National Research Program on Global Air Pollution and Climate Change): 1993, *Report of*

International Review Meeting IMAGE 2.0, Amsterdam, NRP Report 00-09, NRP, Bilthoven, the Netherlands.

Olson, J. , J.A. Watts and L.J. Allison: 1985, *Major World Ecosystem Complexes Ranked by Carbon in Live Vegetation: A Database*, NDP-017, Oak Ridge National Laboratory, Oak Ridge, Tennessee.

Oort, A.H.: 1983, *Global Atmospheric Circulation Statistics, 1958-1973*. NOAA Prof. Pap. 14., U.S. Dept. of Commerce, Rockville, Md. U.S.A.

Peng, L., M.-D. Chou and A. Arking: 1982, Climate studies with a multi-layer energy balance model. Part I: model description and sensitivity to the solar constant, *J. Atm. Sci.*, **39**(12): 2639-2656.

Prentice, I.C., W. Cramer, S.P. Harrison, R. Leemans, R.A. Monserud and A. Solomon: 1992, A global biome model based on plant physiology and dominance, soil properties, and climate, *J. Biogeography.*, *19*: 117-134.

Rotmans, J.: 1990, *IMAGE: an integrated model to assess the greenhouse effect*, Kluwer.

Rotmans, J., H. de Boois and R.J. Swart: 1990, An integrated model for the assessment of the greenhouse effect, *Clim. Change*, **16**: 331-356.

Sedjo, R.A.: 1992, Temperate forest ecosystems in the global carbon cycle, *Ambio*, **21**: 274-277.

Tans, P.P., I.Y. Fung and T. Takahashi: 1990, Observational constraints on the global atmospheric CO_2 budget, *Science*, **247**: 1431-1438.

Thompson, A.M., R.W. Stewart, M.A. Owens and J.A. Herwehe: 1989, Sensitivity of Tropospheric Oxidants to Global Chemical and Climate Change, *Atm. Env.*, **23**(3): 519-532.

de Vries, H.J.M., R.A. van den Wijngaart, G.J.J. Kreileman, J.G.J. Olivier and A.M.C. Toet: 1994, A model for calculating regional energy use and emissions for evaluating global climate scenarios, *Wat. Air Soil Pollut.*, **76** (this volume).

WRI: 1992, *World Resources, 1992-1993. A Guide to the Global Environment. Toward Sustainable Development.*

Zuidema, G., G.J. van den Born, G.J.J. Kreileman and J. Alcamo: 1994, Simulation of global land cover changes as affected by economic factors and climate, *Wat. Air Soil Pollut.*, **76** (this volume).

MODELING THE GLOBAL SOCIETY-BIOSPHERE-CLIMATE SYSTEM: PART 2: COMPUTED SCENARIOS

J. ALCAMO, G.J. VAN DEN BORN, A.F. BOUWMAN, B.J. DE HAAN, K. KLEIN GOLDEWIJK, O. KLEPPER, J. KRABEC, R. LEEMANS, J.G.J. OLIVIER, A.M.C. TOET, H.J.M. DE VRIES, H.J. VAN DER WOERD

National Institute of Public Health and Environmental Protection (RIVM)
P.O. Box 1, 3720 BA, Bilthoven, the Netherlands

Abstract. This paper presents scenarios computed with IMAGE 2.0, an integrated model of the global environment and climate change. Results are presented for selected aspects of the society-biosphere-climate system including primary energy consumption, emissions of various greenhouse gases, atmospheric concentrations of gases, temperature, precipitation, land cover and other indicators. Included are a "Conventional Wisdom" scenario, and three variations of this scenario: (i) the Conventional Wisdom scenario is a reference case which is partly based on the input assumptions of the IPCC's IS92a scenario; (ii) the "Biofuel Crops" scenario assumes that most biofuels will be derived from new cropland; (iii) the "No Biofuels" scenario examines the sensitivity of the system to the use of biofuels; and (iv) the "Ocean Realignment" scenario investigates the effect of a large-scale change in ocean circulation on the biosphere and climate. Results of the biofuel scenarios illustrate the importance of examining the impact of biofuels on the full range of greenhouse gases, rather than only CO_2. These scenarios also indicate possible side effects of the land requirements for energy crops. The Ocean Realignment scenario shows that an unexpected, low probability event can both enhance the build-up of greenhouse gases, and at the same time cause a temporary cooling of surface air temperatures in the Northern Hemisphere. However, warming of the atmosphere is only delayed, not avoided.

Keywords: climate change, global change, integrated assessment, integrated models, scenario analysis, carbon cycle, biofuels

1. Introduction

Although climate-related research usually centers on one aspect or spatial scale of the climate change issue, in reality climate issues involve interrelated elements of the society, biosphere, and the climate system. This wide range of issues are reflected, for example, in the reports of IPCC's first assessment (IPCC, 1990a,b,c). The purpose of this paper is to present scenarios which capture some of the scope and detail of these interrelated issues. These scenarios are computed by the IMAGE 2.0 model, an integrated model of the global environment and climate change.

IMAGE 2.0 consists of three sub-systems of models -- "Energy-Industry", "Terrestrial Environment", and "Atmosphere-Ocean". The model gives roughly equal weight to each of these sub-systems. The Energy-Industry models compute the emissions of greenhouse gases in 13 world regions as a function of energy consumption and industrial production. End use energy consumption is computed from various economic/demographic driving forces. The Terrestrial Environment models simulate grid-scale changes in global land cover based on climatic and socio-economic factors, and the flux of CO_2 and other

Water, Air, and Soil Pollution **76**: 37–78, 1994.

greenhouse gases between the biosphere and atmosphere. The Atmosphere-Ocean models compute the buildup of greenhouse gases in the atmosphere and the resulting zonal-average temperature and precipitation patterns.

The time horizon of model calculations extends from 1970 to 2100, and many terrestrial calculations are performed on a grid of 0.5^0 latitude by 0.5^0 longitude; economic-based calculations (relating to energy, industrial production, and agricultural demand) are performed for 13 world regions rather than on a global grid. Climate calculations are performed on a two-dimensional grid of 10^0 latitudinal bands and nine or more vertical layers in both the atmosphere and ocean.

The fully linked model has been tested against data from 1970 to 1990, and after calibration, can reproduce the following observed trends: regional energy consumption and energy-related emissions, terrestrial flux of carbon dioxide and emissions of other greenhouse gases, concentrations of greenhouse gases in the atmosphere, and transformation of land cover. The model can also simulate the observed latitudinal variation of annual average atmospheric temperatures (averaged over a climatologic period).

An overview of the content and testing of IMAGE 2.0 is given in a companion paper (Alcamo et al., 1994). The Energy-Industry system is described in de Vries et al. (1994); the Terrestrial Environment sub-system in Klein Goldewijk et al. (1994), Kreileman and Bouwman (1994), Leemans and van den Born (1994), and Zuidema et al. (1994); and the Atmosphere-Ocean sub-system in Krol and van der Woerd (1994) and de Haan et al. (1994).

The following scenarios are described in this paper:

(i) Conventional Wisdom Scenario -- This scenario makes conventional assumptions about future demographic, economic, and technological driving forces. This is a reference scenario in that it makes no assumptions about climate-related policies. Input data for the main driving forces are based partly on the assumptions of the IS92a scenario of the Intergovernmental Panel on Climate Change (IPCC, 1992).

(ii) Biofuel Crops Scenario -- The Conventional Wisdom and Biofuel Crops scenarios have identical amounts of biofuel usage in the global energy system, and only differ in their assumptions about how/where modern biofuels are grown. In the Conventional Wisdom scenario, biofuels are assumed to come from sources that do not require new cropland, whereas the Biofuel Crops scenario assumes that a substantial amount of biofuels come from new cropland. [In this paper, the terms biofuels and biomass are used interchangeably and both refer to modern biofuels from crop residues, energy crops, plantations and similar sources rather than traditional biofuels such as fuelwood. The IMAGE 2.0 energy model (de Vries et al., 1994) has separate categories for modern biomass, fuelwood and renewables (excluding modern biomass and fuelwood) such as hydroelectric, solar and wind power.]

TABLE 1

Overview of Scenarios.

name of scenario	global population	global economic growth	biofuels included ?	biofuels require new cropland ?	changed ocean circulation
Conventional Wisdom	11.5 B by 2100	1990-2025: 2.9 % a^{-1} 2025-2100: 2.3 % a^{-1}	YES	NO	NO
Biofuel Crops	As Conventional Wisdom	As Conventional Wisdom	YES	YES	NO
No Biofuels	As Conventional Wisdom	As Conventional Wisdom	NO	n.a.	NO
Ocean Realignment	As Conventional Wisdom	As Conventional Wisdom	YES	NO	YES

Note: only global totals given in this table; see Tables 3-6 for selected regional values.

(iii) No Biofuels Scenario -- In this other variation of the Conventional Wisdom scenario, the sensitivity of the global climate system to modern biofuel use is investigated. For this scenario biofuels are removed from the Conventional Wisdom scenario and are replaced by oil.

(iv) Ocean Realignment -- This "surprise" scenario investigates the consequences on the global society-biosphere-climate system of a major change of the ocean's circulation pattern.

Table 1 compares the key assumptions of the four scenarios, and Table 2 summarizes the input data needed to produce these scenarios. The world regions included in IMAGE 2.0 are listed in Table 3. The input data of the scenarios are reviewed in section 2 of this paper, and scenario results are reported in section 3. We note that scenarios ii, iii, and iv can also be viewed as sensitivity studies (although not sensitivity *analysis* in the conventional use of the term) of the Conventional Wisdom scenario.

2. Conventional Wisdom Scenario

2.1 ASSUMPTIONS OF "CONVENTIONAL WISDOM" SCENARIO

The "Conventional Wisdom" scenario is based on conventional assumptions about socio-economic trends. Estimates of future population and economic growth are taken from the IS92a scenario of the Intergovernmental Panel on Climate Change (IPCC, 1992). The IS92a scenario is an intermediate case out of six "reference" scenarios developed by the IPCC. Although the IS92a scenario is an intermediate scenario, the IPCC did not propose

TABLE 2

Main scenario-dependent variables in IMAGE 2.0

Category of Inputs	Variables
Socio-economic	Population (total and urban) GNP
Energy-related	Value added of industrial output Value added of commercial services Private consumption Number of passenger vehicles Fuel mix Fuel prices Efficiency of primary energy conversion Autonomous efficiency improvements Emission factors
Agriculture-related	Food trade, export/import N fertilizer use Technology-related crop yield increase Animal production coefficient Ratio of concentrate:roughage for livestock Fraction of animal feed provided by a particular crop
Related to Atmosphere-Ocean	Ocean circulation pattern

TABLE 3

Regional Population Assumptions for Conventional Wisdom Scenario. Source: IPCC (1992), Scenario: "IS92a". Units: Millions.

Region	1990	2000	year 2025	2050	2100
Canada	27	28	29	28	27
USA	250	270	302	298	295
Latin America	448	534	715	824	877
Africa	642	844	1540	2208	2875
OECD Europe	378	393	407	395	388
Eastern Europe	123	131	143	149	148
CIS	289	306	335	350	347
Middle East	203	272	508	730	937
India + S. Asia	1171	1412	1970	2375	2644
China + C.P. Asia	1248	1431	1756	1896	1963
East Asia	371	447	624	752	837
Oceania	23	24	25	24	24
Japan	124	131	136	132	130
World	5297	6223	8490	10161	11492

TABLE 4

Regional Urban Population Assumptions for Conventional Wisdom Scenario.
Source: 1990 data -- WRI (1990)
Units: Percent of Total Population

Region		year	
	1990	2025	2100
Canada	77	80	85
USA	75	78	84
Latin America	71	85	85
Africa	34	54	85
OECD Europe	78	85	85
Eastern Europe	71	85	85
CIS	66	80	85
Middle East	57	85	85
India + S. Asia	26	39	66
China + C.P. Asia	33	63	85
East Asia	36	61	85
Oceania	80	81	83
Japan	77	85	85

it to be the most likely scenario. We have made the decision to interpret its assumptions (for example, population and GNP) as the "conventional wisdom". At the same time we do not pass judgement on the feasibility or desirability of the scenario's assumptions.

2.1.1 Population Assumptions

The regional population assumptions of Scenario IS92a (Table 3) are based on World Bank estimates, which are close to UN's medium projection (IPCC, 1992). According to this scenario world population will more than double by the year 2100, reaching 11.5 billion people.

The scenario of urban vs. rural population in each region (Table 4) is based on extrapolating the urbanization trend between 1970 and 1990 in each region up to a maximum of 85% urbanization. The future linear increase is consistent with UN estimates for Africa and Asia up to year 2025 (WRI, 1990); the assumed maximum 85% urbanization is the UN estimate for Latin America in 2025 and corresponds to the current percentage of urbanization in some northern European countries, where a maximum may have been reached.

2.1.2 Economic Growth Assumptions

Economic growth assumptions (Table 5) follow those of Scenario IS92a of the IPCC (1992) and take into account recent changes in Eastern Europe and the Commonwealth of Independent States (CIS), as well as consequences of the Persian Gulf war. The IPCC (1992) reports that the GNP growth assumptions of this scenario are at the low end or below the recent range of World Bank forecasts. Nevertheless, the IS92a scenario implies a rapid increase in income per capita in the developing world, although a large income gap

TABLE 5

Regional Economic Growth Assumptions for Conventional
Wisdom Scenario. Source: "IS92a" scenario (IPCC ,1992).
Units: GNP Annual Percentage Growth

Region	year	
	1990-2025	2025-2100
Canada	2.06	1.31
USA	2.09	1.25
Latin America	1.85	2.20
Africa	1.57	2.39
OECD Europe	2.06	1.31
Eastern Europe	1.87	1.18
CIS	1.87	1.18
Middle East	1.36	1.98
India + S. Asia	2.97	2.84
China + C.P. Asia	4.23	3.07
East Asia	2.97	2.84
Oceania	2.71	1.28
Japan	2.71	1.28

remains in the year 2100 between developed and developing regions.

2.1.3 Energy-related Assumptions

The energy-related variables for an IMAGE 2.0 scenario are listed in Table 2. We briefly describe them in this section, while more details are given in de Vries, *et al.* (1994).

To compute future end use consumption of energy, assumptions are required about future levels of "activity" in each end use sector. The measures of activity are: value-added of industrial output (industry sector), value-added of services (commercial sector), private consumption (residential sector), number of passenger vehicles (transport sector), and GNP ("other" sector). These data are summarized in Table 6.

Industrial Output and Services. For OECD regions, it is assumed that the value-added of industrial output and services remains at their current fraction of GNP. Therefore, as GNP in a region increases according to the Conventional Wisdom scenario, the value-added of industrial output and commercial services in this region proportionately increases. As for non-OECD regions, they are assumed to follow the historical pattern of structural change of OECD economies, i.e., as GNP rises, industrial output initially increases, then peaks and declines; meanwhile the decline of industrial output is paralleled by an increase in commercial services (Maddison, 1991). In this scenario, non-OECD regions are assumed to follow this pattern. The fraction of GNP devoted to industrial output increases, peaks, and then declines while the fraction of GNP devoted to commercial services increases when the industrial output fraction decreases.

Private Consumption. Private consumption in OECD regions remains fixed at its current

TABLE 6
Assumed activity levels for energy end-use sectors

Region	Value added industrial output ($ cap⁻¹ a⁻¹)					Value added commercial services ($ cap⁻¹ a⁻¹)				
	1970	1990	2025	2050	2100	1970	1990	2025	2050	2100
Canada	3141	4282	8746	12114	23240	3686	7114	14529	20124	38607
USA	3635	4629	9555	13020	24172	6338	9210	19011	25903	48091
Latin America	535	619	1222	2099	6199	723	915	1796	3190	10070
Africa	335	268	491	848	2531	241	295	520	1036	4112
OECD Europe	3025	4189	8556	11851	22736	4251	6791	13869	19210	36854
Eastern Europe	646	1122	2058	2637	4330	861	1138	2509	3537	7033
CIS	1850	2560	4210	5640	10123	617	1341	6094	8164	14653
Middle East	1524	1299	2357	3523	7873	641	1094	2221	3981	12789
India + S. Asia	45	82	353	688	2619	45	111	388	862	4254
China + C.P. Asia	93	375	1249	2413	9008	38	131	1062	2547	14633
East Asia	159	490	1519	2819	9713	207	513	1451	3214	15779
Oceania	2768	3470	8853	12164	22965	4571	6370	16251	22329	42155
Japan	3032	6330	16147	22186	41887	3895	7409	18900	25969	49029

Region	Private consumption ($ cap⁻¹ a⁻¹)					Number of passenger vehicles (vehicles per 1000 cap)				
	1970	1990	2025	2050	2100	1970	1990	2025	2050	2100
Canada	5024	6969	14234	19715	37824	327	472	600	605	615
USA	6528	9341	19281	26272	48774	450	566	600	604	611
Latin America	1022	1211	2244	3718	10202	30	73	132	135	140
Africa	455	456	770	1367	4305	12	15	43	55	61
OECD Europe	4440	6882	14056	19469	37350	197	375	420	420	420
Eastern Europe	1077	1305	2732	3761	7126	34	145	215	222	237
CIS	2503	4388	6928	9282	16661	18	59	215	222	237
Middle East	1058	1396	2362	4165	12958	12	41	75	75	75
India + S. Asia	175	241	672	1249	4310	1	3	16	36	59
China + C.P. Asia	157	323	1496	3213	14826	0	3	15	35	58
East Asia	321	658	1944	3924	15987	5	16	43	56	61
Oceania	4645	6083	15516	21320	40251	307	413	485	485	485
Japan	4259	7860	20050	27549	52011	102	254	315	315	315

fraction of GNP. This means that private consumption increases proportionately to GNP. By comparison, in developing regions this fraction is not fixed, but is assumed to increase to the current average fraction in OECD countries, as GNP in the developing region approaches the current average GNP of OECD countries.

Passenger Vehicles. Studies of historical trends in transportation have shown that the number of vehicles in a society are proportionately related to wealth, but are also constrained by the availability of roads, the density of populations, and other country-specific factors (Grübler and Nakicenovic, 1991). As a result it is probably not wise to assume that there is a universal relationship between income and vehicles per person, nor that there is a typical time period by which each region will reach the saturation number of vehicles. For our estimates we use technological diffusion data from different countries which indirectly take into account constraints to number of vehicles (Grübler and Nakicenovic, 1991). We use these data to estimate the saturation value of vehicles per capita for each region; we further assume that saturation will be reached in year 2100. For the year 2025, we use vehicle estimates from the U.S. EPA (1990) for different regions, and interpolate for years in-between. An exception is made for the four world regions currently having very low levels of vehicle usage (Africa, India plus South Asia, China plus Centrally Planned Asia, and East Asia). For these regions we assume that the current global average (61 vehicles/ 1000 cap) will be reached in year 2100. For intermediate years, we assume that the increase in vehicles in these four regions will follow a typical "S curve" trajectory, as proposed by Grübler and Nakicenovic (1991).

Fuel Mix and Prices. The trend of greenhouse gas emissions from each region's energy economy is closely related to the amount and mix of fuels consumed. The IMAGE 2.0 model endogenously computes the amount of energy consumed in each of five end-use sectors (industry, commercial, residential, transport, and "other") of each region, based on the activity levels just described. However, the fuel mix in each sector is prescribed (although version 2.1 of IMAGE will endogenously compute the fuel mix in each region.) For the Conventional Wisdom scenario, the fuel mix for each sector (i.e. the fraction of total end use energy consumption delivered by each energy carrier) has been estimated from results of the model used to generate the IS92a scenario (IPCC, 1992; Pepper *et al.*, 1992).

The computation of end use energy consumption in IMAGE 2.0 also depends on a scenario of future fuel prices which are used to determine the level of energy conservation. Future trends in prices of coal, gas and oil are the same for each region and are taken from the Edmonds-Reilly model (Edmonds and Reilly, 1985). For coal, the price index (scaled to 1975) is 1.55 in 2050 and 2.37 in 2100. For gas, the index is 4.10 in 2050 and 7.71 in 2100, and for oil 2.46 in 2050 and 2.38 in 2100. Prices of fuelwood are held constant. Prices of biomass are held constant until 2025 and are then assumed to be 10% higher than current prices in 2050, and 20% higher in 2100.

Energy Conversion Efficiency. Emissions of greenhouse gases also depend on the energy used to convert primary to secondary energy, which in turn depends on the assumed efficiency of electricity and heat generation. For the Conventional Wisdom scenario, it was assumed that efficiency of converting coal, gas, and oil to electricity increases linearly with time from its 1990 value (which varies from region to region) to the value of 0.50 in 2100 in OECD regions, Eastern Europe, CIS and Middle East; and to 0.45 in 2100 in other regions. Other assumed conversion efficiencies are presented in de Vries *et al.* (1994).

Autonomous Efficiency Improvements. Another important variable affecting end use consumption of electricity and heat are so-called "autonomous" factors that lead to improvements in end use energy efficiency. By definition, these are improvements that are *not directly related to increases in fuel prices.* For electricity, autonomous improvements of the energy intensity are assumed to arise from technological development rather than from higher fuel prices. For energy in the form of heat, we assume that technologies for delivering heat become cheaper, making price-driven energy conservation more attractive. The assumed rate of improvement is region-specific, ranging from 0 to 2% a^{-1} for heat and 0 to 5.5% a^{-1} for electricity (de Vries *et al.*, 1994). The higher rates of improvement are assigned to developing regions under the assumption that they can realize large gains in their currently inefficient energy systems.

Emission Factors. In order to compute future emissions of greenhouse gases, it is also necessary to assign future emission factors to these gases. For the Conventional Wisdom scenario, it is assumed that emission controls lead to decreases of emission factors of NO_x in all sectors, CH_4 in fuel production, and CO and VOC in transport. Emission factors for N_2O in transport are assumed to increase as a side effect of catalyst-type emission controls on vehicles. All other emission factors are held constant. More information about these assumptions is available in de Vries *et al.* (1994).

2.1.4 Agriculture-related Assumptions

The types of agriculture-related assumptions required for the Conventional Wisdom scenario are presented in Table 2. Details are given in Zuidema *et al.* (1994).

Food Trade. Future agricultural demand will strongly affect land cover patterns and these will affect the flux of CO_2 and other greenhouse gases from the terrestrial environment. Agricultural demand, of course, depends on the global trade of agricultural commodities. However, since IMAGE 2.0 does not compute world food trade (this is planned for version 2.1), a very simple approach is taken. We assume current exports of food products from the developed world increase by 50% from 1990 to 2050 and level off afterwards. Net exports from developing regions double their 1990 level by 2100. Export of animal products stay constant at their 1990 level, while sugar export is assumed to be zero. The allocation of crop exports to importing regions is weighted according to the crop consumption of importing regions.

Crop Yield. Future land cover patterns will greatly depend on the need for agricultural

TABLE 7

Summary of scenario results. These are global average or total results unless otherwise specified.

YEAR+ SCENARIO	CO2-emissions from energy /industry	Carbon Cycle (Pg C a⁻¹) net biosphere flux*	ocean flux*	Methane emissions (Tg CH₄ a⁻¹)	Atmospheric Concentrations CO₂ (ppm)	CH₄ (ppm)	trop. O₃	Change of agri- cultural area (10³ km²)	Change of forest area (10³ km²)	Average surface temperature (°C) Northern Hemisphere	Southern Hemisphere
1990	6.1	-1.2	-1.6	492	358	1.7	-	26.7	47.2	14.2	13.0
2050							**	**	**	***	***
Conv. Wisdom	15.2	-7.2	-3.0	688	522	2.5	+11.6%	+9%	-26%	+1.4	+1.0
Biofuel Crops	15.2	-6.0	-3.1	692	534	2.6	+12.6%	+30%	-32%	+1.5	+1.0
No Biofuels	17.0	-7.5	-3.2	677	539	2.4	+ 9.0%	+9%	-26%	+1.4	+1.0
Ocean Realign.	15.2	-4.5	-3.1	686	563	2.6	+12.7%	+12%	-27%	+0.0	+1.1
2100											
Conv. Wisdom	24.0	-8.2	-4.2	778	777	2.3	+10.0%	+14%	-27%	+2.4	+1.8
Biofuel Crops	24.0	-6.7	-4.5	793	821	2.4	+12.0%	+65%	-31%	+2.7	+2.0
No Biofuels	29.2	-8.9	-4.8	746	857	1.7	+0.2%	+15%	-27%	+2.4	+1.9
Ocean Realign.	24.0	-6.7	-4.1	778	863	2.4	+11.7%	+18%	-28%	+1.2	+2.0

*   Minus sign indicates net sink.

**   Percentages are relative to 1990.

*** 2050 and 2100 are absolute changes in °C relative to 1990.

land, and this in turn will depend on the potential yields of crops. There are two aspects to these yields -- The first is the potential yield resulting from local climate and unmanaged soil conditions; this is computed by the Terrestrial Vegetation submodel of IMAGE 2.0 (see Leemans and van den Born, 1994) and is not a scenario variable. The second is the influence of fertilizer and other technological inputs (tractors, management know-how) on yield. These variables must be prescribed for each scenario. The yield increase due to nitrogen fertilizer is based on a representative yield response curve for cereals from Addiscot (1991). Future nitrogen fertilizer use is also a scenario variable and is derived from the IS92a scenario of the IPCC. The effect of other technological inputs (tractors, management know-how) on yield after 1990 is based on trends in each region between 1970 and 1990 (see Zuidema et al., 1994).

Animal Productivity and Other Variables. The future trend of animal productivity (ratio of non-productive animals versus productive animals, production of meat and dairy products per cow) can influence future land requirements in developing regions because improved productivity can lead to smaller grassland requirements per unit animal product. This scenario assumes that animal efficiencies in developing regions will linearly approach the current (1990) efficiencies in OECD Europe as incomes of these regions approach the current income of OECD Europe. Two other scenario variables also affect future land requirements for animals: the assumed composition of feed (roughage or concentrate) influences the amount of range land required, while the type of crop used to provide feed determines how feed will compete with human requirements for crops. Both of these variables were fixed at their 1990 values. Quantitative information about these assumptions is available in Zuidema et al. (1994).

2.2   RESULTS OF THE CONVENTIONAL WISDOM SCENARIO

Because of the large volume of output generated by the IMAGE 2.0 model for each scenario, we present detailed results for only two of thirteen world regions -- OECD Europe from the developed regions, and Africa from the developing regions. Selected results from other regions are presented when they are of particular interest for the scenario. In addition, global calculations are given. Results for the four scenarios are summarized in Table 7.

2.2.1 Results from Energy-Industry
For the Conventional Wisdom scenario, the trend of primary energy consumption in OECD Europe is quite different from Africa (Figures 1 and 2). After 2000, energy demand in OECD Europe slowly increases due to slowly increasing population. Also, at the level of economic activity in this scenario (Table 6), the increase in end use energy consumption for each unit increase in economic activity is small compared to Africa and other developing regions because consumption is near saturation. The slowly increasing trend in primary energy consumption is actually outweighed by improvements in energy efficiency spurred by higher fuel prices and technological developments. The result is a stabilization

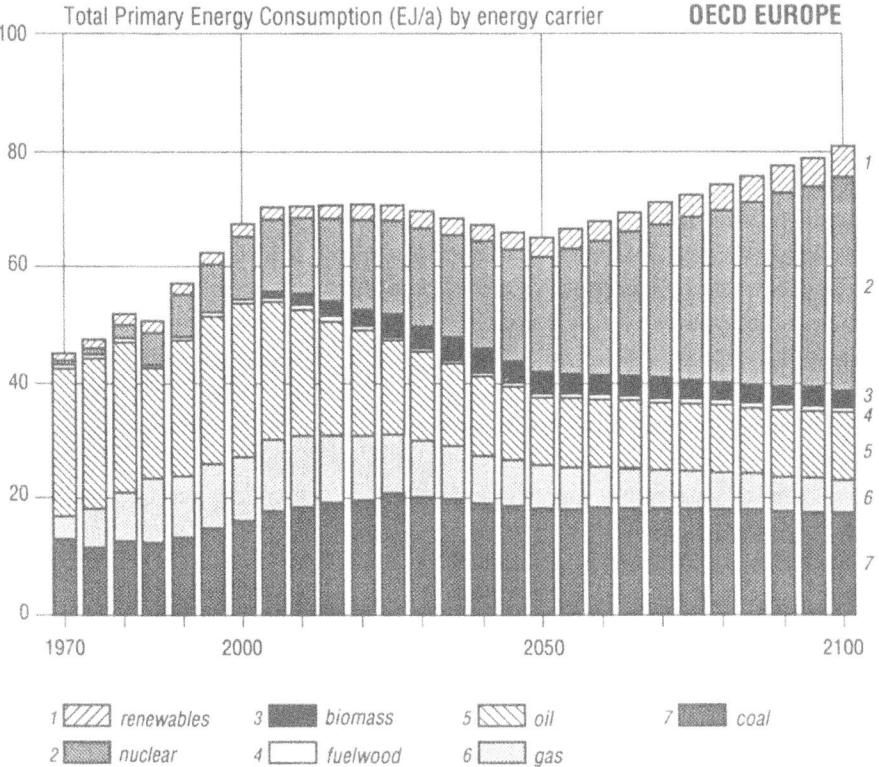

Figure 1: Total primary energy consumption by energy carrier in OECD Europe. (EJ a^{-1}).

of primary energy consumption in the first half of the century (Figure 1). In the second half of the century, energy consumption slightly increases because growth in economic activity outpaces the rate of energy conservation stimulated by energy price increases. We remind the reader that this is meant to be a climate-policy-free scenario, and therefore does not consider that future taxes or other economic policy instruments could boost energy prices still higher and stimulate further conservation.

By contrast to OECD Europe, large increases in population and income in Africa lead to tremendous increases in consumption of all fuels in the second half of next century (Figure 2). Moreover, the increase in end use energy consumption for each unit increase in economic activity remains relatively high in the next century because Africa has a high energy intensity relative to its level of economic activity at the start of the simulation (1970).

Since we will be analysing different biofuel-related scenarios later, we note here that modern biofuels (excluding fuelwood, dung, and other "traditional biofuels") account for 14.4 EJ a^{-1} of Africa's primary energy consumption in 2050 and 57.7 EJ a^{-1} in 2100. For OECD Europe, these figures are 4.1 EJ a^{-1} in 2050 and 3.3 EJ a^{-1} in 2100, and for the world, 74.1 EJ a^{-1} in 2050 and 208.0 EJ a^{-1} in 2100. By comparison, the "Renewables-

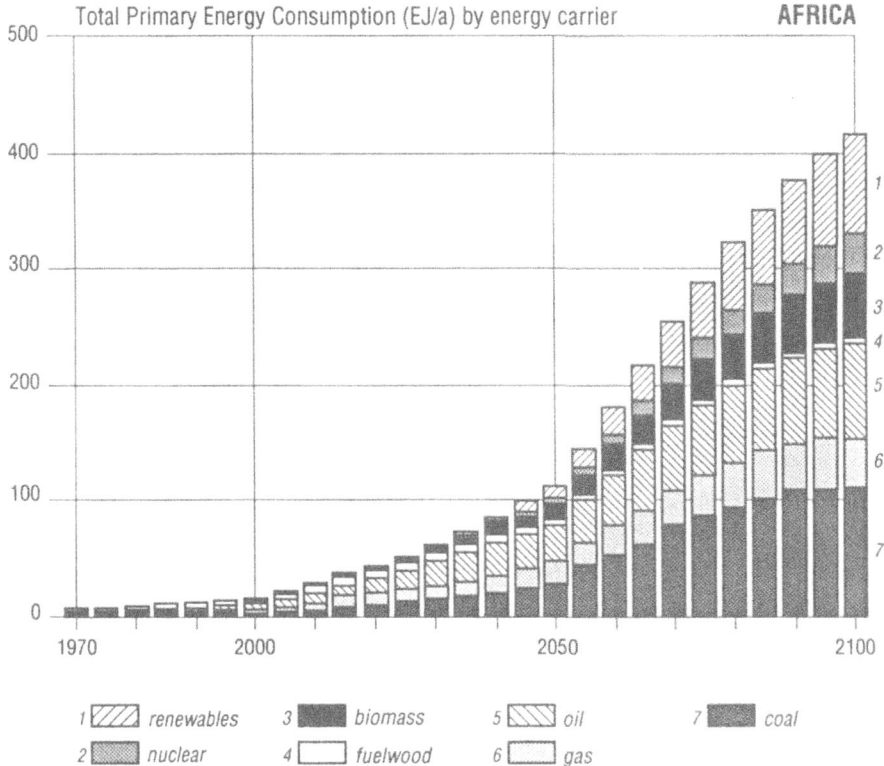

Figure 2: Total primary energy consumption by energy carrier in Africa. (EJ a^{-1}).

Intensive Global Energy Scenario" of Johansson *et al.* (1993b) includes 206 EJ a^{-1} of world biomass consumption in year 2050.

The trend in European emissions is also quite different from Africa's trend (Figures 3 and 4). Emissions of CO_2 from the energy system in OECD Europe decline due to the combined effect of slowly increasing energy consumption and a shift to low or non-CO_2 fuels (Figure 3). Currently, the main source of CO_2 in OECD Europe is power generation, followed by the sectors of industry and transport (Figure 3). According to this scenario, future power generation will account for an even greater share of total emissions from energy in OECD Europe.

After an initial increase, emissions of O_3 precursors (CO, NO_x, and VOC) and N_2O decrease because the consumption of end use energy decreases in the transport sector, one of the main sources of these emissions. The emissions of O_3 precursors also decrease because of assumed air pollution controls in various energy sectors. Emissions of CH_4 are reduced because of a shift from fossil fuels to nuclear power in OECD Europe (according to this scenario) and increased efficiencies.

Emissions of CO_2 and other gases from Africa spiral upwards following increased energy consumption and industrial activity (Figure 4). By 2030, Africa's CO_2 emissions

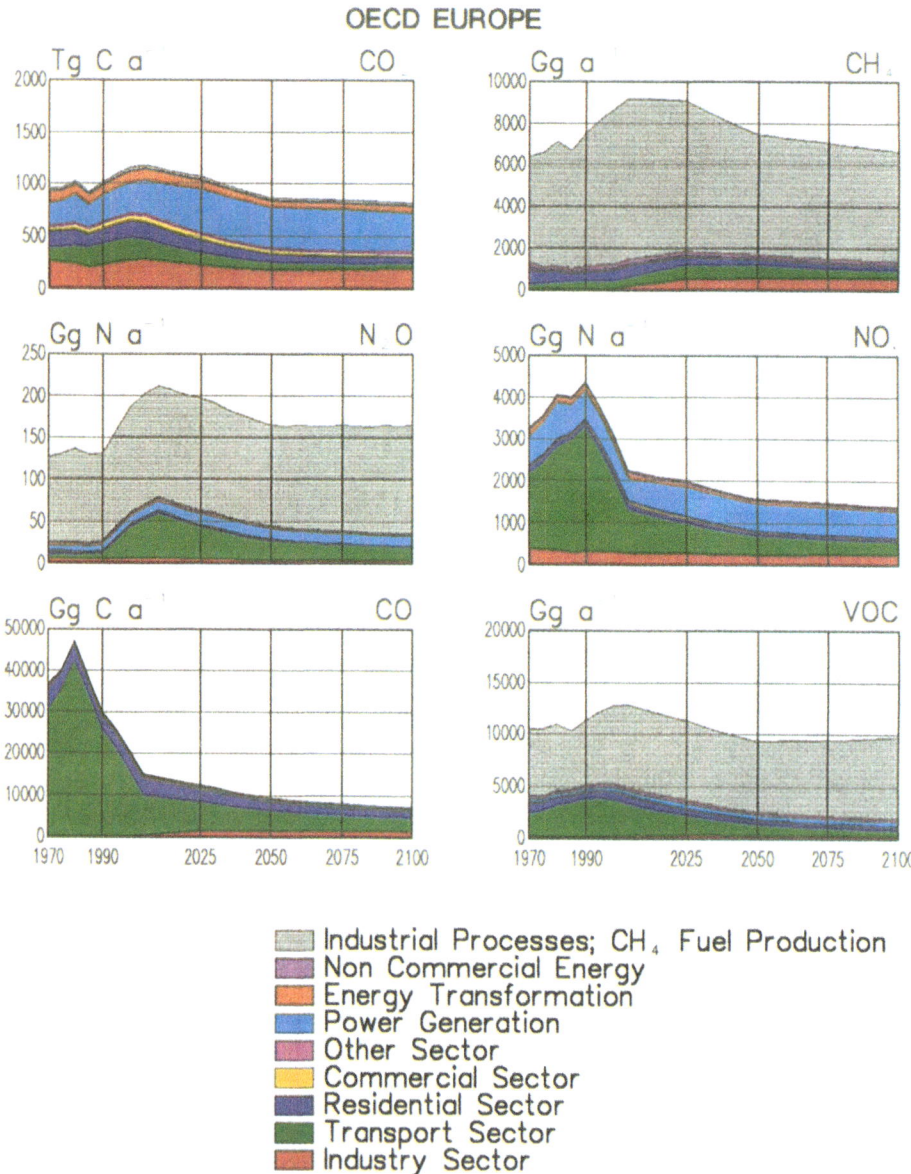

Figure 3: Annual emissions from energy consumption and industrial processes in OECD Europe.

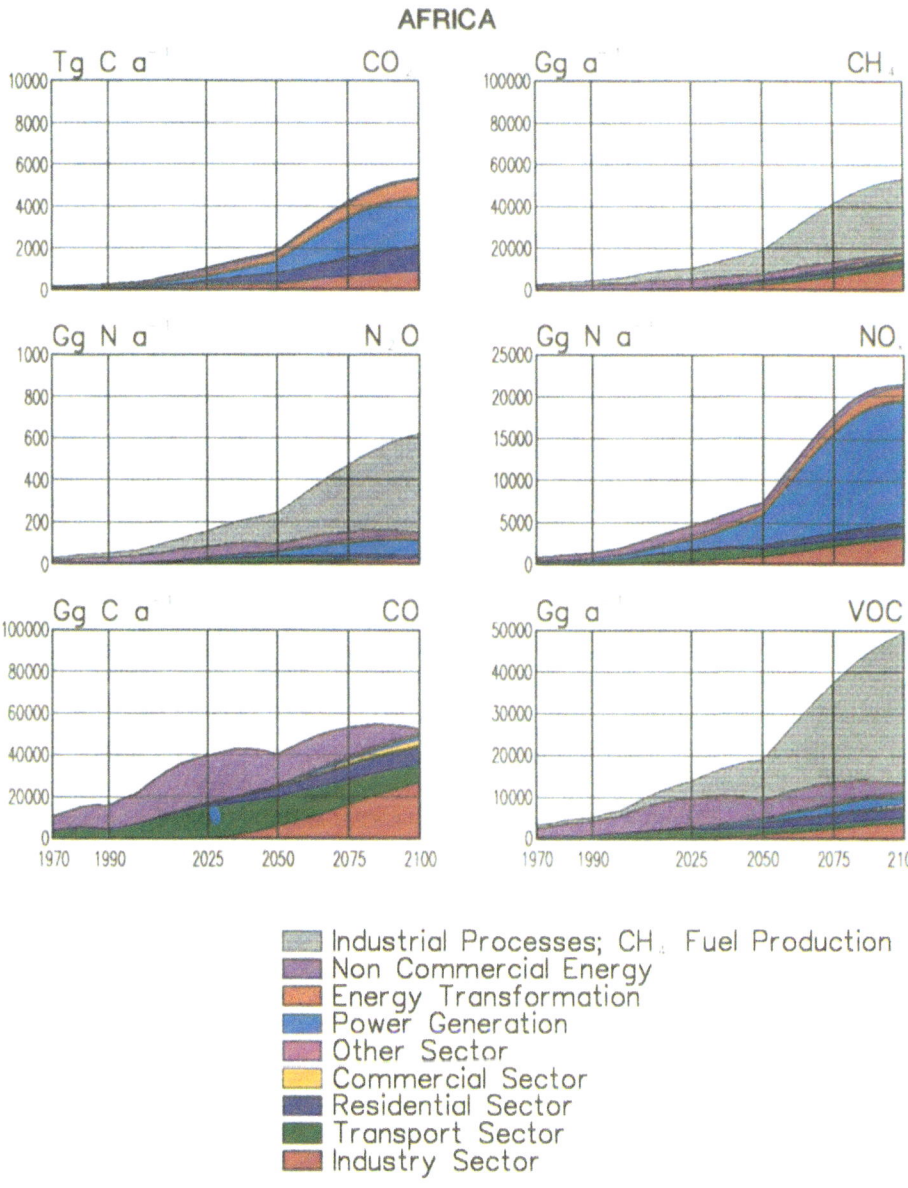

Figure 4: Annual emissions from energy consumption and industrial processes in Africa.

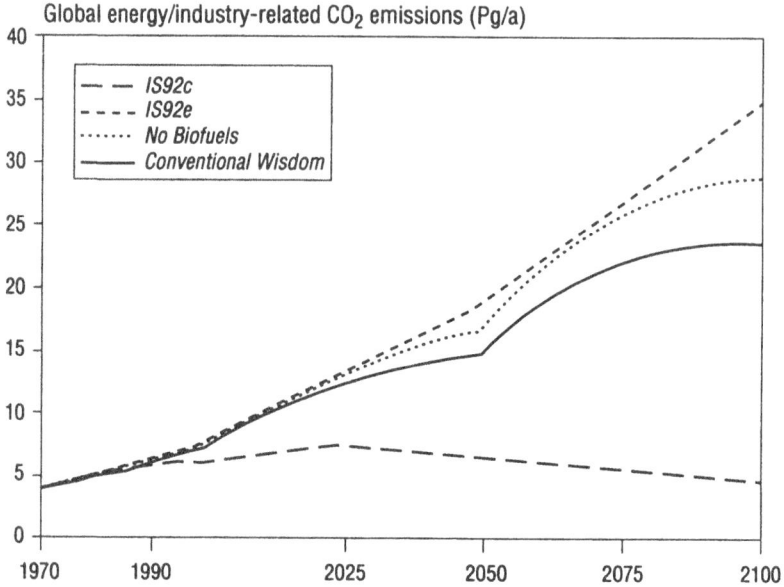

Figure 5: Global energy/industry-related CO_2 emissions for IMAGE 2.0 scenarios and IS92 scenarios. (Pg a^{-1}).

surpass OECD Europe's emissions. For this scenario, the main source of energy-related CO_2 emissions in Africa is power generation (as in OECD Europe), but the second most important source is the residential sector. Future emissions of CH_4 mainly stem from losses in the gas distribution system, while large increases in N_2O arise from increasing industrial activity. Most of the increase in NO_x emissions comes from power generation, while increases in VOC emissions can be attributed to increased industrial production for which no emission controls are assumed. The increase in CO emissions in the second half of the next century stems from energy consumed by industry.

The trend of CO_2 emissions from other developed regions (U.S., Canada, Eastern Europe, CIS, Oceania, and Japan) resembles OECD Europe's trend (although trends in the CIS are somewhat anomalous in showing a strong increase up to 2025) while regions in the developing world are closer to Africa's trends. The pattern of global CO_2 emissions shows that increasing emissions from developing regions prevail over the stabilization of emissions in OECD regions (Figure 5). As a result, global CO_2 emissions increase from 6.1 Pg C a^{-1} in 1990 to 24.0 Pg C a^{-1} in 2100. Emissions of other energy-related greenhouse gases show similar increases over the simulation period.

Estimates of emissions of CO_2 from the energy/industry system for this scenario fall within the range of the minimum (IS92c) and maximum (IS92e) scenarios of the IPCC (1992). This is not too surprising since some of the inputs of the intermediate scenario, IS92a, are also used as inputs to the Conventional Wisdom scenario. Emissions of the Conventional Wisdom scenario fall in the upper range of the IPCC scenarios because we compute a somewhat higher total primary energy consumption (1815 EJ a^{-1} in 2100) than the IS92a scenario (1453 EJ a^{-1} in 2100), and because different emission factors are assumed.

2.2.2 Results from Agriculture and Land Cover

Greenhouse gas emissions from land use / land cover are related to the type and extent of land cover and the intensity of different types of land use. In the IMAGE 2.0 model, shifts in land cover are computed from changes in demand for agricultural commodities and fuelwood which in turn stem from population and economic growth. The model also takes into account technological improvements in crop yield and animal productivity, as well as changes in the potential vegetation and crop productivity related to climate and current soil conditions, and the current location of different types of land cover (Zuidema *et al.*, 1994). Estimation of agricultural demand begins with computation of per capita intake of different commodities according to 8 categories of crops and 5 types of animal products. Most of the non-meat calories consumed by inhabitants of OECD Europe consist of temperate cereals, but per capita consumption of this commodity declines over the simulation period (Figure 6a). Overall consumption of both vegetable and meat products level off by 2025 because per capita consumption of most commodities is at or near saturation.

The leveling off of demand for animal products leads to stable numbers of most types of livestock after 2025. This leads to a stabilization of the amount of feed required for these animals, which together with a decrease in human consumption of cereals and other crops, leads to a leveling off or decline of total crop demands (Figure 6b). Less area is needed in Europe to grow crops as the total demand for crops levels off and crop yields increase per hectare because of more favorable future climate, increased fertilizer use, and technological crop improvements (Figures 7, and compare Figures 8a and b). As a consequence, some agricultural land reverts to its climate-potential land cover. In the case of Europe this is mostly deciduous forests (Figure 9a).

By contrast, increased income in Africa leads to an increase in per capita consumption of most agricultural commodities (Figure 6a). The larger per capita meat demand and increase in population leads to a large increase in the number of animals. Rising per capita food consumption is multiplied by increased population so that total crop demand in Africa rises steeply between 1990 to 2100 (Figure 6b). To satisfy the demand for crops and meat products, the model computes that extensive new areas will be needed for agriculture and grassland (Figure 7b, and compare Figures 8a and b), even though fertilizers and other inputs are assumed to enhance yield per hectare. The amount of agricultural land increases from 325 to 980 Mha between 1990 and 2100 (Figure 7). The expansion of agricultural land and grassland (Figure 9b) is mainly at the expense of savanna and tropical forested areas (Figure 8a). By the year 2060 the demand for grassland cannot be met since all savanna and forest have been cleared; subsequently animal densities increase on the available grasslands. We note that this scenario assumes that most food demand in Africa is met by growing crops within the region, rather than by importing food from Europe and other regions with excess agricultural land. Since this scenario assumes that per capita income increases substantially in Africa, it can be argued

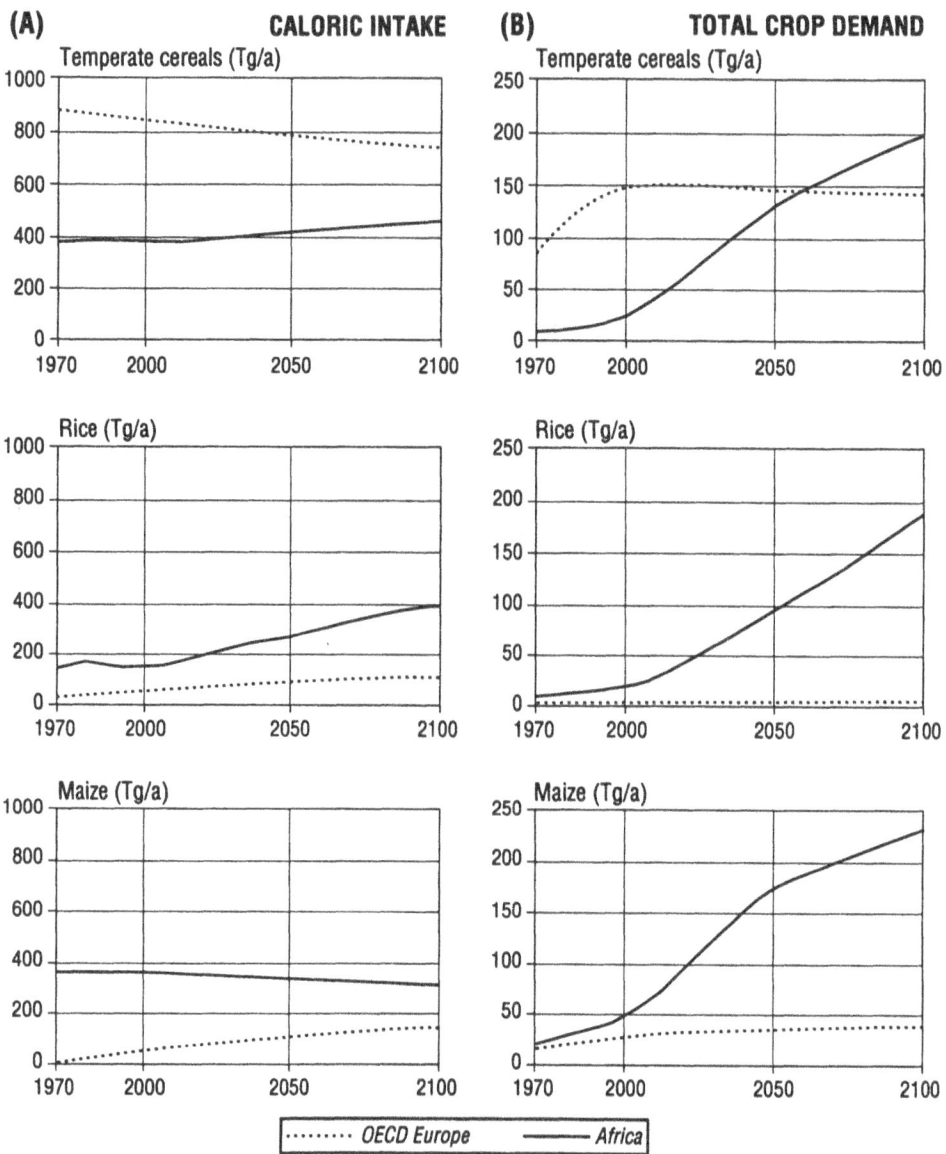

Figure 6: Trends in agricultural demand for OECD Europe and Africa: (a) Caloric intake (kcal cap⁻¹ day⁻¹), (b) Total crop demand (Tg a⁻¹).

that Africans may grow less of their own food in the future, and import more.

Globally, the amount of agricultural land expands to the end of next century, when it begins to level off (Figure 7c and Table 7). Not only Africa experiences a huge expansion of grassland and agricultural land, but also Asia, and the Middle East (Figure 8). In Asia this leads to extensive deforestation. At the same time, the trends discussed above for

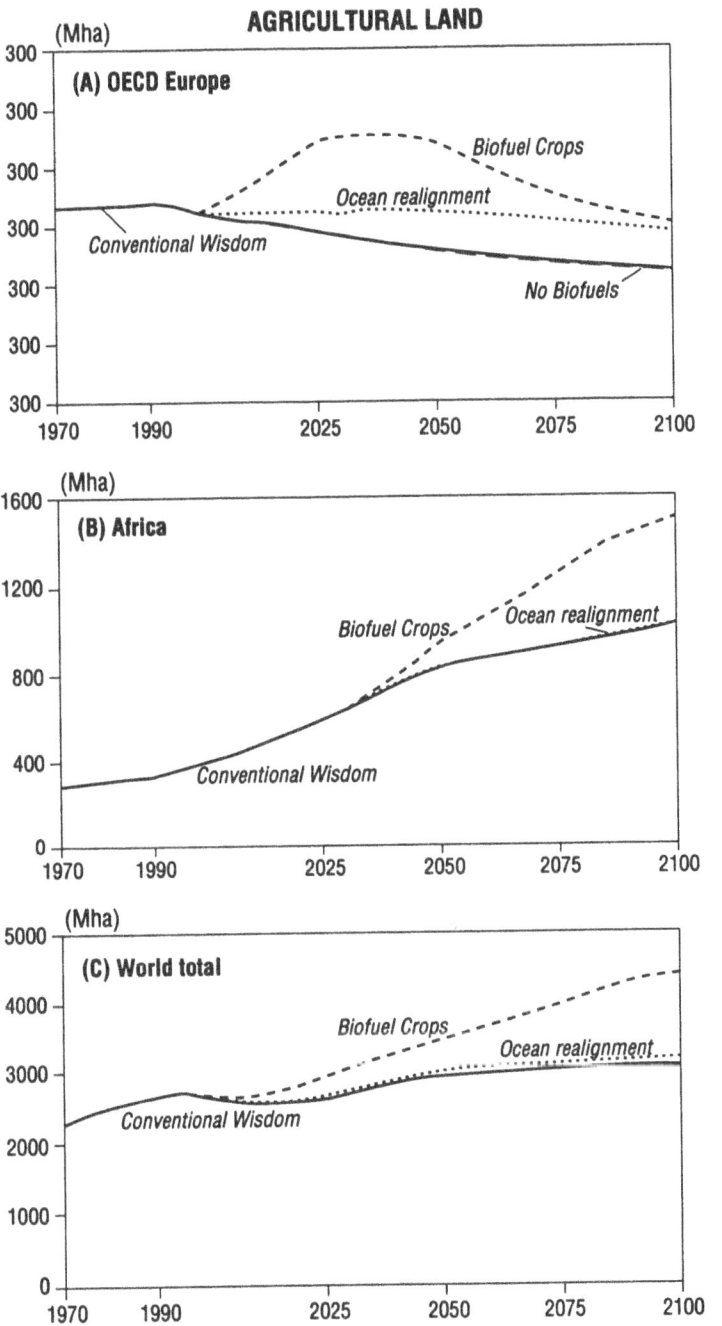

Figure 7: Trends in amount of agricultural land for Conventional Wisdom scenario: (a) OECD Europe, (b) Africa, (c) world.

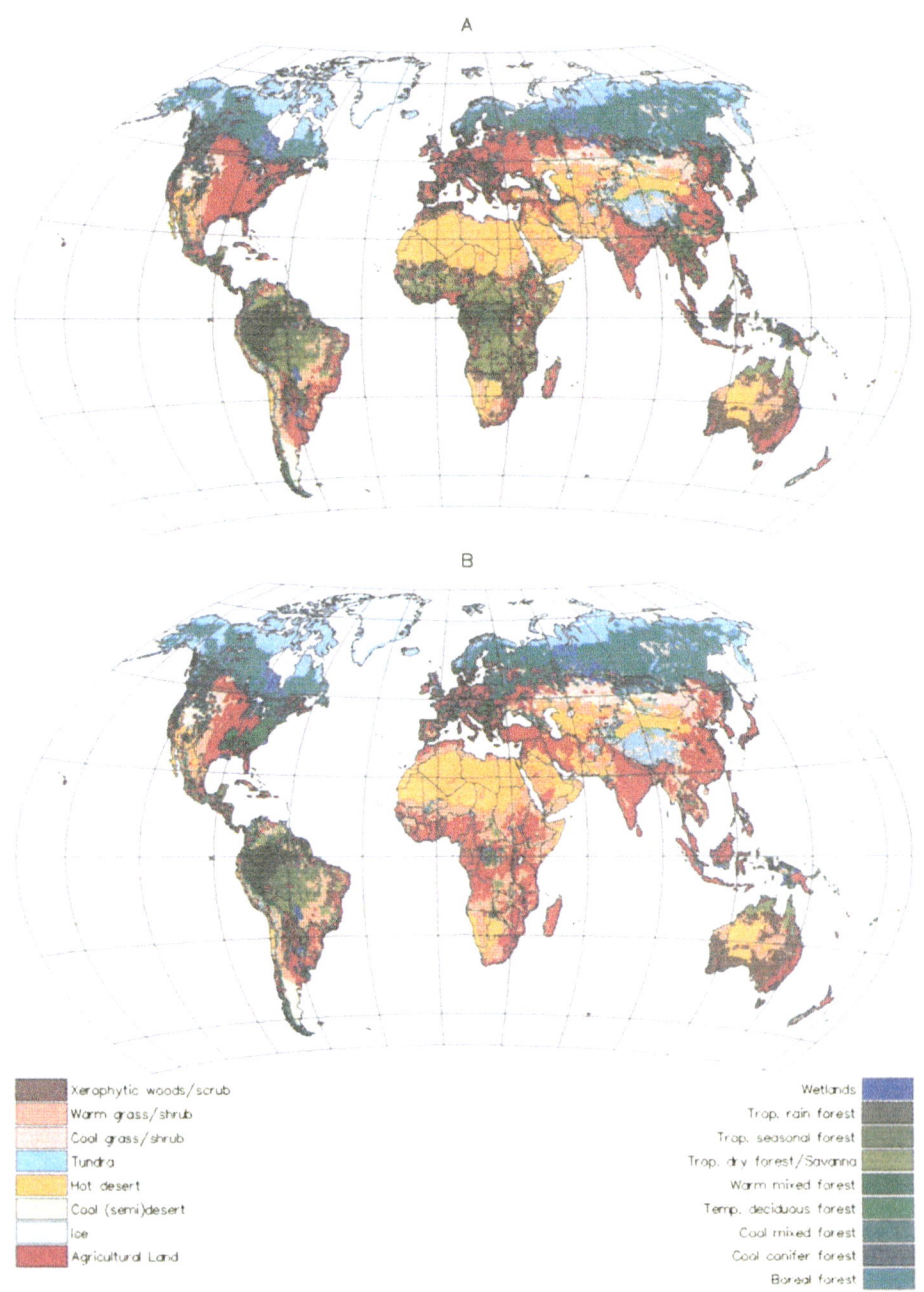

Figure 8: Global land cover: (a) 1990, (b) Conventional Wisdom scenario, 2050.

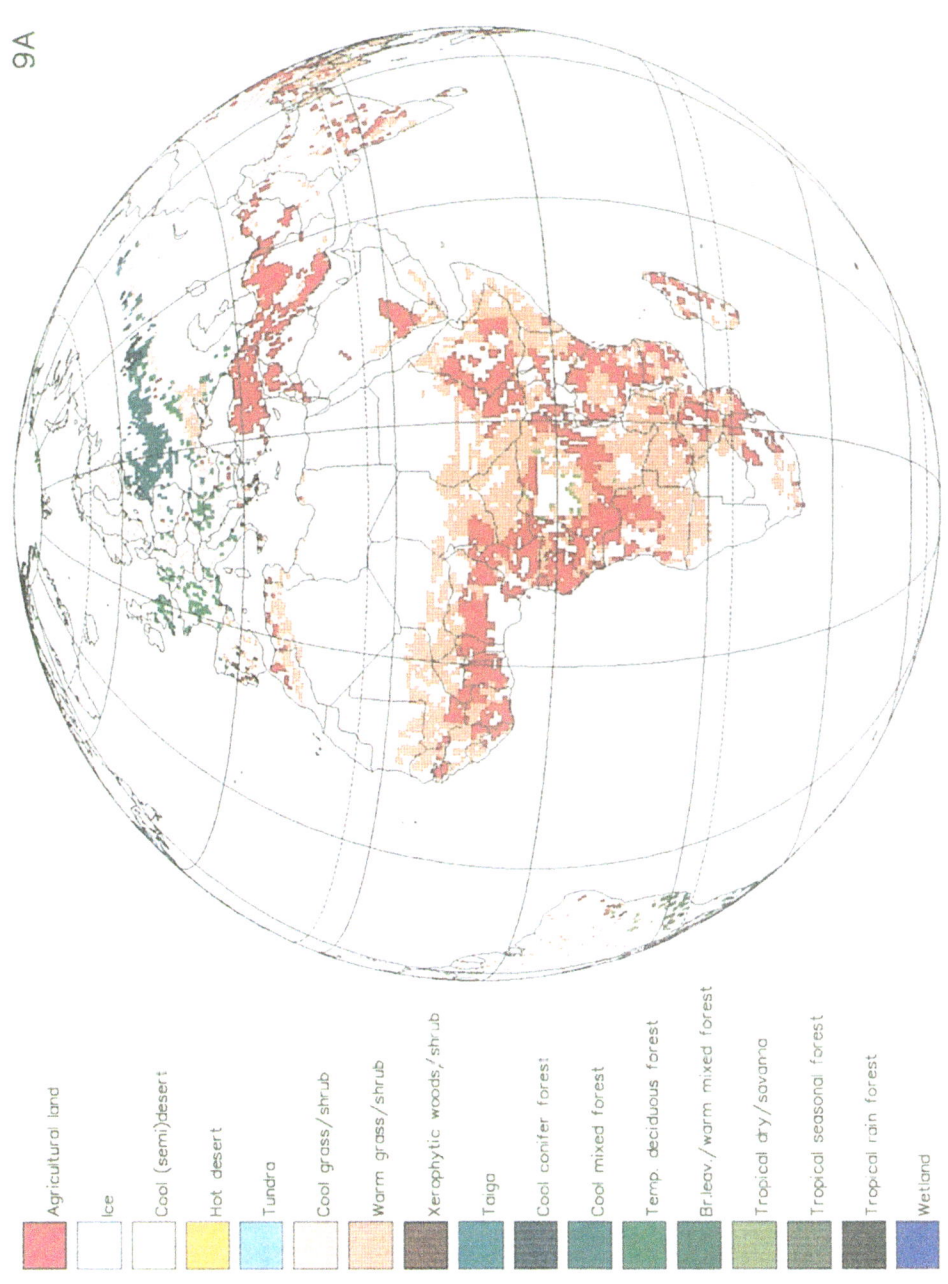

9A

Agricultural land
Ice
Cool (semi)desert
Hot desert
Tundra
Cool grass/shrub
Warm grass/shrub
Xerophytic woods/shrub
Taiga
Cool conifer forest
Cool mixed forest
Temp. deciduous forest
Brleav/warm mixed forest
Tropical dry/savanna
Tropical seasonal forest
Tropical rain forest
Wetland

Figure 9(a): Land cover changes, *Conventional Wisdom scenario, year 2050* compared to year *1990*.

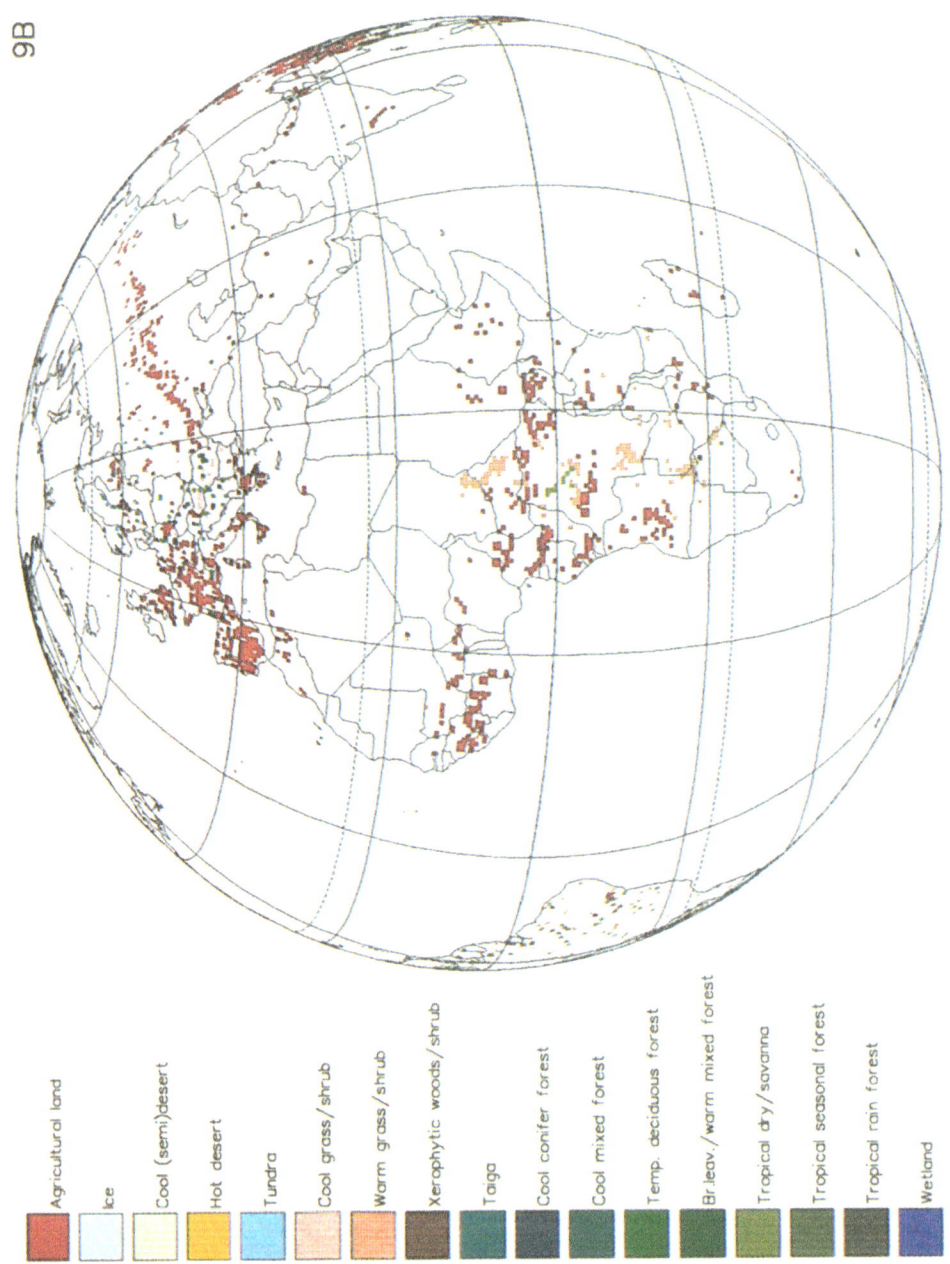

Figure 9(b): Land cover changes, *Biofuel Crops scenario, year 2050*, compared to *Conventional Wisdom scenario, year 2050*. These are land cover types which change *in addition* to the changes portrayed in Figure 9a.

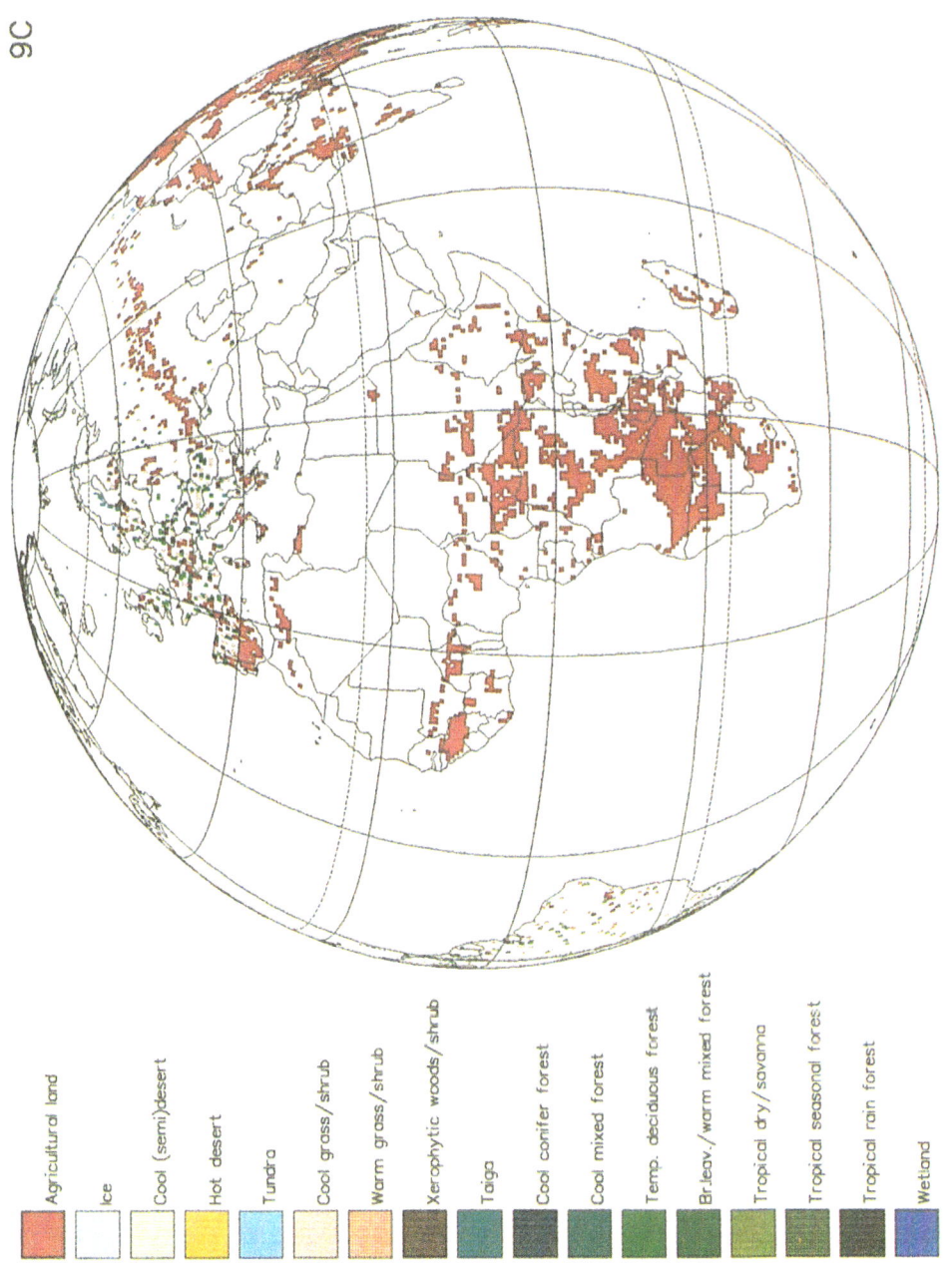

Figure 9(c): Land cover changes, *Biofuel Crops scenario, year 2100*, compared to *Conventional Wisdom scenario, year 2100*.

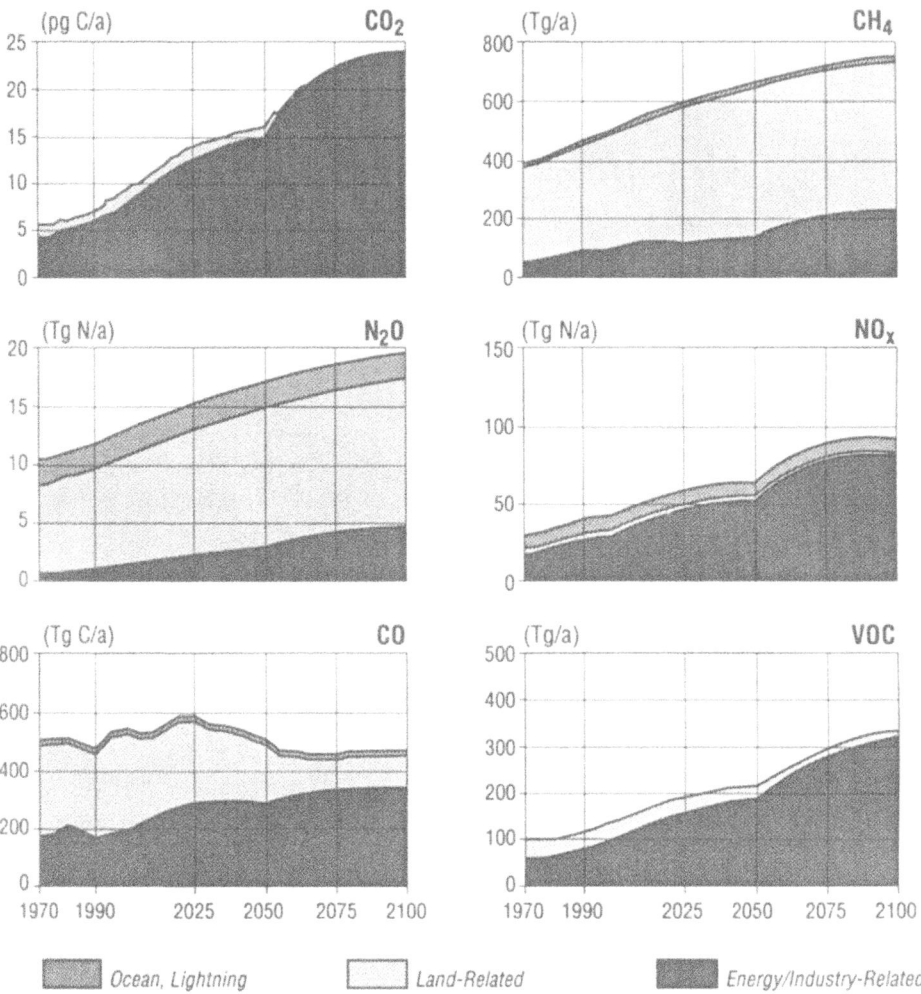

Figure 10: Global emissions of greenhouse gases and ozone precursors from natural, land-related, and energy/industry-related sources.

Europe also apply to North America and all of the CIS; for example, forests replace abandoned agricultural land in Siberia and on the East Coast of USA and Canada (Figure 8).

2.2.3 Comparing Emissions from Energy/Industry and Land.

Global emissions coming from natural sources, land activity, and energy/industry are compared in Figure 10. Land-related emissions of CO_2 are connected mostly with the process of deforestation. According to the Conventional Wisdom scenario, the global rate of deforestation dwindles in the second half of next century. This occurs because either forests have disappeared (for example, in Africa and India plus South Asia) or because stabilizing food demand and increasing crop yield slow the expansion of agricultural and range land (in the case of Latin America). A consequence of declining deforestation rates is that land-related CO_2 emissions are unimportant in the second half of next century (Figure 10). A related issue, the role of the biosphere in the carbon cycle, is taken up below.

Methane is emitted mostly from land-related sources (e.g. wetlands and animals). Since agricultural activity expands in the next century, CH_4 emissions continue to increase (Figure 10). Emissions of N_2O are also chiefly land-related (from natural soils in particular) and increase because of moisture and temperature feedbacks to soil.

Land-related emissions of ozone precursors (NO_x, CO, and VOC) stem from seasonal savanna burning, biomass burning following deforestation, and agricultural waste burning. Emissions from savanna and biomass burning will decrease steadily because of the declining rate of deforestation rates and dwindling extent of savanna lands. By comparison, the main sources of NO_x and VOC emissions are related to energy and industry rather than land use, and these sources continue to rise throughout the next century, especially because of expanded economic activity in developing regions (Figure 10). The net result is that total NO_x and VOC emissions continue to increase. The situation is similar for CO, except that land-related emissions make up a much larger part of total emissions. Consequently, the decrease in land-related emissions outweighs the increase in energy/industry emissions, and total emissions decline after 2025 (Figure 10). More information about land-related emissions for the Conventional Wisdom scenario is given by Kreileman and Bouwman (1994).

2.2.4 Results Concerning Total Carbon Fluxes

In this section we focus on the total flux of carbon because of its important contribution to radiative forcing. To compute the flux of carbon between the biosphere and atmosphere, the IMAGE 2.0 model takes into account plant primary productivity, soil respiration and burning of biomass from cleared land (Klein Goldewijk et al., 1994). In addition, the rate of soil respiration depends on moisture availability and temperature, and the rate of plant productivity depends on CO_2 air concentration, temperature, and the length of the growing season. For reference, the calculated 1990 carbon fluxes are presented in Figure 11a and discussed in Alcamo et al. (1994) and Klein Goldewijk et al. (1994). The year 2050 fluxes (Figure 11b) show increased uptake in the northern boreal forests due to increased temperature and CO_2, and in the USA and Europe also because of reversion of agricultural land to forest. At the same time, many areas of Africa and Asia become new sources of CO_2 owing to expansion of grassland and agricultural land and the resulting burning of biomass and increased soil respiration. In Latin America, forests continue to act as sinks because of assumed climate feedbacks (see Klein Goldewijk et al., 1994). The net effect of these different trends is that the biosphere acts as a larger and larger *sink* of atmospheric CO_2, increasing from 1.2 Pg C a^{-1} in 1990 to 8.2 Pg C a^{-1} in 2100 (Figure 12a). The ocean also behaves as a net sink of CO_2 according to IMAGE 2.0 calculations, increasing from 1.6 Pg C a^{-1} in 1990 to 4.2 Pg C a^{-1} in 2100. This is mostly due to higher atmospheric concentrations of CO_2.

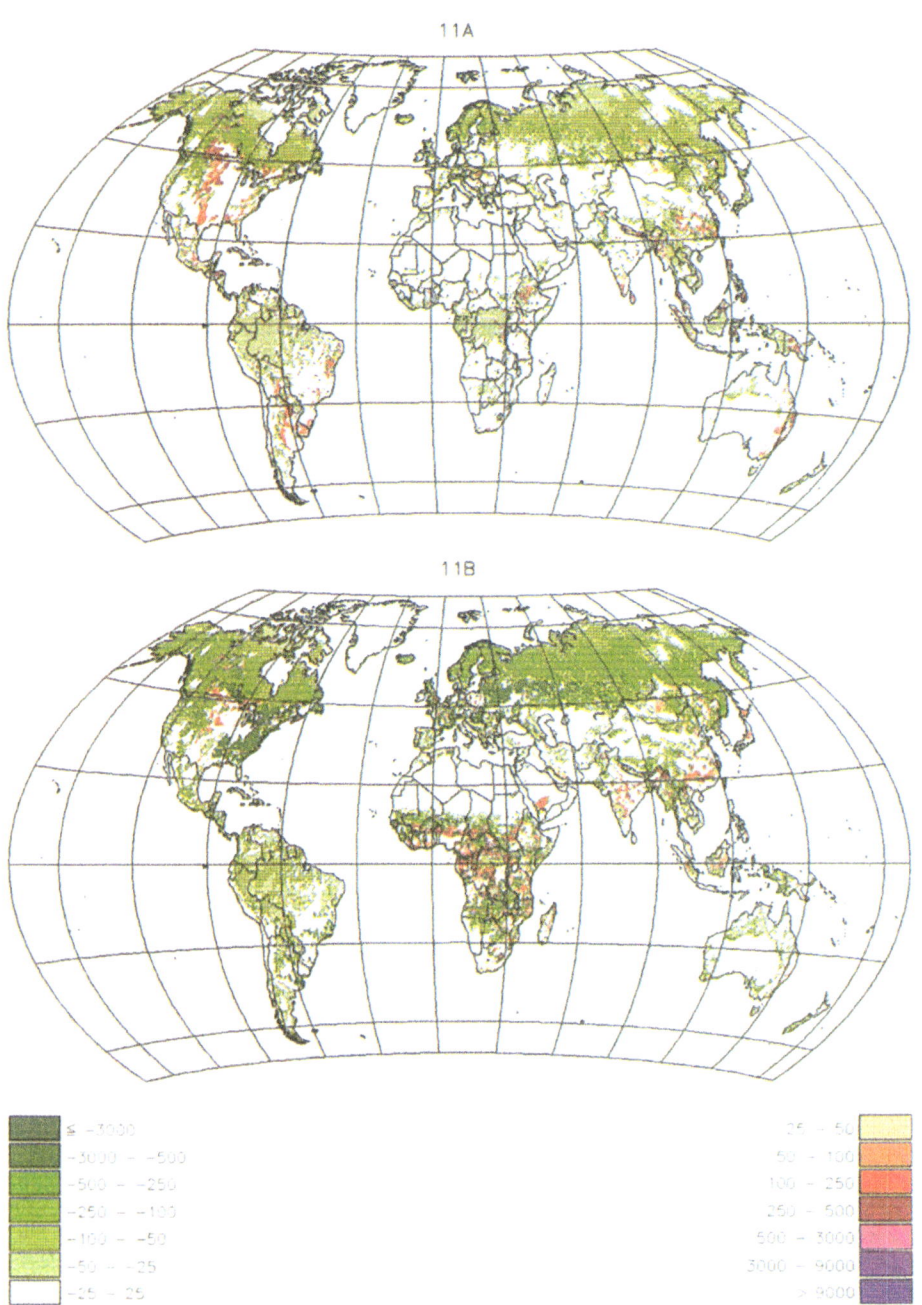

Figure 11a-b: Flux of C between the biosphere and atmosphere, (a) *year 1990*, (b) *Conventional Wisdom scenario, year 2050*. Negative numbers indicate a net biospheric sink of C (t C km^{-2} a^{-1}).

11C

11D

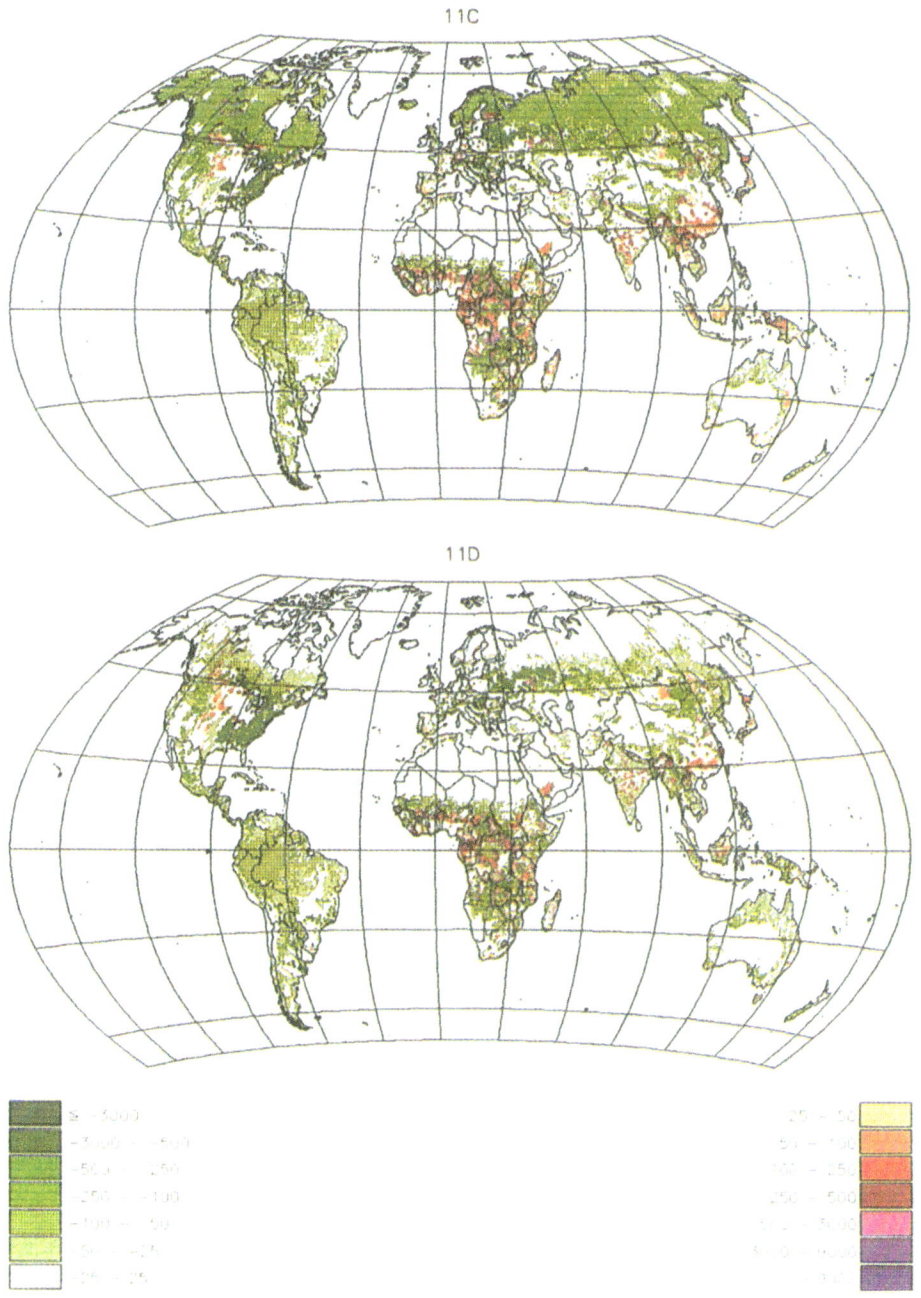

Figure 11c-d: Flux of C between the biosphere and atmosphere, (c) *Biofuel Crops scenario, year 2050*, (d) *Ocean Realignment scenario, year 2050*. Negative numbers indicate a net biospheric sink of C (t C km^{-2} a^{-1}).

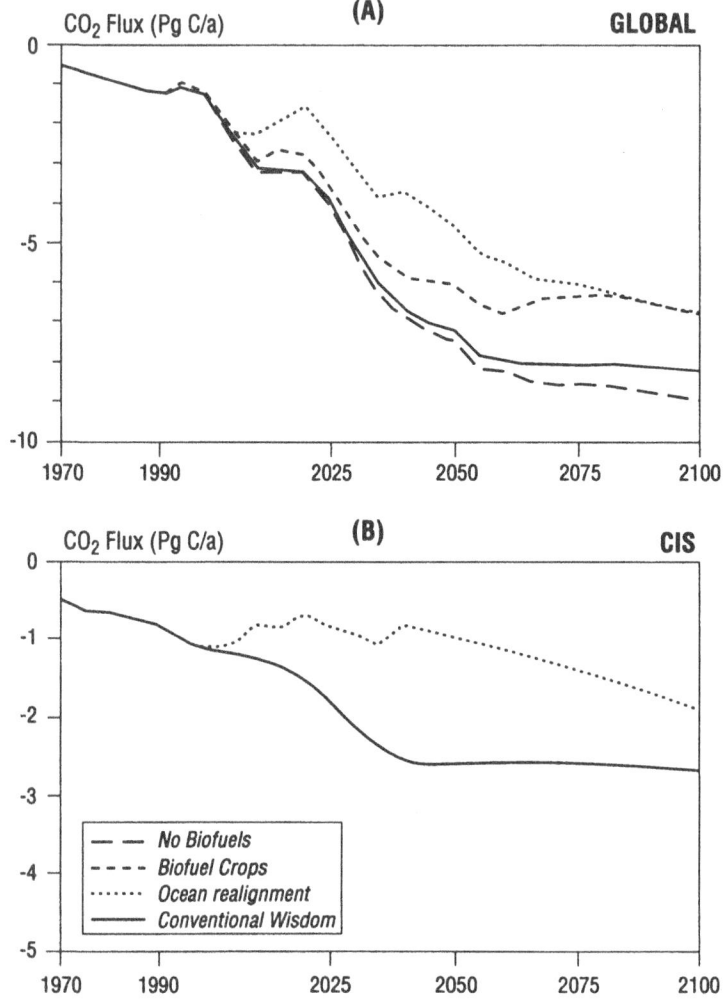

Figure 12: Net carbon flux between the biosphere and atmosphere: (a) Global, (b) CIS.

As noted above, the source of CO_2 from the world's energy/industrial system increases from 6.1 in 1990 to 24.0 Pg C a^{-1} between 1990 and 2100. The sum of these fluxes in 2100 result in a net build-up of 11.6 Pg C a^{-1} in the atmosphere.

2.2.5 Results from Atmosphere and Climate
Because of the above described changes in global carbon flux, CO_2 in the atmosphere increases from 358 ppm (slightly above measured concentrations) to 777 ppm between 1990 and 2100 (Figure 13a). At the same time, methane concentrations rise steadily from 1.7 ppm in 1990 to 2.5 ppm in 2050, but slowly decrease afterwards (Figure 13b). The initial increase is due to increasing emissions, as well as depletion of its atmospheric sink,

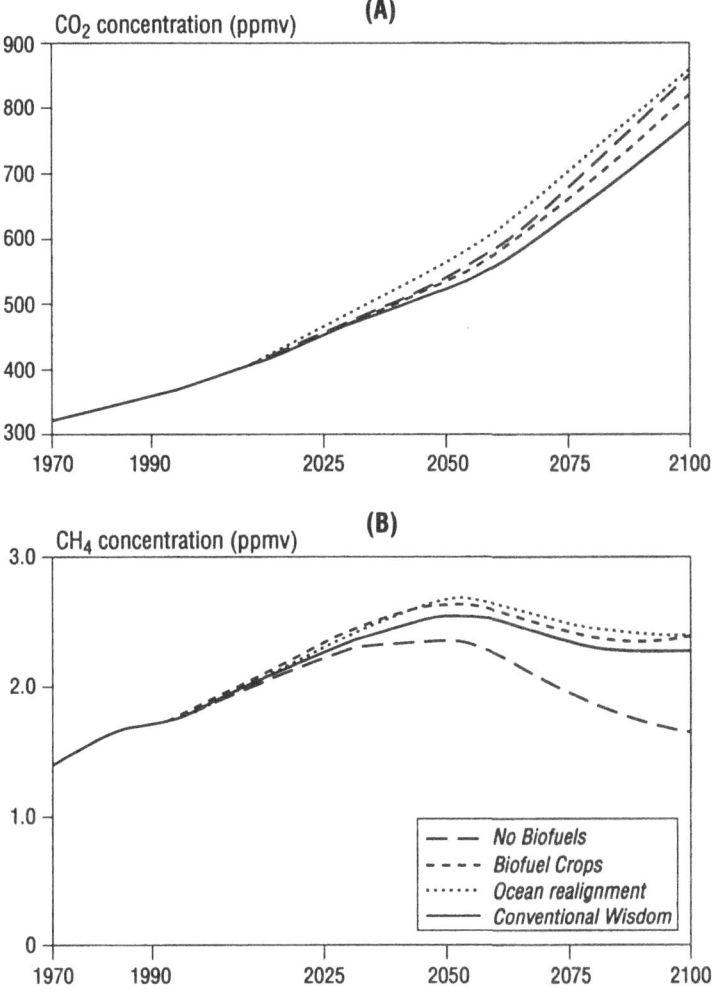

Figure 13: Atmospheric concentrations of greenhouse gases: (a) CO_2, (b) CH_4.

hydroxyl radical. The atmospheric concentration of hydroxyl recovers when CO emissions decline after 2025, and begins to serve as a more effective sink for CH_4. Consequently, CH_4 in the atmosphere slowly declines after 2050 although its global emissions continue to increase (Figure 10). The concentration of N_2O follows its upward emissions trend, increasing from 305 to 430 ppb between 1990 and 2100, while CFCs decline over this period due to the assumption of a partial compliance to the London Amendments of the Montreal Protocol, leading to a phase out of all CFCs by 2075.

The net effect of the changes in greenhouse gas concentrations is a substantial increase in surface temperature in both the Southern and Northern Hemispheres (Figure 14a). Model results show the zonal pattern of temperature change that is typical of more

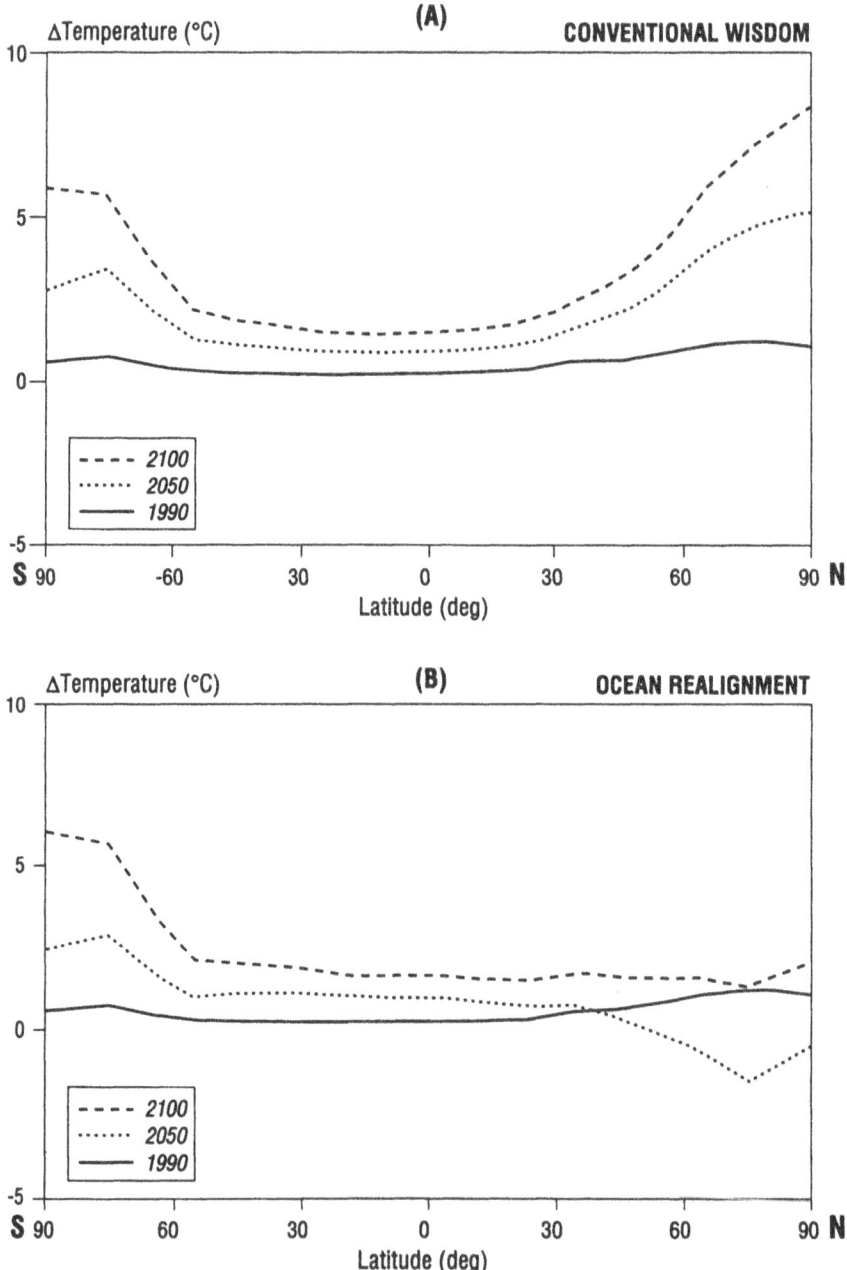

Figure 14: Latitudinal profile of change of atmospheric surface temperature relative to 1970: (a) Conventional Wisdom scenario, (b) Ocean Realignment scenario.

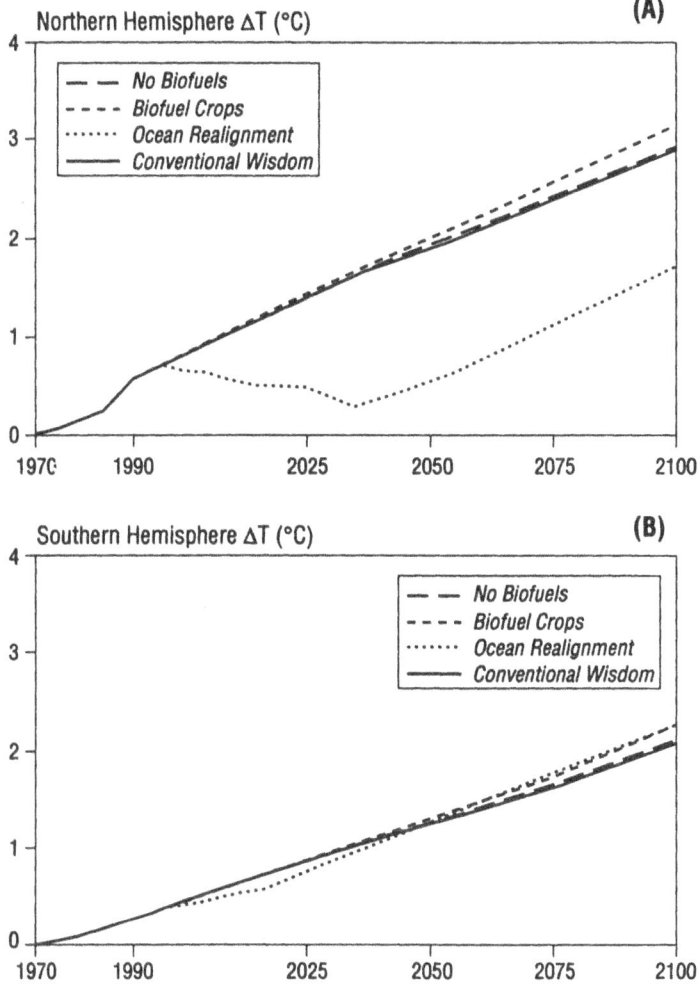

Figure 15: Trend of change of atmospheric surface temperature: (a) Northern Hemisphere, (b) Southern Hemisphere.

complicated general circulation models, namely a lower temperature increase in the tropics because of extensive heat flux from this region, with a substantially higher increase in temperate regions (Figure 14a). Around the equator, surface temperatures between 1970 and 2100 increase about 1.5°C, whereas in the middle northern latitudes the increase is around 3 to 5°C (Figure 14a). Temperature changes in the Southern Hemisphere are smaller than in the Northern Hemisphere because of the modifying effects of the South's larger surface area of ocean (Figures 14a and 15).

Since calculations of society, biosphere, and climate are coupled in IMAGE 2.0, increases in surface temperature affect potential crop productivity, productivity of existing vegetation, and the rates of emissions of different greenhouse gases (e.g. N_2O from soils).

These factors profoundly affect atmospheric levels of greenhouse gases, which in turn affect surface temperatures and other aspects of climate, which again feed back to potential crop productivity, productivity of vegetation, and so on, until the loop is closed in the society-biosphere-climate system for each model time step.

2.2.6 Synopsis of Conventional Wisdom results

The Conventional Wisdom scenario provides a comprehensive (but incomplete) picture of the chain of consequences following "conventional wisdom" driving forces. Energy use and industrial activity slows down in OECD regions, while it rapidly expands in other regions. CO_2 and other emissions related to energy/industry follow this pattern. At the same time the leveling off of population and a small marginal increase in per capita consumption leads to a stabilization of per capita food demand. This, and improved crop yields per hectare due to technology and improved climate, leads to a decline in total crop and animal demands. As a result, agricultural area shrinks, and the resulting forestation leads to greater uptake of CO_2 by the biosphere in the north.

In developing regions, large increases in population and GNP also increases energy consumption and industrial activity, leading to increased emissions. The demand for food also greatly increases, and results in expanding agricultural and grassland areas, depleting forests and savanna in Africa and Asia, and increasing flux of CO_2 between the biosphere and atmosphere. The net global effect of these trends is a rapidly increasing atmospheric level of most greenhouse gases, and significant increase in surface temperatures.

3. Biofuel Crops Scenario

3.1 ASSUMPTIONS "BIOFUEL CROPS" SCENARIO

The Conventional Wisdom scenario assumes that biofuels used in the world's energy system are derived from crop residues and other sources that do not require new cropland. Consequently, the use of biofuels does not lead to an increase in agricultural area. The assumptions of the "Biofuel Crops" scenario are the same as the Conventional Wisdom scenario except that it assumes that a large fraction of biofuels will be provided by energy crops grown on additional cropland. Specifically:

* Biofuels used in the transport sector of tropical or partly tropical regions (Latin America, Africa, and East Asia) are derived from sugar cane, and in other regions from maize.
* Of the biofuels used in other energy sectors, 60% are assumed to come from crop residues and other sources not requiring new cropland. The remaining 40% of the demand comes from elephant grass (*miscanthus sp.*) grown on new cropland. This is a C4 species with a potential maximum yield of about 50 t ha^{-1} a^{-1}, and a plausible worldwide range because of its relatively high potential productivity in both temperate and tropical climates (Figure 16). Elephant grass is also a good indicator species because its growing properties resemble that of eucalyptus, poplar, willows, and other

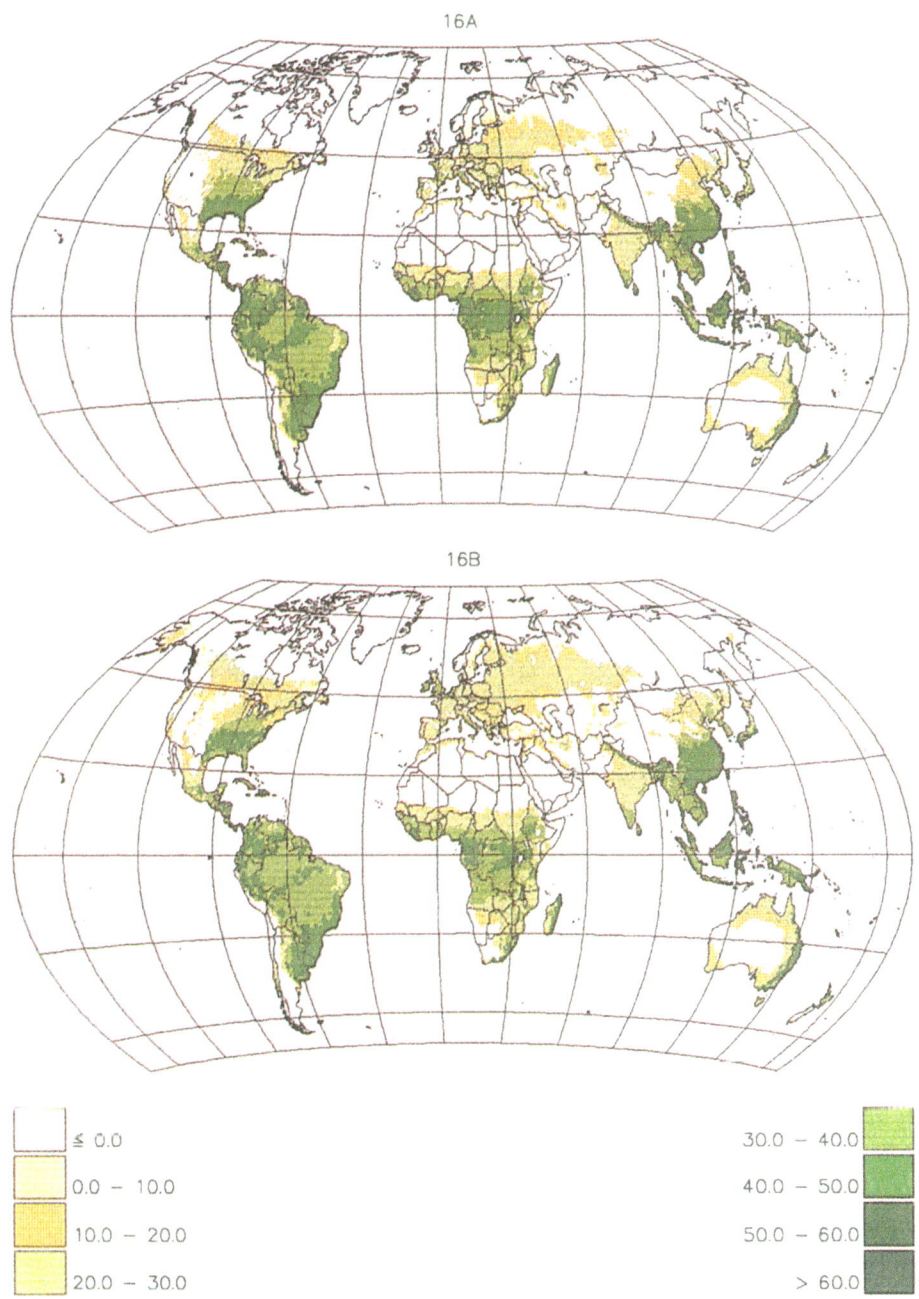

Figure 16: Potential productivity of elephant grass: (a) 1970, (b) Biofuels Crops scenario, 2050. (t dry weight ha^{-1} over the growing period).

crops that may be used for biofuels.

The preceding assumptions can be compared to those of Johansson *et al.* (1993a) who assume in the Renewables-Intensive Energy Scenario (RIGES) that about 55% of world biomass supplies in the year 2025 are provided by energy crops grown mainly on "excess" cropland in industrialized countries.

It is emphasized that land requirements for biofuels are likely to be overestimated because we assume *ad hoc* that a large percentage of biofuels must come from energy crops on new cropland. Some studies contend that large quantities of biofuels can be provided on marginal lands outside of prime cropland (see, for example, Swisher, 1993 and Woods and Hall, 1993). In addition, we only take into account three energy crops, whereas there are many other crops that might be better suited to a particular climate and soil and consequently have higher local yields. Moreover, we do not consider the costs of growing the assumed energy crops, which in reality should lead to efficient use of land.

All other assumptions in this scenario are the same as in the Conventional Wisdom scenario.

3.2   RESULTS OF THE "BIOFUEL CROPS" SCENARIO

The energy-related assumptions for this scenario are the same as the Conventional Wisdom scenario, so the computed energy-related emissions are also identical (Table 7).

IMAGE 2.0 takes into account the growing characteristics of the energy crops and estimates their change in potential productivity as climate changes. As an example, in this scenario the potential productivity of elephant grass increases between the years 1970 and 2100 especially in Canada and Russia due to changes in temperature and precipitation (Figure 16). Changes in productivity, together with the change in demand for these biofuels, leads to the allocation of additional agricultural land for these crops. The reader is referred to Leemans and van den Born (1994) and Zuidema *et al.* (1994) for descriptions of the methodology for calculations of potential crop productivity and land cover changes.

As to changes in land cover, we again focus on Europe and Africa as examples. Figure 9a depicts the new land cover types that appear between 1990 and 2050 according to the Conventional Wisdom scenario. As noted previously, large new areas of grassland and agricultural land are needed in Africa to satisfy increased food demand, whereas forested areas reappear in Europe because of stabilizing food demand and increased crop yield. Figure 9b shows the additional agricultural areas required in year 2050 for energy crops according to the Biofuels Crops scenario (over and above the new agricultural areas shown in Figure 9a).

In year 2050, 14% more agricultural area is required for biofuel crops in Africa and 71% more for OECD Europe as compared to the Conventional Wisdom scenario (Figures 7a and b). In OECD Europe this leads to deforestation instead of the forestation computed in the Conventional Wisdom scenario. The relatively small increase of agricultural land in Africa for year 2050 can be explained by the relatively small absolute amount of modern biofuels used in the first half of the 21st century. The use of biofuels in Africa becomes more substantial in the second half of the century, and this is reflected in the land cover simulation which indicates large new areas of agricultural land necessary for energy crops (Figure 7b, and compare 9b and 9c). By comparison, the use of biofuels declines in the second half of the century in Europe, and the yield per hectare of energy crops increases because of technology. Consequently, less area is required for biofuels in 2100 than in 2050 (Figure 7a, and compare 9b and 9c).

Not only will Africa require substantial new agricultural areas for biofuels in this scenario, but globally a 20% increase is required for 2050 and 45% for 2100 (Figure 7c, Table 7). Since agriculture replaces forests and other land cover types capable of assimilating more carbon, there is a substantial reduction in the carbon assimilated by the biosphere (Figures 11c and 12a). What follows is an increase in atmospheric CO_2 in year 2100 from 777 ppm in the Conventional Wisdom scenario to 821 ppm in this scenario (Figure 13a). This results in a slight increase in temperature for the Northern and Southern Hemispheres, as compared to the Conventional Wisdom scenario (Figure 15, Table 7).

Synopsis of Results of Biofuel Crops Scenario
Summing up, following the assumptions of this scenario, the need for biofuels may take up large amounts of new agricultural land in the world. A consequence of expansion of agricultural land is a reduction of the CO_2 assimilated by the biosphere, a small increase in atmospheric CO_2 as compared to the Conventional Wisdom scenario, and somewhat larger global warming. However, we reiterate that the energy cropland requirements assumed in this scenario may be exaggerated since it may be possible to provide a much larger fraction of biofuels from agricultural wastes, plantations on marginal land, and other non-cropland sources (see, for example, Woods and Hall, 1993; Johansson *et al.* (1993a and b). Moreover, the requirements for land would not have been as large, nor the reduction in C uptake by the biosphere so great, if energy crops/trees had been selected that were better suited to local climate and soil. Perhaps this scenario provides a useful estimate of the upper range of land requirements of biofuels.

4. "No Biofuels" Scenario

4.1 Assumptions of "No Biofuels" Scenario

As described earlier, the Conventional Wisdom scenario assumes that biofuels will make a significant contribution to the world's future energy consumption. Indeed, this is the conventional wisdom of current energy studies (see, for example, World Energy Council, 1993). In the "No Biofuels" scenario, we investigate the sensitivity of the climate system to biofuel use. For this scenario, we remove the biofuels specified in the Conventional Wisdom scenario (other than fuelwood). We further assume that oil will be used if biofuels are not available. This is a fairly good assumption for the transport sector where oil and other liquid fuels are the major energy carriers. For other sectors, however, it is rather difficult to decide on the fuel that would be used in place of biofuels. For example, coal can be used as well as oil in power generation. Consequently, the use of oil is simply a default assumption for this sensitivity study. We note that the total supply of oil, required by this scenario over the next century does not exhaust the presently known oil reserves.

All other assumptions are the same as the Conventional Wisdom scenario.

4.2   RESULTS OF THE "NO BIOFUELS" SCENARIO

Removing biofuels from the energy system results in an increase in CO_2 emissions of 1.8 Pg C a^{-1} in year 2050, and 5.2 Pg C a^{-1} in year 2100 over the Conventional Wisdom scenario (Table 7). This is because biofuel combustion is assumed to have zero net C emissions (because an equal amount of CO_2 is assumed to be assimilated by regrown biomass). The difference is relatively small in 2050 as compared to 2100 because the Conventional Wisdom scenario assumes that biofuel use will increase greatly in Africa and Asia in the second half of next century. Following the rise in emissions, atmospheric concentrations of CO_2 are also larger in this scenario as compared to the Conventional Wisdom scenario; the concentration is 17 ppm greater in 2050, and 80 ppm in 2100.

Methane emissions and atmospheric concentrations, on the other hand, decrease relative to the Conventional Wisdom scenario because unit emissions of CH_4 from biofuels are higher than from oil (Figure 13). This is a crucial result and depends on the implicit assumption of how biofuels are burned. If they are gasified, for example, most of the CH_4 would be utilized rather than emitted to the atmosphere. However, scenario assumptions imply that it is combusted without gasification. Lower emissions of CH_4 lead to lower concentrations of this substance in the atmosphere. This also applies to CO and NO_x, two other important precursors of tropospheric ozone. The lower concentrations of O_3 precursors leads to lower concentrations of tropospheric ozone.

The net effect of these changes on radiative forcing are important. The increase in CO_2 concentration tends to increase radiative forcing, while the decrease in CH_4 and tropospheric O_3 tends to decrease it. The net effect is a very small difference in the change of surface temperature between this and the Conventional Wisdom scenario (Figure 15 and Table 7).

Synopsis of Results of No Biofuels Scenario
Summing up, emissions from biofuels result in lower atmospheric levels of CO_2, but higher levels of CH_4 and O_3. The net result is a small difference in climate change between scenarios with and without biofuels. These results point out the importance of taking into account all emissions as well as the composition of the atmosphere. However, as noted above, these conclusions also depend on assumptions about biomass utilization.

These results also raise interesting questions -- How sensitive are scenario results to the assumed mix of fuels that are used instead of biofuels? How does the effect of biofuels on tropospheric ozone depend on the background atmospheric concentration of ozone precursors? What influence does the timing of introduction of biofuels have on the rate of climate change?

5. "Ocean Realignment" Scenario

5.1   ASSUMPTIONS OF "OCEAN REALIGNMENT" SCENARIO

The previous scenarios have examined the effect of human driving forces on the society-biosphere-climate system. In this scenario we examine the influence of an unexpected change of natural driving forces on the system, namely, the slowing down of ocean circulation and reduction in the downwelling rates in the North Atlantic and Antarctic Circumpolar Ocean. We base our assumptions on the model experiments of Mikolajewicz *et al.* (1990) who examined the effects of an increase of surface air temperatures due to a doubling of CO_2 on the thermohaline circulation of the ocean. We adapted their results by modifying the fixed two-dimensional circulation scheme contained in the IMAGE 2.0 ocean model so that deep water formation assumed in the model: (1) decreases to 30% of its original volume in the North Atlantic in the period 1990 to 2040, and (2) reduces to 55 % of its original volume in the Antarctic Circumpolar Ocean over the same period. Afterwards the circulation is taken to remain constant.

We emphasize that this scenario is meant to illustrate a low probability, "surprise" occurrence. Indeed, the rapidness of the changes in ocean circulation found by Mikolojewicz *et al.* are likely to be due to the nature of their modeling exercise, namely, that they omitted the effect of ocean feedbacks on the atmospheric energy balance. The IPCC notes that such rapid changes in ocean circulation are not computed by models that take ocean feedbacks into account (IPCC, 1992).

Other assumptions are the same as in the Conventional Wisdom scenario.

5.2   RESULTS OF THE "OCEAN REALIGNMENT" SCENARIO

The decrease of downwelling and slower ocean circulation has the important effect of reducing the northward transport of heat in the Atlantic (see de Haan *et al.*, 1994 for more details). What follows is a net cooling north of 40^0 N up to the year 2050 (Figures 14b and 15a). This scenario has less of an effect on the Southern Hemisphere because ocean circulation is not as significantly modified there (Figures 14b and 15b).

Cooling in the Northern Hemisphere has an important influence on the global build-up of greenhouse gases. For example, carbon uptake is especially reduced in northern boreal forests because of their extensive area, the cooling they are exposed to, and the assumed relationship between net primary productivity and temperature (Figure 11d). (See Klein Goldewijk *et al.*, 1994 for a description of this relationship.) This effect is particularly pronounced in CIS where C uptake in year 2050 decreases from 2.6 Pg C a^{-1} to 1.0 Pg C a^{-1} between the Conventional Wisdom and Ocean Realignment scenarios (Figure 12b). The global biospheric uptake of the two scenarios differs by about 2.7 Pg C a^{-1} in year 2050 (Figure 12a). With reduced C uptake, atmospheric CO_2 reaches 90 ppm higher than the Conventional Wisdom scenario in year 2100 (Figure 13a).

Another effect of the cooler temperatures in this scenario is a lower mixing ratio of water vapor in the atmosphere. This results in lower production of hydroxyl radical, which is the main atmospheric sink of CH_4. Consequently, CH_4 concentrations are higher in this scenario than in the Conventional Wisdom scenario (Figure 13b). Although higher levels of greenhouse gases increase radiative forcing, this is compensated by the reduced transport of heat from the tropics. Nevertheless, the trend of declining surface temperatures in the Northern Hemisphere is reversed after 2035 because of the increase in radiative forcing (Figures 14b and 15a). However, by the end of the century the temperature gain in the middle latitudes is only 1.5 ^0C as compared to 3 to 5^0C in the Conventional Wisdom scenario (Figures 14a,b).

Cooler temperatures also reduce potential crop productivity which leads to larger land requirements for the same amount of agricultural demand. This is especially important in the northern temperate regions such as the CIS where the area of agricultural land in year 2100 increases from 137 Mha in the Conventional Wisdom scenario to 164 Mha in this scenario. In Eastern Europe, agricultural area in 2100 increases from 61 to 70 Mha, and in OECD Europe from 111 to 146 Mha (Figure 7). The larger area of agricultural land comes partly at the expense of forest land; in year 2100 there is 60 Mha less global forest area in the Ocean Realignment scenario than in the Conventional Wisdom scenario (see Table 7).

Synopsis of the Ocean Realignment Scenario
A change in the circulation of the ocean can result in a temporary cooling rather than warming of the Northern Hemisphere. This cooling would reduce uptake of carbon in the northern boreal forests and other areas, and leads to a greater build-up of CO_2 in the atmosphere than in the Conventional Wisdom scenario. The build-up of CO_2 and other gases will eventually reverse the cooling trend, although temperatures will remain substantially cooler in the Northern Hemisphere as compared to the Conventional Wisdom scenario up to 2100 and beyond. One outcome of the cooler temperatures is the need for more land to produce the same amount of food in the North (assuming no change in trade patterns), and subsequently a lower rate of forestation of abandoned land. We repeat, however, that this is a low probability scenario and is most useful in illustrating the large differences between the Conventional Wisdom scenario, and an unexpected "surprise" scenario. These differences underscore the need to test the robustness of climate policies against different kinds of uncertainties and "surprises" (Clark, 1986).

6. Discussion and Conclusions

Although the foregoing scenarios are fairly comprehensive, they omit many factors that could alter their outcome and conclusions, and that require further study. For example, the land cover simulations assume that the expansion of agricultural land and grassland will lead to the elimination of all forest and savanna areas in some regions without considering that society's intervention will probably prevent it from disappearing altogether. Related to this, the IMAGE 2.0 model does not take into account land costs or other economic factors that would slow the depletion of land and energy resources. Also of relevance to assessing agriculture and other impacts, the simulation does not include extreme climate events, such as extended cold or dry periods, which could have long term effects on agriculture if frequently occurring.

As a general comment, because of the omissions of the model, it is best to view the scenarios in this paper as a type of sensitivity study that can provide insight into couplings and linkages in the society-biosphere-climate system. With this in mind, we review some of the scenarios' main conclusions:

6.1 CONVENTIONAL WISDOM SCENARIO

- The slowing of population and economic growth in *developed* regions, together with increased conservation, leads to stabilization of energy- and industry-related emissions. This also results in shrinking agricultural area, and forestation of abandoned cropland. This forestation, and the effect of climate feedbacks on vegetation, enhances the uptake of CO_2 to the northern terrestrial biosphere.
- Rapid population and economic growth in *developing regions* increases emissions from energy and industry, and leads to expansion of agricultural land and initially grassland at the expense of forests and savannas. Later, the expansion of agricultural land encroaches on grasslands which implies increasing animal densities on the remaining grasslands and perhaps accelerated land degradation. Rapidly changing land cover also increases greenhouse gas flux from the terrestrial environment in the first half of the next century.
- Globally, emissions of all greenhouse gas emissions and ozone precursors increase, with the exception of CFCs and CO. Emissions of CFCs decrease because of assumed partial compliance with the Montreal Protocol, and CO emissions because of diminishment of its land-based sources -- deforestation and savanna burning.
- The increase of CO_2 emissions from developing regions outweighs the enhanced carbon sink in the Northern Hemisphere. Consequently, there is a considerable increase of CO_2 in the atmosphere (along with most other greenhouse gases), and a significant increase in surface temperatures.

6.2 BIOFUEL CROPS SCENARIO

- According to the assumptions of this scenario, a substantial amount of new agricultural area will be required to deliver biofuels in the next century.
- Trends of land requirements for biofuels are quite different in different regions. In Africa, land requirements are fairly small in 2050 but are much greater in 2100, whereas in Europe, land requirements are proportionately larger in 2050, but decrease towards 2100. These results arise from different regional trends in energy consumption, market share of biofuels, crop yield improvements, and other factors.
- One consequence of the biofuel assumptions in this scenario is that the expanded agricultural areas take up less C than the land cover they replace. Therefore, overall C uptake of the biosphere decreases, resulting in a small increase in atmospheric CO_2 as compared to the Conventional Wisdom scenario.

6.3 NO BIOFUELS SCENARIO

- Not using biofuels in the global energy system results in higher atmospheric concentrations of CO_2, but lower concentrations of CH_4 and O_3.
- The net result of changes in atmospheric composition is that there is little net difference in radiative forcing between the Conventional Wisdom scenario which contains biofuels and the No Biofuels scenario.
- The preceding results depend on assumptions about how biofuels are combusted. For example, the build-up of CH_4, and perhaps tropospheric O_3 associated with biofuels use, could be averted by gasifying biofuels.

6.4 OVERALL CONCLUSIONS ABOUT BIOFUELS

Based on the two biofuel scenarios we conclude:
- In order to maximize the benefits (and minimize the costs) of biofuels on the global environment, it is important to give attention to the type of biomass species, to the processes by which they are combusted and delivered (utility boilers, gasification units, etc.), and to their associated emission factors.
- To properly assess the impacts of biofuels, it is also important to analyze their influence on the complete range of greenhouse gases.
- Land requirements for biofuels can have important side impacts, e.g. on the uptake of carbon by the terrestrial biosphere, that should be taken into account in impact assessments.
- We emphasize that these scenarios probably overestimate land requirements for biofuels because: (i) only three possible energy crops were taken into account, whereas there are many other crops that are likely to be better suited to a particular local climate and soil, (ii) it may be possible to provide a much larger fraction of biofuels from agricultural wastes, plantations on marginal land, and other non-cropland sources than was assumed in these scenarios.

The authors note here that the "top-down" analysis of biofuels as presented in this paper cannot substitute for "bottom-up" analyses of biofuels which focus on selecting the optimum energy crop for different locations (see, e.g., Swisher, 1993; Woods and Hall, 1993). Many factors having to do with local crop suitability, cultural values, and institutional factors can be included in a bottom-up analysis in order to optimize the production/delivery of biofuels and minimize their impacts. But only some of these factors have been incorporated in our approach. On the other hand, it is difficult for a bottom-up analysis to link local/regional biofuel development with the global biosphere/climate system as is done in IMAGE 2.0. It is our view that the two approaches are complementary by providing different and useful types of information for evaluating biofuel development.

6.5   OCEAN REALIGNMENT SCENARIO

- This scenario illustrates that an unexpected, low probability natural event -- in this case a slowing down of the Atlantic's circulation -- can both enhance the build-up of greenhouse gases, and at the same time cool surface air temperatures in the Northern Hemisphere.
- An increase in surface temperature in the Northern Hemisphere is not avoided, only postponed.
- One side effect of the hemispheric cooling is the lowering of crop productivity in the Northern Hemisphere relative to the Conventional Wisdom scenario, and the larger requirements for agricultural land.
- The large differences between this scenario and the Conventional Wisdom scenario emphasize the importance of testing the robustness of climate policies against different kinds of uncertainties and "surprises" (Clark, 1986).

In summing up, the Conventional Wisdom and biofuel scenarios illustrate the many cross impacts that can ensue from human-related driving forces. The Ocean Realignment scenario makes the same point about unexpected changes in natural driving forces. Both emphasize the importance of simulating as comprehensively as possible the complete chain of processes in the global society-biosphere-climate system, both in time and in space.

Acknowledgements

The authors acknowledge the key contributions of Eric Kreileman, Maarten Krol, and Gé Zuidema to the development and applications of the IMAGE 2.0 model, as well as to the completion of this paper. We also recognize Rob Swart and Fred Langeweg for their support of the development of IMAGE 2.0. The authors thank Coos Battjes and Jelle van Minnen for their comments on this manuscript. This research has been supported by MAP Grant 481507 of the Dutch Ministry of Environment, Housing, and Physical Planning, and the following grants of the Dutch National Research Programme on Global Air Pollution and Climate Change: NOP Numbers 851037, 851040, 851042, 851044, and 851045. Some results in this paper are contained in "Integrated Modeling as Input to Assessment of Climate Change Mitigation and its Impacts" presented at the IIASA International Workshop on "Integrative Assessment of Mitigation, Impacts, and Adaption to Climate Change", 13-15 October, 1993.

References

Addiscot, T.M., Whitemore, A.P. and D.S. Powlson: 1991, *Farming, Fertilizers and the Nitrate Problem*, CAB International, 170 pp.

Alcamo, J., G.J.J. Kreileman, M.S. Krol and G. Zuidema: 1994, Modeling the global society-biosphere-climate system, Part 1. Model description and testing, *Wat. Air Soil Pollut.*, **76** (this volume).

Clark, W.C.: 1986, Sustainable development of the biosphere: themes for a research program, in: W.C. Clark and R.E. Munn (eds), *Sustainable Development of the Biosphere*, Cambridge University Press, 491 pp.

Edmonds, J. and J. Reilly: 1985, *Global Energy: Assessing the Future*, Oxford University Press.

de Haan, B.J., M. Jonas, O. Klepper, J. Krabec, M.S. Krol, and K. Olendrzynski: 1994, An atmosphere-ocean model for integrated assessment of global change, *Wat. Air Soil Pollut.*, **76** (this volume).

Grübler, A., Nakicenovic, N., 1991, *Evolution of Transport Systems: Past and Future*, IIASA Research Report 91-8, IIASA, Laxenburg, Austria.

IPCC (Intergovernmental Panel on Climate Change): 1990a, *The IPCC Impacts Assessment*, WMO/UNEP. Australian Government Publishing Service: Canberra.

IPCC (Intergovernmental Panel on Climate Change): 1990b, *The IPCC Response Strategies*, WMO/UNEP.

IPCC (Intergovernmental Panel on Climate Change): 1990c, *The IPCC Scientific Assessment*, WMO/UNEP, Cambridge University Press.

IPCC (Intergovernmental Panel on Climate Change): 1992, J.T. Houghton, B.A. Callendar and S.K. Varney (eds), *Climate Change 1992. The Supplementary Report to the IPCC Scientific Assessment*, Cambridge Univ. Press.

Johansson, T.B., H. Kelly, A.K.N. Reddy, and R.H. Williams: 1993a. Renewable fuels and electricity for a growing World economy, in: Johansson, T.B., H. Kelly, A.K.N. Reddy, and R.H. Williams (eds), *Renewable Energy*, Island Press: Washington D.C. pp. 1-72.

Johansson, T.B., H. Kelly, A.K.N. Reddy, and R.H. Williams,: 1993b. A renewables-intensive global energy scenario, in: Johansson, T.B., H. Kelly, A.K.N. Reddy, and R.H. Williams (eds), *Renewable Energy*, Island Press: Washington D.C., pp. 1071-1142.

Klein Goldewijk, K., J.G. van Minnen, G.J.J. Kreileman, M. Vloedbeld, and R. Leemans: 1994, Simulating the C flux between the terrestrial environment and the atmosphere, *Wat. Air Soil Pollut.*, **76** (this volume).

Kreileman, G.J.J. and A.F. Bouwman: 1994, Computing land use emissions of greenhouse gases, *Wat. Air Soil Pollut.*, **76** (this volume).

Krol, M.S. and H.J. van der Woerd: 1994, Atmospheric composition calculations for evaluation of climate scenarios, *Wat. Air Soil Pollut.*, **76** (this volume).

Leemans, R. and G.J. van den Born: 1994, Determining the potential distribution of vegetation, crops and agricultural productivity, *Wat. Air Soil Pollut.*, **75** (this volume).

Maddison, A.: 1991, *Dynamic Forces in Capitalist Development - A Long Run Comparative View*, Oxford University Press, 109 pp.

Mikolajewicz, U., B.D. Santer and E. Maier-Reimer: 1990, Ocean response to greenhouse warming, *Nature*, **345**: 589-593.

Pepper, W., Leggett, J., Swart, Wasson, J. Edmonds, J., Mintzner, I.: 1992, *Emission Scenarios for the IPCC, An Update'*, *Unpublished Report* Prepared for IPCC Working Group I.

Swisher, J.: 1993, Bottom-up comparisons of CO_2 storage and costs in forestry and biomass energy projects, *Biomass and Bioenergy*, submitted.

U.S. EPA (U.S. Environmental Protection Agency): 1990, Policy options for stabilizing global climate, Draft Report to Congress.

de Vries, H.J.M, J.G.J. Olivier, R.A. van den Wijngaart, G.J.J. Kreileman, and A.M.C. Toet: 1994, Model for calculating regional energy use, industrial production and greenhouse gas emissions for evaluating global climate scenarios, *Wat. Air Soil Pollut.*, **76** (this volume).

Woods, J. and D.O. Hall: 1993, *Biofuels as a sustainable substitute for fossil fuels: their potential for CO_2 emissions reduction*, A Study for Agriculture and Energy Section of FAO.

World Energy Council: 1993, *Renewable Energy Resources: Opportunities and Constraints, 1990-2020*, World Energy Council.

WRI (World Resources Institute): 1990, *World Resources 1990-91*, Oxford University Press, Oxford, 383 pp.

Zuidema, G., G.J. van den Born, G.J.J. Kreileman and J. Alcamo: 1994, Simulation of global land cover changes as affected by economic factors and climate, *Wat. Air Soil Pollut.*, **76** (this volume).

MODEL FOR CALCULATING REGIONAL ENERGY USE, INDUSTRIAL PRODUCTION AND GREENHOUSE GAS EMISSIONS FOR EVALUATING GLOBAL CLIMATE SCENARIOS

H.J.M. de Vries, J.G.J. Olivier, R.A. van den Wijngaart,
G.J.J. Kreileman and A.M.C. Toet.

National Institute for Public Health and Environmental Protection (RIVM),
, P.O. Box 1, NL-3720 BA Bilthoven, The Netherlands

Abstract: In the integrated IMAGE 2.0 model the "Energy-Industry System" is implemented as a set of models to develop global scenarios for energy use and industrial processes and for the related emissions of greenhouse gases on a region specific basis. The Energy-Economy model computes total energy use, with a focus on final energy consumption in end-use sectors, based on economic activity levels and the energy conservation potential ("end-use approach"). The Industrial Production and Consumption model computes the future levels of activities other than energy use, which lead to greenhouse gas emissions, based on relations with activities defined in the Energy-Economy model. These two models are complemented by two emissions models, to compute the associated emissions by using emission factors per compound and per activity defined. For investigating energy conservation and emissions control strategy scenarios various techno-economic coefficients in the model can be modified. In this paper the methodology and implementation of the "Energy-Industry System" models is described as well as results from their testing against data for the period 1970-1990. In addition, the application of the models is presented for a specific scenario calculation. Future extensions of the models are in preparation.

Keywords: energy modeling; greenhouse gas emissions; climate change

1. Introduction

Energy combustion and industrial processes are major sources of greenhouse gas emissions (Houghton *et al.*, 1992). Modifications in the global energy and industrial systems will play a key role in strategies to reduce greenhouse gases. In this paper we describe a set of four models for developing long-term scenarios of global energy use and industrial output together with their expected greenhouse gas emissions. These four models make up the "Energy-Industry System" of the IMAGE 2.0 model. In this paper we also present results from their testing against data from 1970 to 1990 for 13 world regions, and the application of this set of models to a test scenario up to year 2100.

The IMAGE 2.0 model is an integrated model of global climate change which includes three systems of models - "Energy-Industry", "Terrestrial Environment", and "Atmosphere-Ocean" (Alcamo *et al.*, 1994a). Emissions from the Energy-Industry System are added to emissions and fluxes from the Terrestrial Environment System to calculate the build-up of greenhouse gases in the atmosphere. This build-up of gases leads to global climate change which is computed by the Atmosphere-Ocean System. The computed climate change is fed back to the Terrestrial Environment System to compute its impacts on the world's biosphere, and the change in emissions from land use activity. These

Water, Air, and Soil Pollution **76**: 79–131, 1994.
© 1994 *Kluwer Academic Publishers.*

IMAGE 2.0
Framework of Models and Linkages

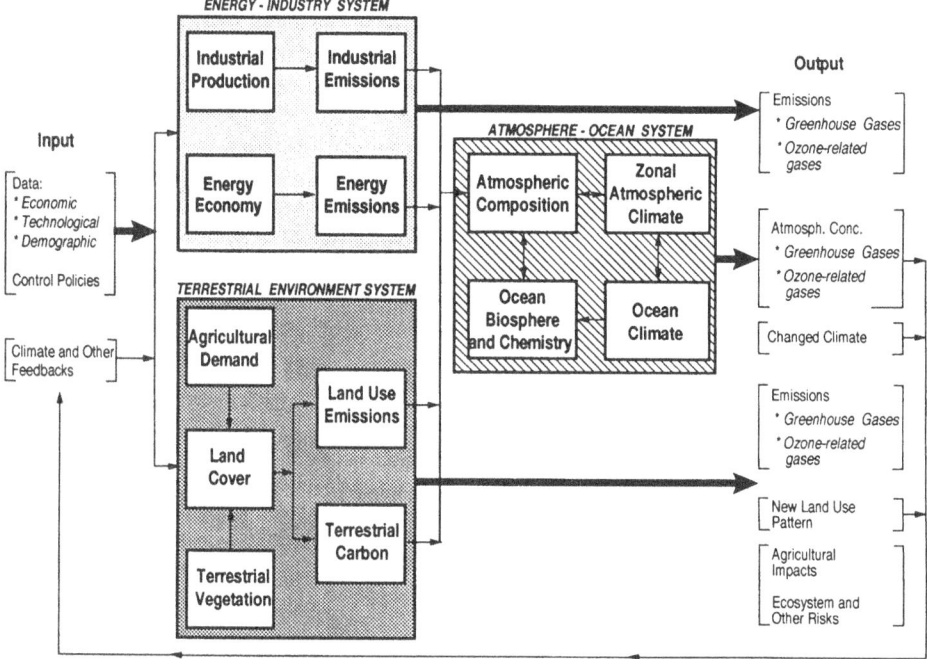

Figure 1: Schematic diagram of IMAGE 2.0 showing its framework of models and linkages.

emissions are added to emissions from the Energy-Industry System to compute new atmospheric concentrations of greenhouse gases in the next time step, and so on. Linked together, these three systems of models provide a quantitative and geographic overview of the global society-biosphere-climate system, which can be used for scientific and policy studies (Figure 1).

2. Overview: Energy-Industry System

The objective of the Energy-Industry System model in IMAGE 2.0 is to compute the emissions of greenhouse gases in 13 world regions as a function of energy consumption and industrial production and consumption. The 13 world regions are depicted in Figure 2. The models are designed especially for investigating the effect of improved energy efficiency, of shifts between energy carriers and of technological developments on future emissions in each region. It can also be used to assess the consequences of different

World regions in IMAGE 2

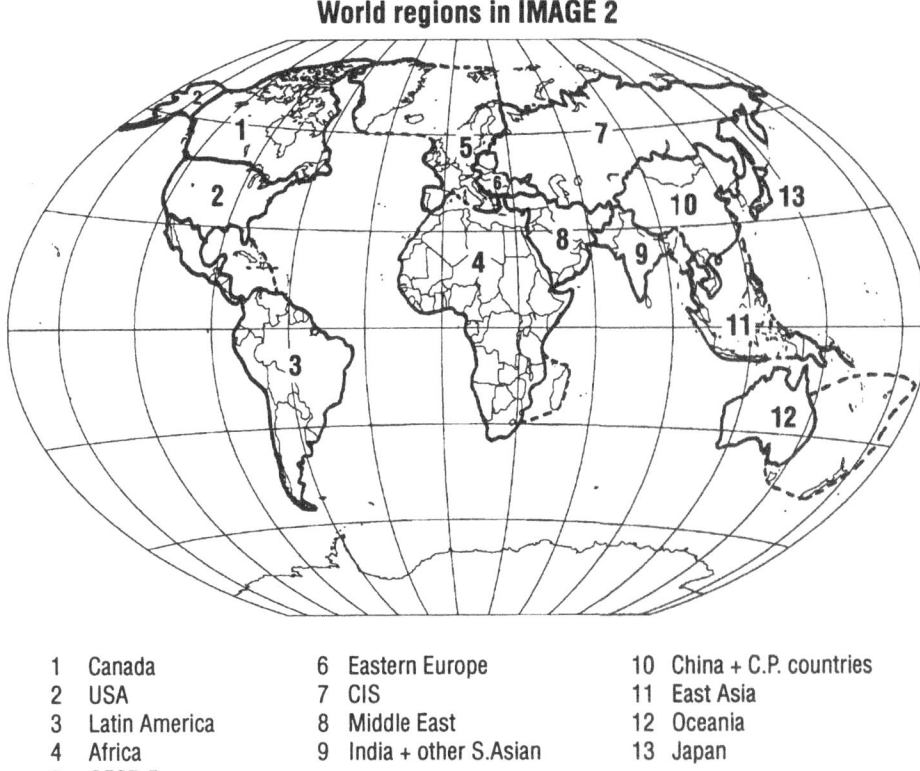

1	Canada	6	Eastern Europe	10	China + C.P. countries
2	USA	7	CIS	11	East Asia
3	Latin America	8	Middle East	12	Oceania
4	Africa	9	India + other S.Asian	13	Japan
5	OECD Europe				

Figure 2: World regions used in IMAGE 2.0.

policies and socio-economic trends on future emissions. The system of models consists of four individual models:

(1) Energy-Economy. The purpose of this model is to compute total energy consumption by end-use and energy supply sectors, with a focus on final energy consumption in each of five economic end-use sectors for each world region. The calculations are based on a change in level of activity in these sectors as well as the economic motivation and technical potential for conservation. Energy conservation scenarios can be investigated by modifying various techno-economic coefficients in the model.

(2) Energy-Emissions. This model computes the emissions of CO_2 and other greenhouse gases, resulting from production, transport and combustion of energy carriers in each world region. Estimates of energy use are taken from the Energy-Economy model. The effectiveness of different "source control" strategies for controlling emissions can be investigated by changing various emission factors in the model.

(3) Industrial Production and Consumption. This model computes the future level of economic activities other than energy use, that lead to greenhouse gas emissions. Future output levels for relevant processes are based on economic

production/consumption data.

(4) *Industrial Emissions.* This model computes the emissions of greenhouse gases resulting from economic activities other than energy use, e.g. CO_2 from cement manufacturing. Estimates of future emissions are based on activity data from the previous model and application of abatement factors.

Each of these models are now described in more detail.

3. Energy-Economy model: Methodology

The conceptual design of the IMAGE 2.0 Energy-Economy model has been done as part of the ESCAPE-project for the Commission of the European Communities (CEC, 1992). In addition to the purpose of the overall Energy-Industry model system, the goal of the Energy-Economy model is to allow the construction of energy use scenarios in a transparent and interactive way. The starting point is the representation of energy use as a chain of energy carrier handling processes from energy supply to end-use of energy: primary fuel production (fossil, renewable), transmission, energy transformation from primary fuels to secondary energy carriers (electric power generation and others such as oil refining), distribution, and final energy consumption in end-use sectors (industry, commercial, residential, transport and other); and finally the conversion into useful energy to provide energy services. We distinguish a set of driving forces, i.e. activities, that are assumed to correlate well with the demand for end-use energy functions[1]. This demand is converted to the input of fossil fuels and other energy sources.

The model is based on the following design considerations :

. calculations cover a *long time horizon* (up to the year 2100);

. explicit introduction of *energy functions* for end-use energy in the key sectors of the economy, allowing efficiency and conservation cost functions to be related to economic activity and engineering analyses;

. flexible, explicit inclusion of *conversion techniques* to satisfy the demand for end-use energy with secondary energy carriers as inputs; optionally, competition among these conversion techniques and the corresponding energy carriers driven by relative costs of end-use energy (substitution);

. conversion of *secondary energy carrier* use into primary energy inputs; primary energy inputs are modeled as separate exploitation cycles with depletion, learning and demand-supply dynamics, which relate their costs c.q. prices to cumulated production.

The focus on end-use has been applied earlier by other modellers (e.g. Johansson *et al.*, 1989; Alcamo and De Vries, 1992; Schipper *et al.*, 1992).

A major advantage of this approach is that it focuses on the basic human needs for

---

[1] We will use the terms 'energy functions' and 'energy services' interchangeably to denote those functions (heating, cooling, mechanical drive etc.) which are actually demanded by energy users. The energy which is used to provide these functions c.q. services are denoted by 'useful energy' or 'end-use energy'.

energy by examining how energy functions (services) are related to economic or human activity. It also enables the incorporation of "bottom-up" analyses. The approach also has limitations. There is the problem of having to estimate long term trends for phenomena with much shorter evolution cycles. Also, this approach requires data which are not readily available because most official statistics are still at the level of primary and secondary fuels. Finally, the trends of large numbers of input parameters and variables are not easily checked for consistency because macro-economic calculations are not performed. The solution to this latter problem is to use well-defined and tested economic reference scenarios.

The present Energy-Economy model is a first version, which does not yet have all the above features. Notably, we have energy functions to heat and electricity, have assumed one homogeneous conversion technique and have used exogenous time-paths for the market shares of fuels c.q. electricity generating techniques. Also, prices of primary fuels are exogenous and resource availability is not explicitly dealt with.

3.1   FROM SECTOR ACTIVITY TO USEFUL ENERGY DEMAND

Let us assume there are S sectors, labelled $s = 1,2..S$ in each of the 13 regions, labelled $r = 1,2..13$. Within each sector there is a demand for energy functions, labelled $j = 1,2..J$. The core of the model is the relation *Sector activity * specific useful energy requirement = sectoral useful energy demand*, or in formula form :

$$E_{rst} = \sum_j E_{rsjt} = \sum_j A_{rst} * R_{rsjt} \qquad GJ\ a^{-1} \qquad (1)$$

in any given year t, with A the activity vector, R the specific useful energy matrix and E the useful energy vector. The *activity vector A* gives an exogenous representation over time of those activities which represent the major causal factor behind sectoral energy demand. For the present model simulations we categorize activities according to the sectoral approach of energy and economic statistics[2]: industry, commercial, residential, transportation and other[3] (see Table 1). The activity levels are presently expressed in monetary terms, except for the transportation sector (see Table 3). In the future we intend to make a further disaggregation of activities and to construct activity indicators which are based on e.g. physical output levels (steel, aluminum) or capital stocks (office floor space, number of dwellings).

---

[2]   For an adequate analysis, one would prefer the differentiation between e.g. personal and commercial transport and between market and non-market services (see e.g. Slesser and De Vries (1990)). However, data availability limits such options.

[3]   At present, "industrial value-added" figures include the energy supply sectors (mining, oil refineries etc.); industrial energy use excludes the energy supply sectors and energy conversion losses. Transport energy demand is driven by passenger car per caput; however, energy use in the transportation sector includes freight transport.

TABLE 1

Aggregate sectors implemented in IMAGE 2.0.

Type	Aggregate sector	Definition (examples of sub-sectors)
End-use	Industry	Iron & steel, Non-ferrous, Chemical, Paper & pulp, Building materials
	Commercials	Commercial and public services
	Residentials	Household size classes; dwelling types
	Transportation	Road, rail, water, air transport; passenger, freight transport
	Other end-use	Miscellaneous, e.g. military activities not included in the above
categories		
Energy supply	Electricity generation	Public utilities, autoproducers, cogeneration of heat and power
	Other fuel transform.	Oil refineries, gas works
	Coal production	⎫
	Oil production	⎬ Production, transport, distribution and storage
	Gas production	⎭

Note: The end-use sectors correspond to IEA conventions used in the IEA energy balances.
        Other fuel transformation is introduced to match with total primary energy consumption per region.

The *specific useful energy matrix R* in Equation (1) relates the activity vector to the demand for energy functions. Ideally, there should be enough categories of energy functions to describe the energy service which is actually satisfied: the need for light, hot water, a warm house, cold food storage, extraction work, fluid transport, etc. But the number of categories has to be a compromise between process and engineering features, cross sector similarities and available data. Energy-intensity or specific energy use is defined as the amount of useful energy per unit of activity to satisfy the required energy function[4]. It equals E_{rsjt}/A_{rst} for energy function j in sector s and region r at time t. This quantity changes over time due to a variety of changes in the nature of the activities considered. It may increase as a consequence of changes in product mix and processes, of increasing mechanisation and automatization, and of health and safety regulations. It may decrease because of energy-saving technologies of which the penetration rate is linked to energy prices and capital stock turnover rates (see e.g. Schipper *et al.*, 1992, for a recent analysis).

For dynamic calculations the model requires assumptions about changes in sectoral activity levels and the resulting sectoral energy demand. Therefore, the relation between activity level A_{rst} and specific useful energy matrix R_{rsjt} has to be assessed over time. This complicated relation is condensed into variables such as the stage of industrial development (e.g. GNP/capita), state of technology, fuel prices and others. A conventional approach is to use economic growth and price-elasticities. The price elasticities propose a relation between the change in fuel price and the change in useful energy requirement; the

---

[4] The inverse is the energy productivity, in unit of activity per GJ of useful energy.

growth elasticities propose a relation between the change in the activity level and the change in useful energy requirement. Both formulations in fact reflect a lack of detailed understanding of the specific useful energy requirement $R_{sj} = E_{sj}/A_s$ over time.

In our model we use activity growth elasticities to calculate energy end-use demand :

$$E_{sjt} = (1 + \sigma_{sj} * \Delta X_{sj}/X_{sj(t-1)}) * E_{sj(t-1)} \qquad GJ\ a^{-1} \qquad (2)$$

with $\sigma_{sj} = (\Delta E/E)/(\Delta X/X)$ the elasticity coefficient, and X the explaining exogenous activity indicator (e.g. sectoral value-added per capita). It is assumed that these elasticities are time-independent, which is a doubtful assumption for such a condensed parameter, as empirical research has shown (see e.g. Mount et al., 1974). We use it here as a simplified way to represent the process of structural change on the basis of some representative indicators. The growth-elasticities are estimated from cross-region time-series data under the assumption of constant useful energy prices i.e. prices in the reference year T. As an example, Figure 3 shows the value of σ versus GNP/capita for five countries (having a range of incomes) over the period 1960-1987.

Ideally, the thus derived sectoral energy demand is a demand for energy functions, which are then on the basis of technique data converted into a demand for secondary energy carriers. It may well be that for many regions an adequate representation of region-specific demands for energy functions is difficult or impossible, due to lack of data. In that case assumptions have to be made for a representative set of demands per energy function, based on sector studies in other regions which have analyzed the energy demand structure (see e.g. Nørgård and Viegand, 1992; Benders et al., 1993). Also, a fair representation of techniques to perform energy functions, and their development over time, may in some cases be difficult to assess. Therefore, the set-up of the A-vector (i.e. sectors) and of the R-matrix (i.e. energy functions) is important and to some extent an iterative process in which the *goal* of the scenario-designer is a key issue, and in which a compromise is sought between available data and the required level of detail in assessments of demand for energy functions. At present, we distinguish only two energy functions : heat and electricity which are expressed in GJ of useful energy.

The next important consideration is that there are other factors which tend to decrease the energy-intensity i.e. the elements of the matrix R, over time. These are apart from though not unrelated to structural change. Although difficult to disentangle, one of them is autonomous technical improvements which has consistently increased the energy productivity of major energy-intensive processes over the last century although fuel prices were declining. The other are price-induced conservation investments and the improvement in their productivity.

In the Energy-Economy model we have introduced the *end-use energy conservation cost curve* to simulate investments in lowering energy-intensity as stimulated by rising end-use energy costs. The model mechanism works as follows. If the cost of useful energy (see Appendix A) rises, users will take measures on the basis of a required payback time of P years. First, an initial conservation cost curve is assumed, which relates the reduction in

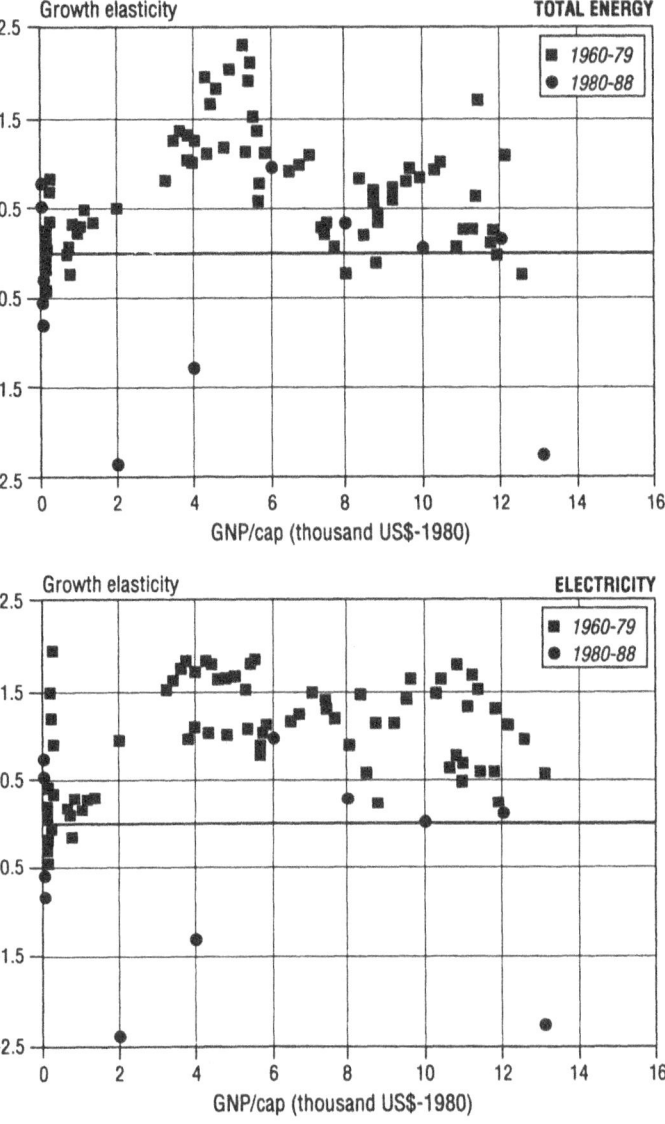

Figure 3: Growth elasticity of total energy and total electricity demand versus GDP/capita for five countries in the period 1960-1988. (sources: World Bank, 1990; IEA, 1992).

energy intensity (β) for sector s, energy function j and reference year T, to the required investments per unit of useful energy saved, I_{rsj}. At time t' the new energy demand will be given by the requirement that:

$$E_{rsjt'} = E_{rsjT} * (1 - \beta_{rsj}) * \prod_{t=T+1}^{t'} (1 + \sigma_{rsj} * \Delta X_{rsj} / X_{sj(t-1)}) \qquad GJ\ a^{-1} \qquad (3)$$

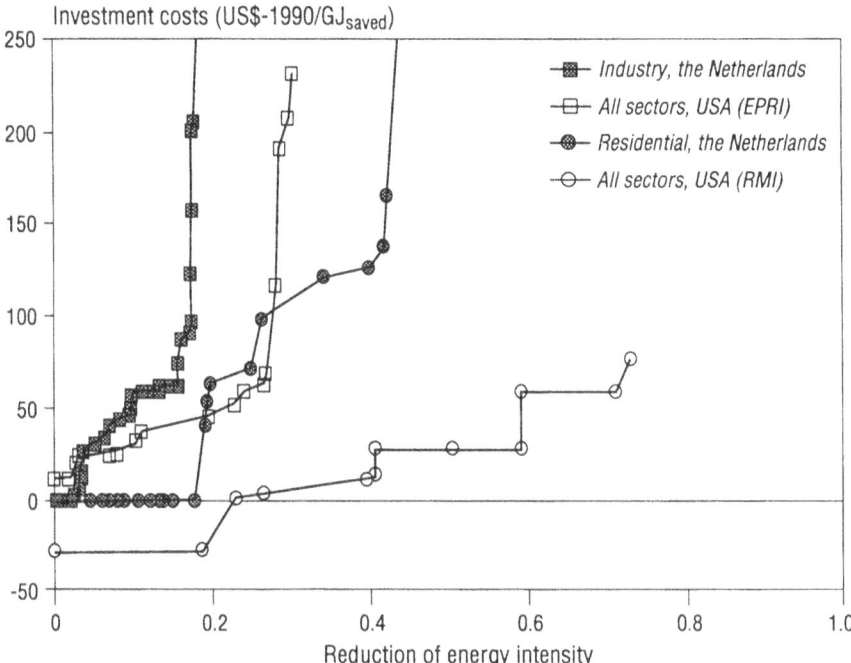

Figure 4: Example of sectoral end-use energy conservation cost curve for electricity for separate sectors in the USA and the Netherlands. (sources: Netherlands: Beer *et al.*, 1992; USA: EPRI and Rocky Mountain Institute (Fickett *et al.*, 1990)). Note: Annualized operation and maintenance costs are included in the investment costs.

with β_{rsj} determined in an iterative process from the requirement that the [marginal] investments per unit of useful energy saved equal the product of annual costs of end-use energy saved and required payback time:

$$I_{rsj} = P * \beta_{rsj} * c_{rsjt} \qquad \$ \, GJ^{-1} \qquad\qquad (4)$$

with c_{rsjt} the cost of useful energy for function j in sector s and region r at year t. The initial year (1970) is chosen as the reference year T. The value of β_{rsj} is limited to a maximum value. In the simulation it is assumed that investments in energy conservation take place only if costs increase; when costs decrease, the former level of energy-intensity is maintained. The actual decline in energy-intensity occurs with a time-delay in the simulation. The conservation cost curves have been constructed for a variety of countries over the last decade. Cross country surveys show a great variety in conservation costs and potentials (Tol, 1993), even if the same model is used (Kram, 1993). In most analyses the curves have been aggregated over all sectors. Figure 4 shows a few curves for separate sectors.

3.2   FROM USEFUL ENERGY DEMAND TO PRIMARY FUEL PRODUCTION

3.2.1   End-use sectors

The next step is to account for the conversion losses and investments in delivering useful energy from secondary energy carriers. This is done by specifying for each element of the useful energy matrix E a set of conversion techniques which are characterized by [time-dependent] conversion efficiency, secondary energy carrier type, market share, specific investment costs and specific emission factors. Let there be I secondary energy carriers, labelled $i = 1,2..I$ and M conversion techniques, labelled $\mu=1,2..M$. The useful energy matrix E is converted into the *secondary energy carrier matrix S* according to the formula:

$$S_{rsit} = \sum_j \sum_\mu U(x)\, \alpha_{rsj\mu}\, E_{rsjt} / \varepsilon_{rsj\mu} \qquad GJ\ a^{-1} \qquad (5)$$

with $U(x)=1$ for $x=i$ and $U(x)=0$ for $x\neq i$. Each element S_{rsit} in the matrix S indicates the amount of secondary energy carrier of type i, used in year t in region r to provide useful energy within sector s. Useful energy for function j can be supplied by a number of techniques T_μ, each having a market share $\alpha_{sj\mu}$ and a conversion-efficiency $\varepsilon_{sj\mu}$ from secondary to useful energy in year t, which are exogenously determined.

It is important to apply consistent definitions of secondary energy and useful energy. One extreme is to refer to the specific energy service, e.g. food storage/preservation in the case of a refrigerator. One has to be explicit about the baseline in terms of energy services provided. To avoid conceptual and data problems in this approach, one often goes to the other extreme and equates secondary and useful energy by setting all efficiencies equal to one. The disadvantage is that one has lost the baseline information which represents the expert's view on technical-thermodynamic potential for (future) demand-side efficiency improvements. As some estimates indicate that energy services are provided at average efficiencies of 5-10% even in the OECD, such information is quite relevant for long-term scenario explorations (Nakicenovic, 1989; Ayres, 1978). An intermediate approach is the EC-approach to define useful energy as the product of fuel input and a fuel-dependent average conversion efficiency (EC, 1988). In this approach it is convenient to define the techniques T_μ in a one-to-one correspondence to the type of secondary energy carrier used. This allows a simulation of fuel diversification on the basis of fuel-specific substitution processes of conversion techniques.

In a dynamic simulation one would like to specify the characteristics of the conversion techniques T_μ over time. This includes changes in conversion efficiency, costs and emission factors. Ideally, these have to be region-specific. In the present version, we have made the simplifying assumption that for each combination of an energy function and a secondary energy carrier, there is a single conversion technique with unit conversion efficiency and fixed capital and operation-and-maintenance costs.

The resulting demand for secondary energy carriers is determined by the market shares $\alpha_{sj\mu}$ which should in turn be related to their relative prices. It is difficult to model adequately the substitution between secondary energy carriers or, for that matter, of conversion techniques. Economists often assume that the sum of fixed and variable costs drive a market penetration process (e.g. Fisher and Pry, 1971). However, learning-by-doing, local availability of specific fuels, technological developments and security of resource supply considerations also influence the preferred choice of secondary energy carriers. Two options appear most relevant. The first is *exogenous setting of the market shares*: one simply sets a time-path for the market share of a secondary energy carrier used to meet the demand for useful energy. This can be done because of policy guidelines, scenario exploration or expert knowledge about a sector, or a combination of all three. For example, one may phase out coal c.q. coal-fired steam generation according to some official projection. This is the approach which has been adopted in the present analysis.

The other approach is that the penetration of conversion techniques is driven by differences in end-use energy costs. This process of *substitution driven by market forces* is modeled in such a way that the rate at which of a secondary energy carrier is penetrating into or driven out of the market is a function of the costs of useful energy from that energy carrier. This could for instance be represented with a mutlinomial logit function. The costs per unit of end-use-energy-delivered are calculated from annualized capital costs, fuel-price related variable costs and operation-and-maintenance (O&M) costs (see Appendix A). A description of the mechanism in the ESCAPE-project is given elsewhere (Olivier *et al.*, 1994). It has been tested for the food sector in the Netherlands and France. The results show that simulations of the combination of energy conservation and price-induced fuel substitution can generate values similar to historical data. However, the phase-in of oil and subsequently natural gas as a substitute for coal required the assumption of a shadow price of coal which is some 100-200% higher than the coal price for end-users as recorded in national statistics. This large difference is in agreement with earlier findings (e.g. Yu and Shu-Dong He, 1990).

3.2.2   Energy supply sectors

The next part of the model is the energy supply system: primary fuel production (e.g. of coal, oil, gas) and energy transformation (electric power generation, refineries, etc.). The Energy-Economy model deals with the energy supply system in a simplified way, as compared to other supply oriented models (e.g. Edmonds and Reilly, 1985). Resource depletion is not modeled explicitly; instead primary fuel prices are exogenous time-paths. Conversion losses in electricity generation and other transformation processes are dealt with in an explicit but simple way.

To compute energy use for electricity generation, first the Gross Electricity Demand (GED) needs to be calculated. To this end the demand for electricity in the end-use sectors is multiplied by a factor, $1 + F_{US}$, to include the use in the supply sector, and with a factor to cover transmission and distribution losses, $1 + F_{TR}$:

$$GED_{rt} = \sum_{s,i} S_{rsit} * (1+F_{US}) * (1+F_{TR})   \qquad GJ_e \, a^{-1} \qquad (6)$$

with the summation over all energy functions provided by electricity (i = electricity). Then, gross electricity demand is converted to secondary fuel input on the basis of a regional mix of power generating options. Each option k is characterized by an energy input type (e.g. coal, nuclear, renewable) and a conversion efficiency ε_{kt}, so the resulting input of secondary energy carrier i, called Fuel For Electricity (FFE) is given by :

$$FFE_{rit} = GED_{rt} * \sum_{k} \alpha_{rkt} / \varepsilon_{kt} \qquad GJ \ a^{-1} \qquad (7)$$

with α_{rkt} the share of option k in region r in year t in the electric power generation mix, and the convention that $\varepsilon_k = 0.33$ for nuclear and $\varepsilon_k = 1.0$ for renewable sources.

Conversion losses in other parts of the energy supply system - i.e. in other energy transformation sectors (oil refineries, gasworks, etc.) and in the fuel production sector - are all aggregated into one sector called "Other Fuel Transformation", to account for all other energy combustion resulting in greenhouse gas emissions. This sector covers the remaining statistical data on total primary energy use necessary for emissions calculations; marine bunkers are also included in this sector. The fuel mix (shares) of this sector is assumed to be equal to the fossil fuel split of total final energy demand in end-use sectors in every year. Energy consumption allocated to this supply sector does not include fuel used as chemical feedstock, since that is included in the end-use sector for industry[5].

3.3   IMPLEMENTATION IN IMAGE 2.0

3.3.1   Model summary

The Energy-Economy model as it has been used for the present simulations is a first and simplified version with the following characteristics :
.   there are five end-use sectors (Table 1);
.   there are two energy functions: heat and electricity;
.   there are six secondary energy carriers (Table 2);
     conversion techniques are only considered for heat, one for every secondary energy carrier and with fixed conversion efficiency and capital and operation-and-maintenance costs;
.   the relation between sector activities and useful energy requirements are based on growth-elasticities which are estimated as a function of the activity indicator (Table 3); it incorporates structural change and part of the autonomous productivity increase over time;
.   price-induced increases in energy productivity are for heat derived from sectoral energy end-use conservation cost curves;
.   exponential decay of the specific investment for conservation for heat is the second way of incorporating autonomous productivity increase;
.   for electricity, there is only an autonomous increase in productivity on top of the part which is incorporated in the growth elasticity; this is for reasons described below;

---

[5]   Energy carriers used for other non-energy use such as lubricants, waxes, bitumen are omitted in regional energy demand calculations; i.e. the energy content is not included in regional primary energy demand calculations, used as input for calculations of fuel production emissions. Fuel for combined heat-and-power production is incorporated through the efficiencies of electric power generation.

TABLE 2

Aggregate energy carrier implemented in IMAGE 2.0.

Type	Aggregate energy carrier	Definition (examples)
Fossil	Coal	Hard coal, sub-bituminous coal/brown coal/lignite, peat, coke
	Oil	Crude oil, Diesel, Gasoline, Heavy fuel oil, LPG
	Gas	Natural gas, Coke oven gas, blast furnace gas
Biofuels	Fuelwood	Fuelwood, charcoal, bagasse, dung (traditional biofuels)
	Other biomass	Ethanol, methanol produced from energy crops (e.g. sugar cane as modern biofuel)
Other	Renewables	Solar, wind, hydro, geothermal
	Uranium	Nuclear energy
	Electricity	

Note: The first 4 aggregate fuel types correspond to IEA conventions used in the IEA energy balances. Other biomass is introduced for scenario construction to include modern biofuel as an renewable energy source.

TABLE 3

Activity indicators for end-use sectors

Sector	Indicator	Unit
Industry	Value Added/cap	US$ (1980)
Commercial	Value Added/cap	US$ (1980)
Residential	Private Consumption/cap	US$ (1980)
Transport	Passenger cars/cap	-
Other	GNP/cap	US$ (1980)

. the energy transformation sector is represented by two sectors "Electric Power Generation" and "Other Fuel Transformation"; secondary fuel prices are derived from exogenous time-paths for primary fuel prices (cf. Table 11);

. electricity use of the energy supply sector and transmission losses are covered by multiplication factors;

. conversion of electricity to secondary energy carriers is based on exogenously set assumptions about the power plant mix and the development of conversion efficiencies over time.

In terms of the formulas in the previous sections, these simplifications can be expressed as follows. In all formulas, end-use sectors are indicated with $s = 1,2..5$ (industry .. other), energy function with $j = 1,2$ (heat, electricity) and fuels with $i = 1,2..6$ (coal .. electricity). Tables 1 and 2 summarize the sectoral and fuel type categories chosen.

Demand for useful energy function j is driven by activities:

$$E_{rsjt} = (1 + \sigma_{rsj} * \Delta X_{rsj}/X_{rsj(t-1)}) * E_{rsj(t-1)} \qquad GJ \ a^{-1} \qquad (8)$$

with X_{rsj} a representative indicator of structural change, namely an activity level per capita. The growth-elasticities are assumed to hold for constant useful energy prices i.e. the prices in reference year T. If end-use energy costs rise, conservation investments will reduce energy-intensity with a factor $(1-ß_{rsj})$ (see Eq. (3)), with $ß_{rsj}$ determined by:

$$I_{rsj} = P * ß_{rst} * c_{rsjt} \qquad \$ \ GJ^{-1} \qquad (9)$$

(if $c_{rsjt} > c_{rsj(t-1)}$). Presently P is taken constant at 5 years throughout all regions and all sectors and the capital costs I_μ are assumed to be fixed for all conversion techniques (see Table A.1 in Appendix A). The resulting demand for heat $(j = 1)$ is converted to demand for secondary fuels according to:

$$S_{rsit} = \alpha_{rsit} E_{rsjt} \qquad GJ \ a^{-1} \qquad (10)$$

with the simplifying assumption of a single conversion for each fuel from useful energy to secondary fuel with conversion efficiency of one. The time-paths for the market shares of the fuels under consideration are specified exogenously.

3.3.2   Model calibration

To calibrate the model, we first have collected the time series on sectoral use of secondary energy carriers from available statistics over the period 1970-1990. For some of the 13 regions these data may be rather weak[6]. The next step has been to collect for each sector one or two key activity indicators that have a major influence on changes in useful energy demand. We then hypothesize that end-use energy demand in all 13 IMAGE 2.0 regions can be simulated with a single elasticity curve for each of the ten sector-function combinations. The growth-elasticities σ_{rsj} as a function of the activity indicator X_{rst} have been estimated from time series of sector activities A_{rst} and useful energy E_{rsjt} for heat and electricity. In doing this, we have used only the data for the period 1970-1980, assuming that the useful energy prices have been fairly constant between 1970 and 1980 and that consequently most of the price induced conservation response to the oil crises had not yet taken place. For less-developed regions one would expect a bell-shaped curve for the growth-elasticities, reflecting the transition from an agricultural society to an industrialized one, followed by an increasing share of the commercial sector (cf. Figure 3). Figure 5 shows for two combinations the resulting curves and the empirical estimates on which they are based. The general pattern of Figure 3 can be recognized, but the data have

---

[6] E.g. the IEA-data base for the Netherlands (IEA, 1992) shows some large discrepancies with the official data from the Dutch Central Bureau of Statistics. See also Schipper et al. (1992).

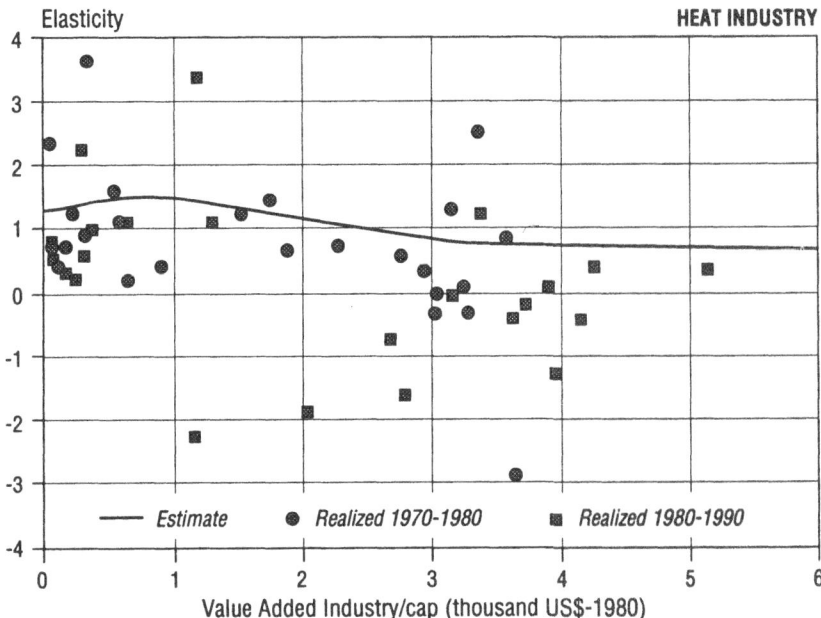

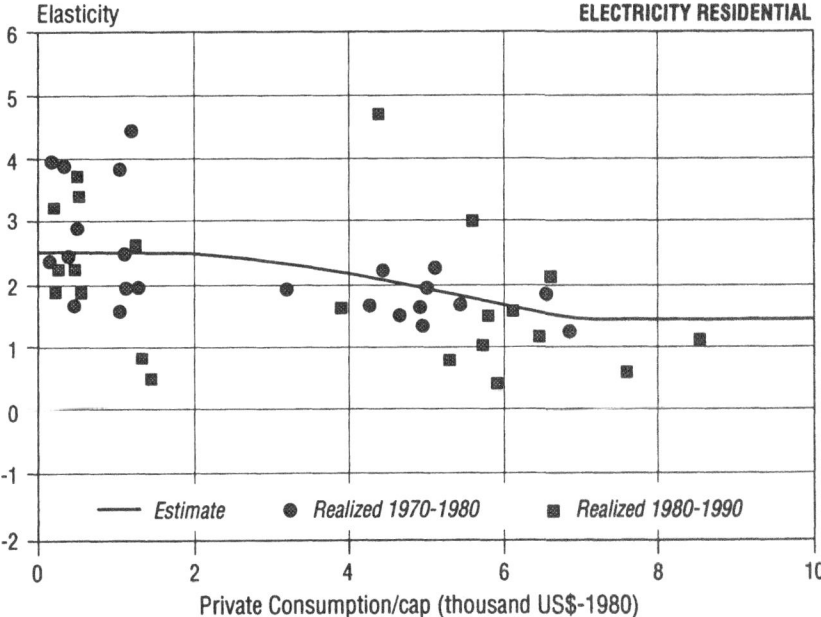

Figure 5: Example of realized and estimated growth-elasticity for industrial heat and residential electricity in five sectors as a function of the activity indicator (cross-region data for the thirteen IMAGE 2.0 regions for the period 1970-1990), calculated with a 5 year time-step. (source: World Bank, 1990; additional sources: for Value Added and private consumption: UNCTAD, 1992; IEA, 1992)

a large spread and the regions are somewhat clustered at low and high activity levels. For residential and commercial heat demand the approach is less satisfactory as e.g. climate differences among the 13 regions are not taken into account yet. For electricity use in the residential and commercial sectors, we obtain much higher values of σ at low income levels than expected, probably reflecting its premium use applications in many Less Developed Countries (e.g. cooling for food storage and air-conditioning). Note that our estimates of the σ versus X curves are chosen somewhat on the high side which reflects the exclusion of price-induced conservation measures taken after 1980.

The next step is to introduce energy conservation. For this purpose we have compiled time series of fuel and electricity prices which are assumed to be representative for the 13 world regions covered by the model. Wherever possible, we have corrected local prices for inflation on the basis of a local GNP-deflator and then converted it to 1990-US dollars. As inflation has surged in the 1980's and there have been significant differences between countries e.g. due to indigenous resource [un]availability and tax policies, it is difficult to construct a reliable indication of how real fuel prices have changed in regions like Latin-America and Africa. For various regions time-series were incomplete and additional assumptions were necessary. Figure 6 gives an example of the price time-series 1970-1990 for steam coal and Heavy Fuel Oil (HFO) for industry in India, used for the IMAGE region

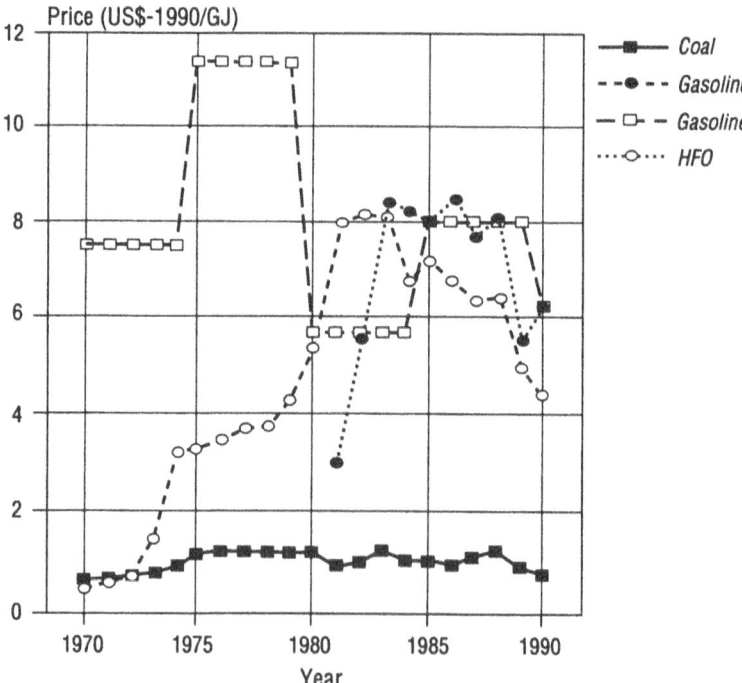

Figure 6: Time series constructed for 1970-1990 of fuel price of steam coal and Heavy Fuel Oil (HFO) for industry in India and gasoline for road transport in Mexico. (sources: World Bank, 1979; OLADE, 1993; IEA, 1992)

"India plus South Asia",which was constructed from a World Bank estimate of real prices for 1970 to 1980, and a combination of local prices from IEA and the US GNP-deflator for 1980 to 1990. For the transport sector in Latin-America we have used gasoline prices in Mexico in local currencies and combined it with the GNP-deflator from the World Bank and estimates made by Sterner (1985). For the former USSR and China the meaning and relevance of the available price time-series are unclear; additional research on these topics is clearly needed.

Given a price time-series for 1970 to 1990, in order to match the historical end-use as closely as possible, we have parameterized a conservation cost curve of the form :

$$I_{rsj} = [(exp(\beta_{max,rsj} - \beta_{opt,rsj}) - 1)^{-1} - (exp(\beta_{max,rsj}) - 1)^{-1}] * \phi_{rsj} * \prod_t (1 - \alpha_{rjt}) \quad \$ \, GJ^{-1} \quad (11)$$

where ϕ_{rsj} is a scaling factor for fitting to historical data, the parameter $\beta_{max,rsj}$ is the maximum achievable reduction (set at 0.9), whereas $\beta_{opt,rsj}$ is the one calculated from the above relation using Eq. (11). Figure 7 shows the shape of the conservation cost curve for three ϕ-values, which is to be compared with the engineering estimates shown in Figure 4. The factor ϕ measures the relative ease in money terms at which specific useful energy can be reduced. The actual reduction $\beta_{act,rsj}$ is calculated using a time-delay of 5 years before reaching $\beta_{opt,rsj}$. On top of this, we incorporated autonomous decrease in energy-intensity by applying a factor $(1-\alpha_{rj}t)$ which effectively makes conservation measures

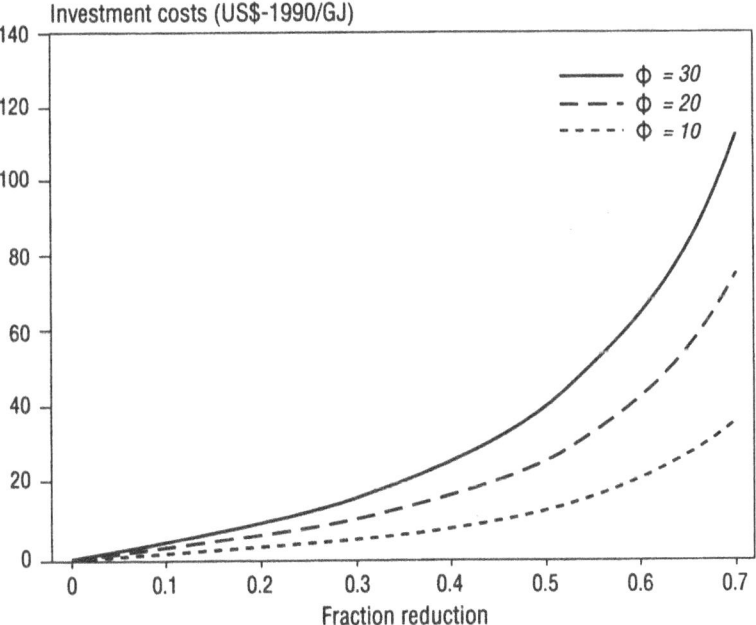

Figure 7: Conservation cost curves for three ϕ-values on a normalized specific energy use axis.

TABLE 4

Results of fuel price induced conservation measures: φ and α-values for heat; historic α-values for electricity.

Region	φ-value for heat by sector					α-value for heat (1990)	α-value for electricity		
	Ind	Trp	Res	Com	Oth	All sectors	1970	1980	1990
Canada	**	91	17	8.5	*(**)	0.010	0.	0.004	0.010
USA	15	113	18	15	*(20)	0.010	0.	0.004	0.010
Latin Am.	13	52	18	15	*(**)	0.006	0.	0.	0.
Africa	**	**	47	*(36)	*(**)	0.006	0.	0.	0.
OECD Eur.	20	170	37	*(46)	17	0.012	0.	0.004	0.010
Eastern Eur.	**	13	8	*(**)	**	0.012	0.	0.006	0.008
CIS	7	27	**	*(7)	7	0.008	0.	0.002	0.008
Middle East	**	**	**	*(**)	*(**)	0.003	0.	0.	0.
India region	16	128	25	*(7)	*(**)	0.010	0.	0.	0.
China region	3	*(21)	12	*(18)	*(**)	0.020	0.010	0.025	0.025
East Asia	23	112	23	*(**)	*(**)	0.010	0.	0.004	0.010
Oceania	19	**	*(26)	*(**)	*(15)	0.010	0.	0.	0.005
Japan	17	107	*(76)	*(45)	*(19)	0.012	0.	0.008	0.012

Note: * Not part of subset of 90 combinations.

        ** Conservation option not applicable because calculated value below historic value.

exponentially cheaper over time (for $\alpha_{rj} > 0$). This factor is in first approximation taken the same for all sectors. Fitting to historical data by adjusting the key parameters is partially a trial-and-error process. On the basis of conservation cost curves, new series of fuel and electricity inputs are calculated and compared with historical data on the basis of historical fuel price fluctuations. Discrepancies are then traced to problems with the price time-series, the consequences of fuel substitution and the estimated growth elasticities. This procedure, which gave unsatisfactory results for various regions, will be repeated in future versions of the model with improved inputs and model assumptions[7].

The results of our preliminary analyses are shown in Table 4, which lists the values of ϕ_{rsj} (j=heat) and α_{rj} (j=heat, electricity) which give the best model fit to historical data. There is a total of 130 combinations (13 regions, 5 sectors, 2 functions). For further analysis we have taken out those sectors for which the share in the regional total was in 1990 less than 10% (heat) c.q. 1% (electricity) unless its value exceeded 0.5% of the world total. Consequently, electricity for transport and for most regions heat in the sectors "Commercial" and "Other" could be omitted. The resulting subset of 90 combinations covers 97% of world final energy demand in 1990. For this subset some conclusions are[8]:

---

[7] For example, the large oil price fluctuations between 1975 and 1985 are a source of discrepancies because we use five-year time intervals. Another cause of discrepancies is that past fuel substitution processes affect energy-intensity, which are not taken into account in our present analysis as we assume all conversion efficiencies to be equal one.

[8] The φ-values are also affected by the absolute energy price levels in the regions, so strict comparison is not possible.

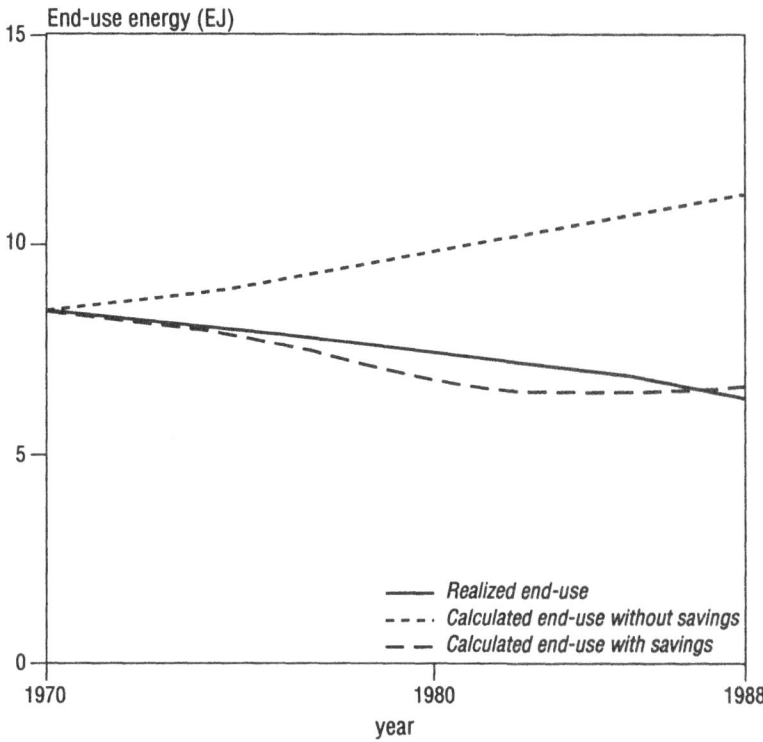

Figure 8: Realized and calculated energy end-use (in PJ) for residential heat in the USA: calculated autonomous trend without savings and with price-induced savings (using φ - 18 and α - 0.1). (sources: IEA, 1992; own calculations)

* Using fuel price changes with values of φ between 13 and 47 gives an adequate simulation of industrial and residential heat demand for most regions (see for example Figure 8). Put in another way, similar and consistent conservation cost curves can explain the observed trend in end-use heat demand. The φ-values are high for heat in the transportation sector, which is mostly gasoline for passenger cars. This may be due to other factors (new designs, comfort) and to longer delays because the consumer has quite limited conservation options in the short term;

* The simulated electricity demand tends to be lower than the historical trend for some sectors/regions. However, for most sectors/regions the conservation cost curves are much steeper with φ-values diverging between 13 and 128 (not shown in Table 4). One reason is that the estimate of the elasticity may not adequately reflect autonomous trends towards higher electricity intensity. Therefore, we decided not to use cost conservation curves for electricity but instead to use only the autonomous decrease in electricity intensity. This is done by modifying Eq. (8) by adding an exogenous time-dependent electricity intensity decrease factor :

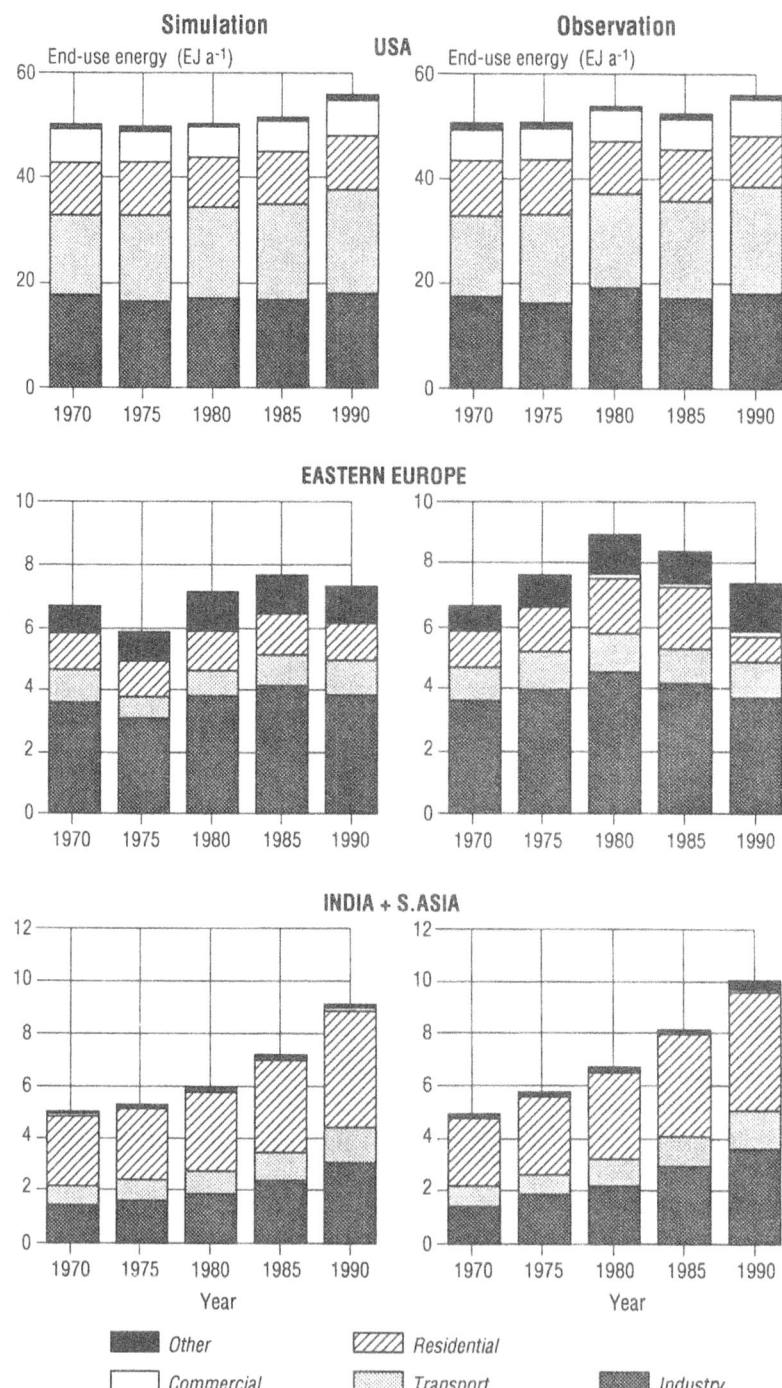

Figure 9: Comparison between model calculations and historical data for 1970-1990 for the USA, Eastern Europe and India plus S.Asia. (sources: IEA, 1992; own calculations)

$$E_{rsjt} = E_{rsj(t-1)} (1 + \sigma_{sj} \Delta X_{rsj}/X_{rsj(t-1)}) * (1-\alpha_{rjt})) \quad GJ \, a^{-1} \tag{12}$$

where α_{rjt} (j=electricity) is the annual decrease of electricity-intensity which is assumed to be equal for all sectors;

* For residential and transport electricity and for heat and electricity in the sector "Other", the discrepancies are large for all regions. Calculated energy use is often smaller than historical energy use even before energy conservation measures are added. Part of the explanation is that the sector "Other" is a residual category which behaves unpredictably. For most regions it is also small;
* For some regions, the discrepancies between historical and simulated time-series are large, especially the Middle East, Africa, Eastern Europe and Oceania; this has to be investigated in more detail but plausible reasons are the abundance of oil (Middle East) and the divergence of fuel prices throughout the region (Africa, Oceania);
* For CIS and the China region, the data on fuel prices are not reliable and many conservation measures may not be significantly influenced by fuel price changes but instead be the result of regulation (see e.g. Levine *et al.* 1992). This may explain the disagreement between calculations and historical data.

Model calculations versus IEA data are presented for 1970-1990 for the USA, Eastern Europe and the India region in Figures 9. Fairly lose agreement has been obtained.

4. Energy-Emissions model: Methodology

The objective of the Energy-Emissions model is to calculate the emissions of the greenhouse gases carbon dioxide (CO_2), methane (CH_4), nitrous oxide (N_2O), nitrogen oxides (NO_x), carbon monoxide (CO), volatile organic compounds (VOC) and sulphur dioxide (SO_2) stemming from the use of energy, as calculated in the Energy-Economy model. Combustion is the main source of greenhouse gas emissions from energy use, of which CO_2 is the most important compound. The model also takes into account emissions, notably CH_4 and VOC, from the coal, oil and gas production and transmission sectors.

4.1   FROM ENERGY USE TO EMISSIONS

The general method used in the Energy-Emissions submodel to estimate combustion emissions is described by the following basic formula for sector s in region r at time t:

$$EM_{Crst} = \sum_{i,j} S_{rsijt} * EF_{Crsij} (1 - r_{Crsijt}) \qquad g\,a^{-1} \tag{13}$$

where EM_{Crst} is the annual emission of compound C, S_{rsijt} is the secondary energy consumption, and EF_{Csij} is the Emission Factor in the reference year T (in g/GJ) for compound C (CO_2, CH_4, N_2O etc.). The index j refers to the energy function, the index i to the type of conversion technique which has a one-to-one correspondence to the fuel type, and r_{Crsijt} is the abatement factor. The effective emission factor at time t equals the emission factor, without any abatement or including the abatement level in the reference year T, multiplied by one minus the abatement factor r in year t.

This equation is applied to each region of the IMAGE model, thus allowing region-specific emission factors to be used. This methodology of applying *emission factors* and *activity levels* is the typical approach to estimating emissions at the national or regional level (see e.g. EPA, 1986; Corinair, 1991; OECD, 1990, 1991). For primary coal, oil and gas production sectors, where emissions are partly due to non-combustion processes such as venting and leakage, the model uses a similar approach with the amount of fuel handled as activity level.

The total annual emissions of compound C in year t and region r of energy function/process types j and techniques/fuel types i, is equal to the sum of combustion and process emissions over all sectors s:

$$EM_{Crt} = \sum_{sij} S^E_{rsijt} * EF^E_{Csijt} + \sum_{sij} S^T_{rsijt} * EF^T_{Crijt} + \sum_{ij} PF^H_{rijt} * EF^H_{Crijt} \quad g\ a^{-1} \quad (14)$$

with S^E the secondary fuel combusted in end-use sectors, S^T the secondary fuel combusted in the transformation sector and PF^H the primary fuel handled (produced, transported/distributed) and EF^E, EF^T and EF^H the corresponding emission factors.

The activity level, used to calculate emissions from the primary production of coal, oil and gas, is taken as the total fuel consumption in the region (end-use sectors plus energy supply sectors) (see Table 1). Non-energy use of fuels, other than as chemical feedstock, is neglected; therefore, the calculated total primary energy consumption (including marine bunkers) is a few per cent less than the actual amount. An additional source of uncertainty is that imports/exports and fuel transformations other than electricity production (e.g. oil refining) are only crudely treated in the Energy-Economy model (see Section 3.2.2). As a consequence, the regional amount of fuel produced/transmitted is somewhat inaccurate.

Delayed emissions from the combustion of carbon containing products, which were produced using fossil fuels as chemical feedstock (so-called non-energy use), are not included in this model. Other minor sources are also omitted: CO_2 from associated gas venting or flaring, VOC from venting, gas and oil leakage, and evaporation of stored oil products. For the coal, oil and gas production sectors (including transmission) only CH_4 emissions are taken into account. Since the uncertainty of emission factors for CH_4 is rather large, the error caused by our assumptions on regional primary energy production, as discussed in the previous paragraph, will not be of significance in estimating CH_4 emissions.

The Energy-Emissions model calculates the actual emissions from fossil fuel use as well as from the preparation and combustion of traditional and modern "biofuels" (e.g. fuelwood and bio-ethanol, respectively). When fuelwood and biofuel crops are produced in a sustainable way - i.e. the carbon fixation rate by regrowth equals the burning rate - no net emission of CO_2 will occur. In the IMAGE 2.0 model this has been assumed for current biofuel production and combustion. While the Energy-Economy model only deals with production/harvesting, transport and final use of fuels, the extent to which actual CO_2

emissions from fuelwood are offset by regrowth is computed in the IMAGE 2.0 Terrestrial Carbon model (Klein Goldewijk *et al.*, 1994)[9].

4.2   EMISSION FACTORS

Since we use highly aggregated sectors and energy functions/processes in the IMAGE 2.0 model (see Tables 1 and 2), we must also use aggregated emission factors. For the 13 world regions, aggregated emission factors have been taken directly from literature. These factors have been calibrated to regional emission estimates from the IPCC 1992 Supplement (Houghton *et al.*, 1992; Pepper *et al.*, 1992). As initial estimates we have used emission factors for combustion in 1990 for CO_2, CO and NO_x reported in IEA (1991), for CH_4 and N_2O from Van Amstel (1993), and for VOC from Corinair (1991). In summary, the emission factors used for combustion are essentially equal for all sectors and regions for CO_2, and for all regions for CH_4, N_2O and VOC, whereas for NO_x, CO and VOC from oil use in transportation a distinction has been made between different world regions, notably OECD, CIS/Eastern Europe, Centrally Planned Asia and Less Developed Countries, distinguished by Pepper *et al.* (1992). For fuel production and transmission, emission factors are used for the total emission of these activities, with a distinction between coal, oil and gas and between the mentioned four world regions. Detailed information on the emission factors adopted is given in Appendix B.

The emission factors for CO_2 were set equal to a fixed global average value for each of the five fuel types considered, with figures from IEA (1991), except for coal products used in Eastern Europe, for which a small correction was made for the high share of brown coal in total solid fuel consumption. For combustion sources, the emission factors for CH_4 were set equal for all regions with values from IEA (1991) and Berdowski *et al.* (1993).

For fuel production sectors (including transmission) we followed the approach applied by Pepper *et al.* (1992) and back-calculated emission factors from global CH_4 emission budgets for coal mining, gas supply and oil production systems separately and distribution of sectoral emissions over the four world regions according to regional fuel production figures. The emission factors derived in this way were then modified according to insights with respect to large regional deviations from the global average emission factor, e.g. for the gas supply in the CIS.

For N_2O from stationary sources, emission factors were set equal for all regions and sectors with figures for fossil fuels from De Soete (1993) and preliminary figures for fuelwood were adjusted in order to be in line with the global estimate of the Intergovernmental Panel for Climate Change (IPCC) by Pepper *et al* (1992). For N_2O

---

[9]   For completeness, we should add that emissions from waste combustion and landfilling are not dealt with in the Energy-Emissions model, although part of the waste will originate from materials, which were produced using fossil fuel as chemical feedstock (e.g. plastics, bitumen, naphtha, lubricants, waxes). The only exception is waste combusted for energetic purposes which is included in this model (methane from landfills is computed in the Landuse emissions model, see Kreileman and Bouwman (1994)).

emissions from mobile sources, which are very uncertain, the latest compilation of default global average emission factors were used (Olivier, 1993a): the same value was used for almost all regions and all commercial fuels, with the exception of the USA and Canada, which were assigned a higher factor because of the high fraction of catalyst-equipped cars in the vehicle fleet and in view of the global emissions estimate from combustion sources by IPCC (Pepper *et al.*, 1992).

For VOC from combustion sources, emission factors were set equal for all regions to preliminary factors proposed by Corinair (1991). An exception was oil use for transportation. Because of the large uncertainty inherently connected with estimates of VOC emissions, these factors have been adjusted so that calculated emissions agree with regional IPCC estimates (Pepper *et al.*, 1992). For NO_x and CO, region- and sector-specific emission factors were taken from IEA (1991), which are sector-specific for coal, oil and gas use, and defined for OECD, CIS/Eastern Europe and Less Developed Countries for different years. The 1990 factors used in IMAGE 2.0 were derived from the 1986 and 1995 data as specified by the IEA. The transportation sector is the determining category in calibrating CO emissions. The aggregated emission factors for the use of oil products in the transportation sector are based on factors for gasoline and diesel and calibrated by accounting for the fraction of gasoline and diesel consumed for road transportation. This is important because emission factors for diesel and gasoline differ greatly. An exception was made for fuelwood and other biofuels, where a different emission factor for CO was used in case of the industrial and transportation sectors because of their large difference with the residential sector (Piccot *et al.*, 1990). The emission factor for NO_x was adjusted upwards to obtain global figures in line with IPCC estimates for biofuels (Pepper *et al.*, 1992).

4.3 COMPARISON OF RESULTS

As noted above, we have calibrated emission factors in the model to regional and global estimates of emissions for 1990 from the IPCC (Pepper *et al.*, 1992). An exception was made for CO_2 emissions, for which emission estimates for the 13 IMAGE regions were also compared to regional totals derived from the Stockholm Environment Institute (Subak *et al.*, 1992). Calibration has been done separately for commercially traded energy carriers and for fuelwood use.

4.3.1.Commercial energy use

For *carbon dioxide emissions* from *commercial energy use*, the calculated global total of CO_2 is in good agreement with IPCC estimates, whereas regional emissions differ up to 20 to 25% (Table 5). The differences are mainly caused by different assumptions for regional primary fossil fuel use. Comparison of emissions from the individual IMAGE regions with Subak *et al.* (1992) shows discrepancies up to 25%, especially for Eastern Europe (see Table 5). These differences may also originate from other factors than the assumed primary energy use: Subak *et al.* use 1988 data based on a mixture of IEA and UN energy

TABLE 5

Comparison of CO_2 emissions from commercial energy for 1990 of IMAGE 2.0 and IPCC estimates.

Region group[*]	Emissions of CO_2 IMAGE (1990)		(Tg CO_2-C) IPCC (1990)	Diff. IMAGE-IPCC	SEI (1988)		Diff. IMAGE-SEI
OECD		**2976**	**2800**	**+ 5%**		**2596**	**+15%**
- Canada	133				110		+21%
- USA	1453				1296		+12%
- OECD Europe	990				863		+15%
- Oceania	87				68		+28%
- Japan	313				259		+12%
CIS/EE		**1284**	**1700**	**-25%**		**1278**	**+ 0%**
- Eastern Europe	287				327		-12%
- CIS	997				951		+ 5%
CPA		**728**	**600**	**+21%**		**651**	**+12%**
- China region	728				651		+12%
OTHER		**1074**	**900**	**+19%**		**903**	**+19%**
- Latin America	273				239		+14%
- Africa	192				161		+19%
- Middle East	240				205		+17%
- India region	189				156		+21%
- East Asia	180				142		+27%
TOTAL		**6062**	**6000**	**+ 1%**		**5428**	**+12%**

[*] Regions may not fully correspond between IMAGE and IPCC and between SEI and IPCC.

source: IMAGE: energy data from IEA (1992); IPCC: Pepper *et al.* (1992); SEI: Subak *et al.* (1992)

data, and they use country or sector-specific emission factors for CO_2. In addition, some difference can be explained by the differences between 1988 and 1990 fuel consumption. A comparison of the regional emissions by Subak *et al.* (1992) with calculations by Marland *et al.* (1989) also showed discrepancies up to 12% (Von Hippel *et al.*, 1993).

For *methane emissions*, the small contributions due to combustion of energy can be neglected. Regional emissions differ up to 20% compared to IPCC estimates, again because of different energy assumptions. For this reason, no further attempt was made to get a better fit with regional estimates of the IPCC. In the case of *nitrous oxide emissions* the most uncertain contribution is from catalyst equipped cars. The emission factors of the USA and Canada were adjusted to arrive at the global level of emissions (see Table 6).

Global total emissions of *carbon monoxide* and *nitrogen oxide* are close to the IPCC estimates, whereas most regional levels differ up to 5% with the exception of a larger difference for the China region (see Table 6). This may be caused by over-estimation of diesel use in the transportation sector, but no data on the sector level were available. The same correction for the use of diesel was made for NO_x emissions in the transportation sector. Final NO_x figures match quite well with IPCC estimates (see Table 6), except for CIS/Eastern Europe which differs about 25%. No further corrections were made in the

TABLE 6

Comparison of N_2O, CO, NO_x and VOC emissions from commercial energy for 1990 of IMAGE 2.0 and IPCC estimates.

Region group[*]	N_2O (Gg N_2O)			CO (Tg CO)			NO_x (Tg NO_2)			VOC (Tg)		
	IMAGE	IPCC	Diff.	IMAGE	IPCC	Diff.	IMAGE	IPCC	Diff.	IMAGE	IPCC	Diff.
OECD	317	283	+12%	56.5	60	- 6%	41.3	39.4	+ 5%	15.9	16	- 1%
CIS/EE	42	126	-67%	24.9	25	0%	17.1	23.0	-26%	4.5	4	+13%
CPA	38	0	-	12.0	6	+100%	9.2	9.9	- 6%	2.7	1	+173%
OTHER	37	0	-	35.6	38	- 6%	16.5	13.1	+26%	6.3	6	+ 5%
TOTAL	**434**	**409**	**+ 6%**	**129.0**	**129**	**0%**	**84.1**	**85.4**	**- 1%**	**29.4**	**27**	**+ 9%**

[*] Region groups: OECD: Canada, USA, OECD-Europe, Oceania, Japan;
                    CIS/EE: CIS, Eastern Europe;
                    CPA: China region;
                    Others: Latin America, Africa, Middle East, India region, East Asia.

source: IMAGE: energy data from IEA (1992); IPCC: Pepper *et al.* (1992)

TABLE 7

Comparison of CH_4, N_2O, CO and NO_x emissions from non-commercial energy for 1990 of IMAGE 2.0 and IPCC estimates.

Region group[*]	CH_4 (Tg CH_4)			N_2O (Gg N_2O)			CO (Tg CO)			NO_x (Tg NO_2)		
	IMAGE	IPCC	Diff.	IMAGE	IPCC	Diff.	IMAGE	IPCC	Diff.	IMAGE	IPCC	Diff.
OECD	0.7	1.5	-55%	13	0	-	4.7	12.1	-61%	0.7	1.3	-50%
CIS/EE	0.6	2.7	-77%	12	0	-	1.1	1.8	-40%	0.6	0.3	+89%
CPA	1.5	2.3	-36%	24	0	-	12.0	18.1	-34%	1.1	1.6	-35%
OTHER	7.3	5.3	+36%	145	157	-8%	38.3	74.8	+10%	7.2	7.6	- 4%
TOTAL	**10.0**	**11.7**	**-15%**	**194**	**157**	**+23%**	**56.1**	**74.8**	**-25%**	**9.6**	**10.8**	**-12%**

[*] Region groups: OECD: Canada, USA, OECD-Europe, Oceania, Japan;
                    CIS/EE: CIS, Eastern Europe;
                    CPA: China region;
                    Others: Latin America, Africa, Middle East, India region, East Asia.

source: IMAGE: energy data from IEA (1992) [non-OECD regions] and Agrostat (1990) [OECD region]; IPCC: Pepper *et al.* (1992)

other sectors, since differences in emission estimates can be explained by differences in fuel assumptions (see discussion on CO_2). For *VOC emissions*, our calculations agree with the regional IPCC estimates (Pepper *et al.*, 1992).

4.3.2   Fuelwood use

Calculations and comparisons of *fuelwood-related emissions* based on fuelwood consumption statistics (in IMAGE 2.0 taken from IEA and Agrostat statistics) are very uncertain (differences of more than 100% are not uncommon). Emission factors for this category also depend very much on local combustion conditions and quality of the fuelwood itself. For CO_2 and VOC, no data were reported by IPCC. The reported regional emissions of CH_4, N_2O, CO and NO_x are not very accurate. Therefore, a very strict calibration to IPCC estimates for 1990 by Pepper *et al.* (1992) was not considered appropriate.

In Table 7 the regional emissions of 1990 are presented as calculated by IMAGE 2.0 and by the IPCC. The emission factors for N_2O and NO_x were adjusted to the IPCC estimates. For CH_4, emission factors taken from Berdowski *et al.* (1993) resulted in a fairly good agreement of the emissions figures. For emission factors of CO due to fuelwood consumption a distinction has been made between industrial/transport sectors and residential/other sectors. Regional differences are reflecting differences in assumptions of energy consumption and emission factors. Global estimates are in line with IPCC estimates. This, however, is not the case for NO_x emissions.

5. Industrial Production/Consumption and Emissions models

Besides energy-related emissions, described in preceding sections, a number of non-combustion activities (also referred to as "industrial" or "process" sources) should be separately accounted for. In IMAGE 2.0 the Industrial Production/Consumption and Emissions models include the emission source categories presented in Table 9. Since there is no clear distinction between combustion and non-combustion processes, we first checked to what extent "non combustion sources" were included in the definition of emission factors of combustion processes. It was concluded that non-combustion sources were implicitly included in estimates of CH_4 and CO emissions from energy use. Halocarbon emissions (including carbon tetrachloride and methyl chloroform) will be largely phased out according to international agreements. Therefore, these emissions have simply been specified directly according to the scenarios of IPCC (Pepper *et al.*, 1992)[10].

The Industrial Emissions model computes emissions from the Industrial Production/Consumption processes similarly to calculations for combustion-related emissions:

$$EM_{Crt} = \sum_s A_{rst} * EF_{Crs} (1 - r_{Crst}) \qquad kg\ a^{-1} \qquad (15)$$

where EM_C is the emission of compound C, A_{rst} is the activity level in each category s (production or consumption level in tonne of product annually produced/consumed), EF_{Crs} is the Emission Factor (in g/kg = kg/ton product), r_{Crst} the abatement factor, and the other

---

[10] In a later version of IMAGE 2.0 halocarbons emission calculation schemes including delayed emissions will be included, such as developed for the ESCAPE-project for the European Commission (CEC, 1992).

TABLE 8

Activity level data used in the Industrial Production & Consumption model.

Process source	Global activity level in 1990	Reference	Disaggregation indicator	Reference
Cement production	1088 Mton	UN (1990)	population[4]	UN (1990)
Adipic Acid production	1800 kton	Olivier (1993b)	FEC-1990[3]	IEA (1992)
Nitric Acid production	60 Mton	Olivier (1993b)	N-fertilizer	Pepper (1992)
Consumptive use of solvents[2]	11500 kton	Pepper et al. (1992)	FEC-1990[3]	IEA (1992)
Miscellaneous VOC sources[5]	11500 kton	Piccot et al. (1992)	FEC-1990[3]	IEA (1992)
Copper smelting	8403 kton	UN (1990)	FEC-1990[3]	IEA (1992)

[1] Indicator used to disaggregate global to regional activity and to estimate future regional activity level.
[2] Global solvent use is assumed to be half of total industry emissions reported by Pepper et al. (1992) [see Piccot et al. (1992); then, using an emission factor of 1 g/g, by definition the activity level of non-solvent VOC sources is equal to the amount of solvent use.
[3] Total final energy consumption of the industry sector in 1990.
[4] National data available for estimating regional activity in 1990. Population used to estimate future regional activity level.
[5] By definition, using an emission factor of 1 g/g and the estimate by Piccot et al. (1992) that industrial non-solvent emissions are about 50% of the total industrial VOC emissions.

TABLE 9

Source categories and global emissions from Industrial Production & Consumption.

Process source	Compound	Reference	Global emissions IMAGE	IPCC	Diff.
Cement production	CO_2	1)	543	550	-1%
Adipic Acid production	N_2O	2)	487	500	- 3%
Nitric Acid production	N_2O	2)	233	250	- 9%
Consumptive use of solvents	VOC	3)	11.5	} 23	} 0%
Miscellaneous VOC sources	VOC	3)	11.5		
Copper smelting	SO_2	4)	8.9	8	+11%
Halocarbon use[*]	NA	5)	NA	NA	NA

Note: * Figures directly taken from IPCC scenarios.
           NA = Not Applicable

source: 1) Marland et al. (1989)
           2) Reimer et al. (1992), including global average abatement factors, if any.
           3) According to Piccot et al. (1992) global solvent use half of total industry emissions.
           4) Spiro et al. (1992)
           5) Pepper et al. (1992), Table 3.5.4.

indices are as before. In the absence of information on regional abatement factors, we used aggregate effective emission factors (including any abatement) to calibrate emissions with other estimates.

For each source category we first obtain the current global total activity level from the literature (Table 8). To apportion global figures to IMAGE regions we use regional cement production figures from Subak *et al.* (1992); for other activities, we use regional secondary energy consumption in industry (IEA, 1992). Next, for each source category a global average emission factor was taken directly from literature, on unabated emission factors and average abatement percentages, or indirectly from the assumed global total emissions for the source category. Default specific emission factors for cement, adipic acid, solvent use and copper smelting are readily available from the literature. The resulting emission factors as well as the resulting global emissions and a comparison with the corresponding IPCC estimates are shown in Table 9.

For scenario calculations the Industrial Production/Consumption model uses the following driving sources to linearly scale future regional activity levels in the various emission source categories:

.   Cement manufacture: population
.   Nitric acid production: nitrogen fertilizer use
.   Other categories: secondary energy consumption in the industry sector.

6. Some preliminary scenario calculations

6.1   CONSTRUCTION OF THE CONVENTIONAL WISDOM SCENARIO

We have constructed a Conventional Wisdom scenario for the period 1990-2100 based on assumptions of the IPCC (Pepper *et al.*, 1992). This scenario allows for considerable growth of GNP and population, especially in the developing countries. These assumptions also include growth of sectoral fuel prices and changes in the sectoral fuel mix (see Table 10). Is was our aim to use the driving forces of the Business-as-Usual scenario of IPCC "IS92a" as demographic and economic inputs for the IMAGE 2.0 simulation, rather than to reproduce this scenario. Because the IMAGE 2.0 model has slightly different regions, has no endogenous market share calculations for the fuel mix, and uses some other driving forces than GNP, it was necessary to make further assumptions on the use of the IPCC data (see Table 10).

Value added in industry and commercial sectors have been based on historic trends of the employment sectoral share in total GNP (Maddison, 1991). Development of consumer expenditures has been related to GNP by assuming for each region a change by 2100 to the average OECD level in the period 1970-1990. Assumptions about the future number of vehicles and rate of urbanization (used by the Terrestrial Environment system of IMAGE 2.0 in order to estimate global land cover) in each region are described in Alcamo *et al.* (1994b). As discussed in Section 3.4 the sectoral elasticities of heat and electricity are assumed to be only dependent on the activity level per caput (cf. Figure 5). The

TABLE 10

Scenario assumptions for sectoral activity indicators in the Conventional Wisdom Scenario (absolute values in 1990; annual average growth rates (%) for 1990-2025 and 2025-2100).

Region	Population			GNP			Passenger cars		
	1990 (millions)	1990-2025	2025-2100	1990 ($/cap a)	1990-2025	2025-2100	1990 (cars/Mcap)	1990-2025	2025-2100
Canada	26.64	0.21	-0.06	12 905	2.06	1.31	472 386	0.69	0.03
USA	249.99	0.54	-0.03	14 324	2.09	1.25	566 108	0.17	0.02
Latin America	448.03	1.34	0.27	1 758	1.85	2.20	72 647	1.72	0.08
Africa	642.11	2.53	0.84	690	1.57	2.39	15 092	3.00	0.47
Western Europe	377.7	0.21	-0.06	11 672	2.06	1.31	375 015	0.32	0.00
Eastern Europe	123.39	0.42	0.04	2 607	1.87	1.18	144 961	1.13	0.13
CIS	289.35	0.42	0.04	6 095	1.87	1.18	58 971	3.77	0.13
Middle East	203.18	2.65	0.82	3 108	1.36	1.98	40 672	1.76	0.00
India Region	1 171.08	1.50	0.39	318	2.97	2.84	2 939	5.01	1.73
China Region	1 248.28	0.98	0.15	604	4.23	3.07	2 653	5.07	1.83
East Asia	370.68	1.50	0.39	1 181	2.97	2.84	15560	2.97	0.45
Oceania	22.79	0.27	-0.06	9 961	2.71	1.28	413 007	0.46	0.00
Japan	123.54	0.27	-0.06	14 187	2.71	1.28	254 101	0.62	0.00

Note: Value added in Industry and in Services (per capita) are derived from GNP per capita using a statistical relationship between Industry resp. Services sector share as function of GNP (Maddison, 1991). Private consumption (per capita) is derived from GNP per capita by interpolating between the regional average share of GNP and the OECD average (1970-1990) share. Development of cars per capita (2025 values) are based on EPA (1990) and saturation values in 2100 were derived from Grübler and Nakicenovic (1991) (see Alcamo *et al.* 1994a).

source: Pepper *et al.* (1992), EPA (1990), Grübler and Nakicenovic (1991).

TABLE 11

Example of fuel price paths for industry in the USA and the China region in the Conventional Wisdom Scenario.

Region	Energy carrier	1990 (US$/GJ)	2000 (indexed to 1990)	2025	2050	2100
USA	coal	1.41	2.10	2.72	3.26	4.98
	gas	2.96	1.06	1.70	2.68	4.84
	oil	2.72	2.13	3.09	3.80	3.68
India	coal	1.27	1.02	1.31	1.57	2.41
	gas	3.80	1.09	1.75	2.76	4.98
	oil	4.71	2.74	3.97	4.88	4.72

Note: Prices of biofuels are assumed to increase very slowly; prices of fuelwood are assumed to stay constant at 1990 levels.

TABLE 12

Assumptions for autonomous cost decrease of heat demand conservation techniques and autonomous decrease of electricity intensity α in the Conventional Wisdom Scenario.

Region	α-value for heat (in %)			α-value for electricity (in %)				
	1990	2025	2100	1990	2010	2025	2050	2100
Canada	1.0	1.4	.9	1.0	1.0	1.0	1.0	1.0
USA	1.0	1.4	.9	1.0	1.5	1.5	1.0	1.0
Latin Am.	.6	1.0	.9	.0	3.0	3.0	3.0	2.5
Africa	.6	1.0	.9	.0	2.0	4.0	2.0	2.0
OECD Eur.	1.2	1.5	.9	1.0	1.0	1.0	1.0	.5
Eastern Eur.	.8	1.5	.9	.8	5.0	4.0	2.0	.5
CIS	1.2	1.5	.9	.8	5.0	4.0	2.0	.5
Middle East	.3	.5	.9	.0	5.5	5.5	5.0	1.0
India region	1.0	1.5	1.2	.0	5.5	3.5	3.2	3.2
China region	2.0	1.9	1.2	2.5	5.5	5.0	4.0	3.0
East Asia	1.0	1.5	1.2	1.0	5.5	3.5	3.2	3.2
Oceania	1.0	1.4	.9	.5	1.0	1.0	1.0	1.0
Japan	1.2	1.5	.9	1.2	1.5	1.0	.8	.5

Note: Values for other years are calculated by interpolation.

assumptions on the autonomous decrease of heat- and electricity-intensity are shown in Table 12. Efficiencies and costs of conversion techniques in end-use and in power generation are kept constant. Fuel market shares of the end-use sectors and power generation have been estimated from the model-based results in the IS92a scenario. Regional 1990-prices have been multiplied by factors derived from the changes in primary energy prices calculated in the Edmonds-Reilly model (Edmonds and Reilly, 1985), both for fossil-fuel based and renewable sources and between 1990 and 2100 (cf. Table 11). The conversion factors from secondary energy carriers to primary energy inputs can only be adequately calculated on the basis of a supply model in which regional production and upgrading (oil refining, gas and coal liquefaction, coking ovens) are explicitly taken into account. This is not yet available in the present model version, so we have used smooth trend extrapolations from the 1970-1990 values. Electricity distribution losses are kept constant at the 1990 fractional value. The future fossil fuel mix in the Other Fuel Transformation sector has been set equal to the regional fossil fuel mix of secondary energy use in the end-use sectors. The activity of industrial (non-energy) emission sources has been related to other scenario variables, as described in Section 4. For halocarbon use the trends defined in the IS92a scenario were used (Table 3.5.4 in Pepper et al. (1992)).

Concerning emission factors we have made the following assumptions (Table 13):

* CO_2 from modern biomass is assumed to be zero (i.e. an equal amount of carbon is assumed to be fixed by re-vegetation ("sustainable production");
*   for NO_x and CO we apply the figures for 2005 in IEA (1992) to the period 2005-2100 (OECD regions only);

TABLE 13

Scenario assumptions for emission factor development in the Conventional Wisdom Scenario.

Compound	Sector	Fuel type	Region	Emission factor assumption
CH_4	Fuel product.	Oil, coal, gas	All	Decreasing to OECD-1990-value by 2025 [see Table B.2]
N_2O	Transport	Oil, gas	OECD	Increasing to USA and Canada-1990-value by 2010
	Transport	Oil, gas	Non-OECD	Increasing to USA and Canada-1990-value by 2050
NO_x	All	Oil, coal, gas	OECD	Decrease by 2005 of 10-40% [see IEA, 1991]
CO	Transport	Oil, coal, gas	OECD	Decrease by 2005 of 20% (except in Western Europe: 60%)
VOC	Transport	Oil	All	Decrease by 2100 of 50%

* for VOC from oil use in transportation processes we assume a 50% linear reduction from 1990 to 2100;
* for N_2O from transport we assume an increase to 10 g N_2O/GJ by 2010 for OECD regions and by 2050 in other regions (and constant from these years onwards).

All other emission factors are assumed to be constant in time.

6.2   RESULTS OF THE CONVENTIONAL WISDOM SCENARIO

Figures 10 and 11 depict, by sector and by energy carrier, the secondary energy consumption for different regions and the global total as calculated for the Conventional Wisdom Scenario. Figure 12 shows the corresponding global primary energy consumption by energy carrier.

The results for the USA show a general three-phase pattern computed in all OECD regions (Figures 10 and 11)[11]. Up to 2000 we see an increase of energy use due to historic momentum of population and economic growth and due to relatively low assumed energy prices. In the second phase from 2000 to 2050 there is a decrease of energy consumption, mainly due to (1) higher energy price assumptions, which induce more conservation, (2) a sustained increase in energy productivity, and (3) a somewhat lower population growth. In the third phase this trend is reversed and energy consumption is again slowly increasing, as we assume a somewhat lower rate of improvement in electricity efficiency and a less steep increase of energy prices. The latter happens partly because of the assumed "backstop" effect of biofuels. In Eastern Europe a similar development is assumed to take place, although delayed in time and with different intensity.

For LDC-regions the results show a somewhat different pattern, such as presented in Figure 10 and 11 for the India region. Up to 2050 the growth of energy consumption is

---

[11] The discontinuities in the scenario results are caused by the fact that we have chosen only the years 2020, 2050 and 2100 and used interpolations in-between.

dominated by the exponential population and economic growth, partially compensated by our assumption of a persistent and high autonomous increase of energy and especially electricity productivity. The net result is an almost linear increase in energy use. After 2050 energy consumption starts to grow faster as industrial output per caput keeps

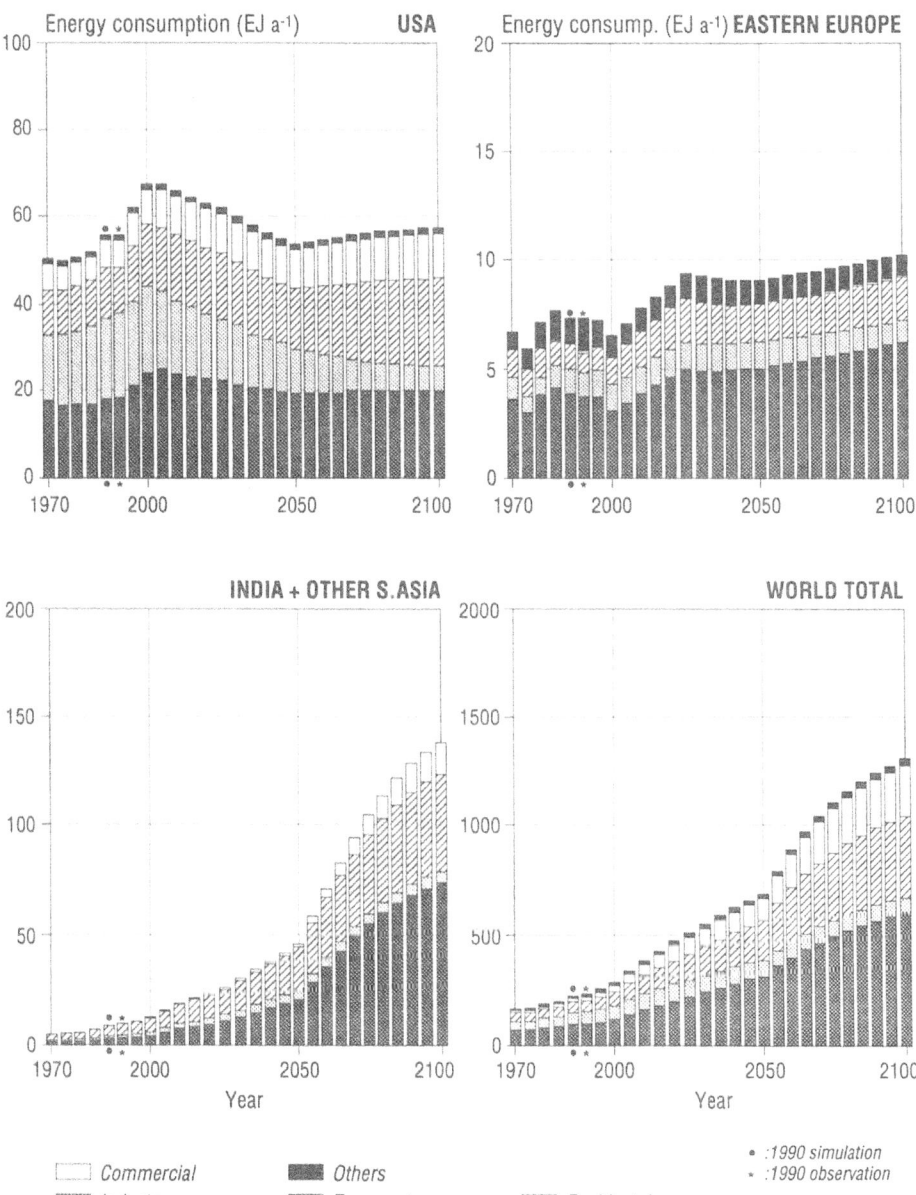

Figure 10: Total final energy consumption by *sector* for the USA, Eastern Europe and the India plus S.Asia and theglobal total in the period 1970-2100 (in EJ a⁻¹) (Conventional Wisdom scenario).

increasing and population growth is still exponential while productivity increases are assumed to slow down. Later on, the emergence of a consumer-oriented service economy and the levelling off of population growth lead to a declining or even zero growth rate in energy use.

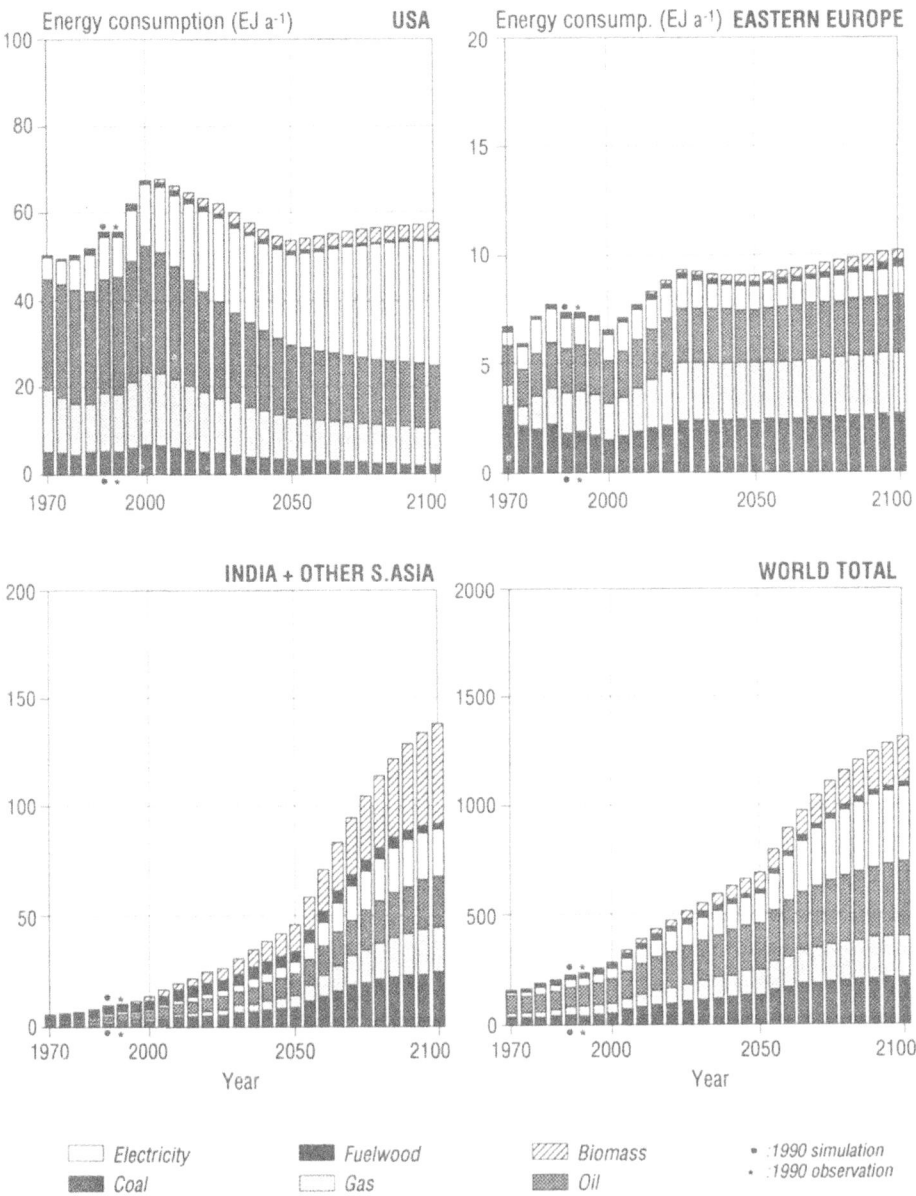

Figure 11: Total final energy supply by *energy carrier* for the USA, Eastern Europe and the India plus S.Asia and the global total in the period 1970-2100 (in EJ a⁻¹) (Conventional Wisdom scenario).

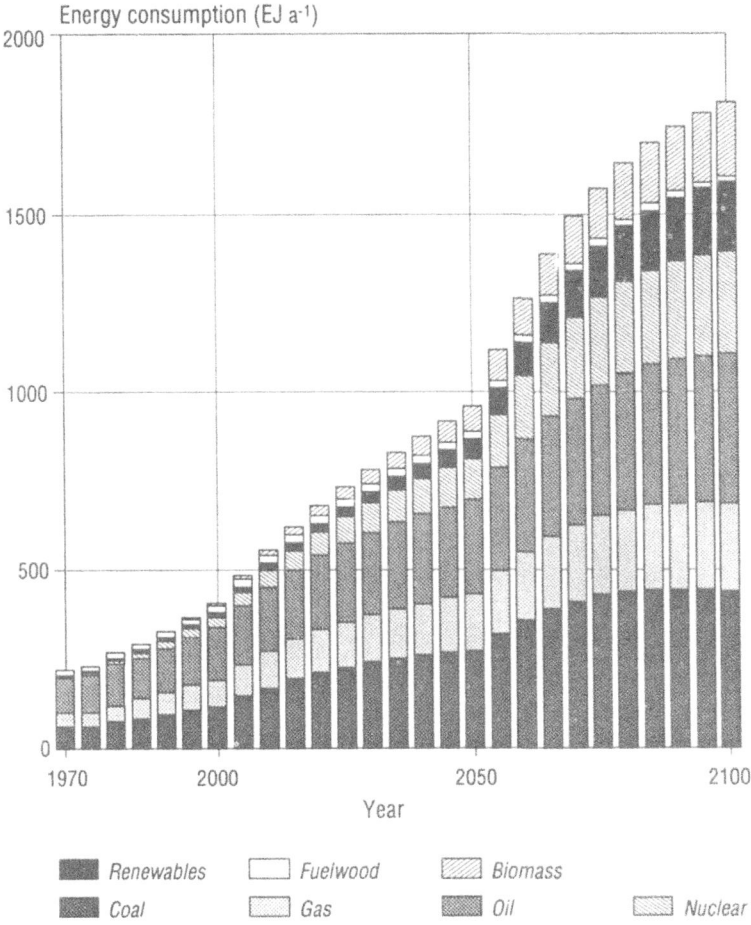

Figure 12: Global total primary energy consumption by energy carrier in the period 1970-2100 (in EJ a^{-1}) (Conventional Wisdom scenario).

In comparison with the IS92a scenario our simulations of energy demand and the corresponding greenhouse gas emissions result in higher figures for electricity use in general and in particular in the industrial sectors in Asia (regions "India", "China" and "South-East Asia"). The main reason is that in IMAGE 2.0 rather large growth elasticities for energy/economic activity are used based on our historical analysis 1970-1990 (see Section 3). Any change in the energy/GNP ratio has to come from declining growth elasticities and declining industry shares in GNP as economic growth and industrialisation proceed, and from autonomous and price-induced heat and electricity conservation. Although substantial for heat, these reductions do not lead to energy demand at a level comparable to the IS92a-scenario values. Another reason for our high estimate as compared to mean IPCC estimates is that we currently assume a constant conversion

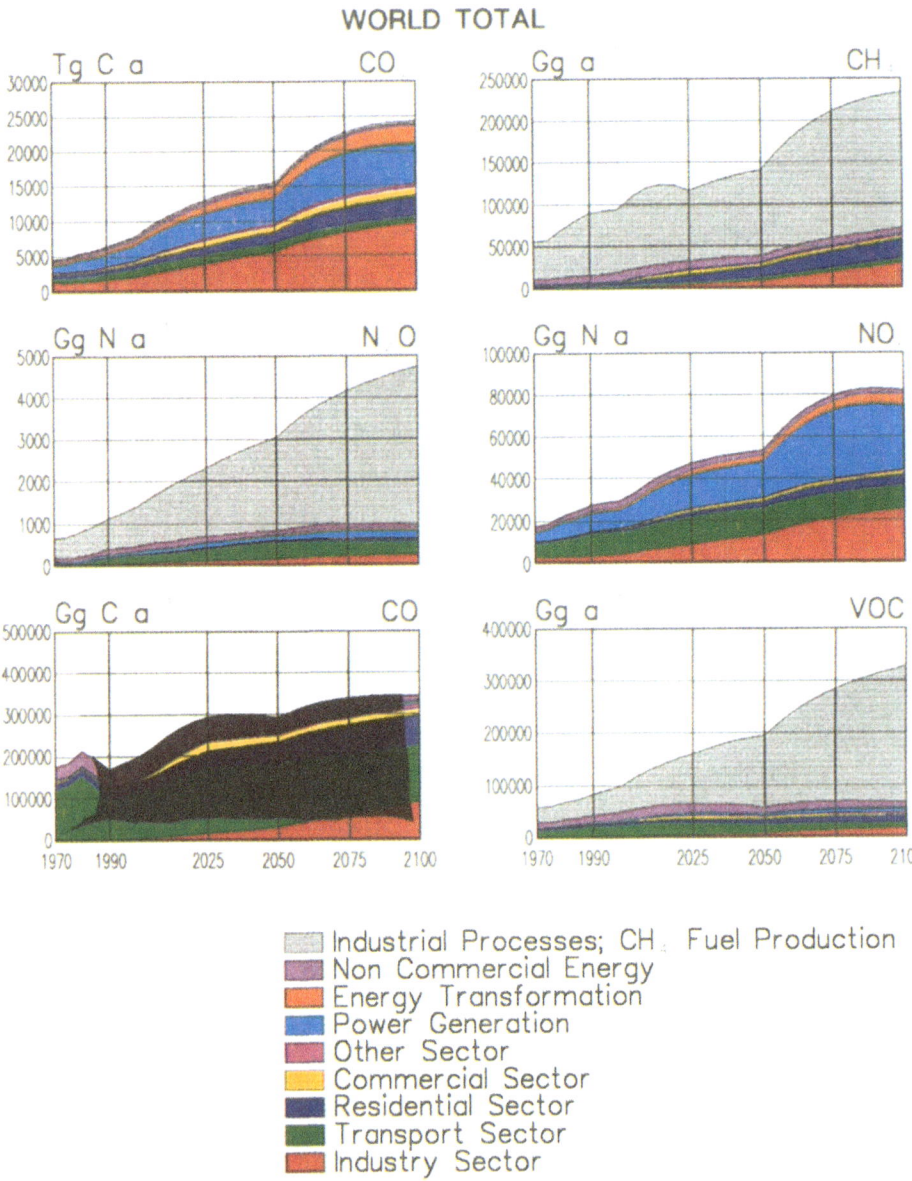

Figure 13: Global emissions from energy use and industrial processes, specified by compound and by sector in the period 1970-2100 (Conventional Wisdom scenario).

efficiency from secondary to end-use energy. As a consequence the large savings from better boiler and furnace equipment, especially feasible in combination with a switch to higher-grade fuels like natural gas, do not enter explicitly in the picture; they are assumedly incorporated in the autonomous energy productivity increase. We intend to do additional research on these model inputs and assumptions.

Figure 13 presents global emission projections. For numerical outputs of regional emissions of CO_2 and N_2O we refer to Alcamo *et al.* (1994b). As expected, the growth of CO_2 emissions follow the growth of global fossil fuel use shown in Figure 12. In Figure 14 regional emissions of CO_2 are shown for the USA, Eastern Europe and the India region. Regarding sectoral emissions, transportation decreases and residential and commercial sectors increase (notably in CIS and Eastern Europe). Industry appears to increase its share particularly in the China and India regions and in Eastern Europe (not shown). Methane emissions are strongly dependent on fossil fuel production, although the impact of abatement assumptions in the period up to 2025 is visible. N_2O and VOC emissions are increasingly dominated by industrial process emissions, since they are coupled to the strongly increasing N-fertilizer use and industrial energy use and no additional abatement was assumed. N_2O emissions from transportation increase their share because of the penetration of catalyst equipped cars worldwide. Finally, CO and NO_x emissions from industry and power generation sector increase, although the transportation sector keeps its high share in CO emissions.

6.3   UNCERTAINTY ANALYSIS OF THE CONVENTIONAL WISDOM SCENARIO

To have an idea of the uncertainties in the CO_2-emission in the Conventional Wisdom scenario, we have made an uncertainty analysis following the methodology described in Janssen *et al.* (1992) and using the accessory software package UNCSAM (UNCertainty / Sensitivity Analysis using Monte carlo sampling). Details of this approach are given in Appendix C.

We first started with identifying the input variables of the model which should be included in the uncertainty analysis. The aim of our analysis is to investigate model output uncertainty due to input variables only, not to scenario variables. In other words, our goal is to assess the influence of the structural parameters of the model on uncertainty of model output, rather than the influence of uncertain scenario assumptions (such as population and economic growth) on model output. All model input variables are listed in Table 14, which also denotes the variables selected for the uncertainty analysis. In view of the present model formulation, we consider population, GNP, number of passenger cars and market shares and prices of energy carriers as scenario variables. Their uncertainty will therefore not show up in our estimate of uncertainty in future CO_2 emissions.

With the selected variables we did an exploratory analysis with an uncertainty range of 25%, uniformly distributed. The variables were varied independently over all the categories indicated in Table 14, with the exception of time. Whenever an input variable has an exogenous time-dependence, we sampled an "uncertainty factor" (e.g. 0.75-1.25 for 25% uncertainty) and multiplied the corresponding variable with this factor throughout the

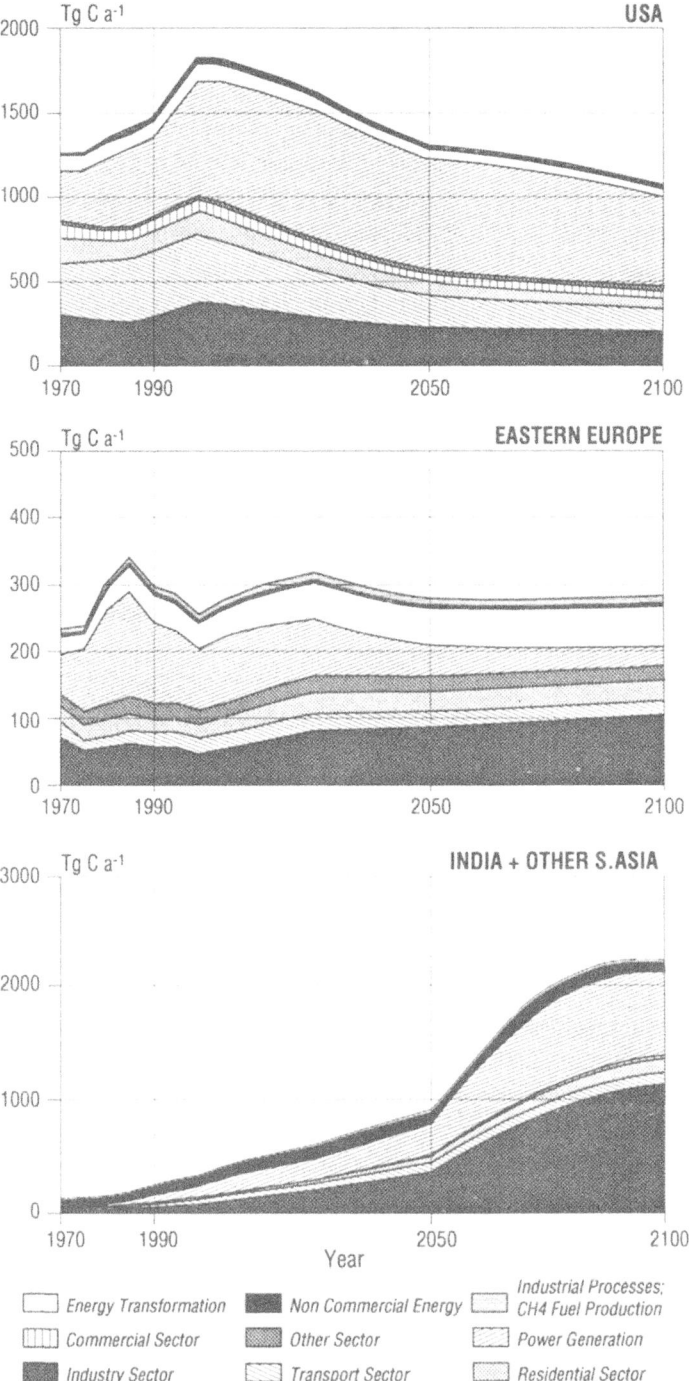

Figure 14: Regional CO_2 emissions from energy use and industrial processes, by sector, for the USA, Eastern Europe and India plus S.Asia in the period 1970-2100 (in Tg CO_2-C a^{-1}) (Conventional Wisdom scenario).

TABLE 14

Input variables for the energy model

VARIABLE	CATEGORIES*						Selected for sensitivity & uncertainty analysis
	R	S	I	K	J	T	
sectoral energy consumption 1970	13	5	-	-	2	-	X
growth elasticities	-	5	-	-	2	X	X
sectoral market shares of energy carriers	13	5	6	-	-	X	
market shares carriers for electricity generation	13	-	-	5	-	X	
conversion efficiency (primary to electricity)	13	-	-	5	-	X	X
own use electricity generation	-	-	-	-	-	X	X
distribution losses electricity generation	13	-	-	-	-	X	X
fraction Other Fuel Transformation	13	-	-	-	-	X	X
end-use prices of energy carriers	13	5	6	-	-	X	
specific investment costs	-	5	6	-	-	-	X
O&M costs	-	5	6	-	-	-	X
load factor	-	-	-	-	-	-	X
technical lifetime	-	-	-	-	-	-	X
interest rate	-	-	-	-	-	-	X
pay-back time	-	-	-	-	-	-	X
scaling factor costs/savings curves	13	5	-	-	2	-	X
autonomous decrease factor costs/savings curves	13	-	-	-	1	X	X
autonomous decrease factor electricity intensity	13	-	-	-	1	X	X
indicator / indicator share of GNP**	13	5	-	-	-	X	X
CO_2-emission factors	13	-	-	5	-	-	X

*   Categories over which the variables range:
    R = region
    S = sector
    I = secondary energy carrier
    K - primary energy carrier
    J = energy function
    T - time

** Industrial value-added for the Industry sector, number of passenger cars for Transport, private consumption for Residential, services value-added for Commercial and GNP for the Others sector.

simulation. We incorporated no correlation between variables into the analysis because we felt we were not able to do a reasonable assessment of possible correlations, though undoubtedly they do exist.

The analysis resulted in an uncertainty distribution of computed CO_2 emissions in 2050. To rank the variables according to their relative influence on computed CO_2 emissions, we also performed a regression for all input variables with the UNCSAM package. Using the Standardized Regression Coefficient (SRC) for the ranking, we discarded the variables that contributed negligibly (SRC > 0.05) to the CO_2 emission uncertainty. We repeated the sensitivity analysis with the remaining variables and found no difference in the output uncertainty distribution. We did not expect the discarded variables to have noticeable influence when their real (instead of standard 25%) uncertainty distribution would be used, so we omitted those variables in further

TABLE 15

Input variables with uncertainty ranges in the uncertainty analysis

Variable	Sector/ Carrier	Energy- function	Region OECD	Others
sectoral energy consumption 1970	all sectors	heat + elec	±10%[1]	±30%
growth elasticities	all sectors	heat + elec	*	*
conversion efficiency electr.generation	all fossil fuels**	electricity	±10%	idem
pay-back time	-	-	±50%	idem
scaling factor c/s curves	industry	heat	±15%	±90%
	transport	heat	±35%	±95%
	residential	heat	±85%	±75%
	commercial	heat	±60%	±95%
	other	heat	±40%	±80%
autonomous decrease factor c/s curves	-	heat	±40%	±50%
autonomous decrease factor electr.int.	-	electricity	±30%	+10%, -50%
indicator shares of GNP	industry,commercial	-	±15%	idem
	residential***	-	±20%	idem
CO_2 emission factors	solid****	-	±4%[1]	idem
	liquid	-	±1%[1]	idem
	gas	-	±3%[1]	idem
	fuelwood	-	±50%	idem

*   For the growth elasticities (which are functions of the sectoral indicator S), the uncertainty has been expressed as: $Elas(S)_{uncert.} = MY * Elas(S)_{standard} + AY$. A band of $±1/3 Elas(max.S)_{standard}$ has been taken and the ranges of the "uncertainty parameters" AY and MY have been chosen in such a way that the $Elas(S)_{uncert.}$ always fitted in this band. Besides, we have imposed the restriction that MY £ 1.

**   Only for fossil fuels; conversion factor by convention is 100% for renewables and 33% for nuclear.

***   Only for VA industry, VA services and Private consumption; the other indicators (GNP and Passenger Cars) are sheer scenario variables and are not taken into account. The indicator shares for industry and commercials have been given correlation -1 to prevent VA industry + VA commercials ≥ GNP.

**** CO_2 emission factor for modern biofuel is zero by convention (assuming sustainable production), so no uncertainty has been applied.

[1]   Source: Marland and Rotty (1984).

uncertainty analysis.

We continued with this reduced set of input variables to do a final uncertainty analysis of the energy model. The variables we did take into account are listed in Table 15, together with the uncertainty ranges we applied. Uncertainty ranges have been assigned after careful consideration and analysis, based on a.o. interregional variation of the data. In general we made a distinction between the OECD regions and the rest of the regions, looked at the variation of data in those groups and did an assessment of the uncertainty ranges for both groups. We made this distinction because we felt that the uncertainty in the non-OECD regions should be larger than in the OECD group. With this final set of input variables and uncertainty ranges, all uniformly distributed, we performed the uncertainty analysis of the energy model looking at the global CO_2 emissions.

Each Monte Carlo sample was passed through the energy model, producing a certain value of the global CO_2 emissions. The resulting ensemble of values, representing the

TABLE 16

Uncertainty distribution of global CO_2 emissions (in Pg CO_2-C)

year	mean	standard-deviation	coefficient of variation	5 percentile	95 percentile
1970	4.4	0.1	3%	4.2	4.
1975	4.4	0.3	6%	4.0	4.8
1980	5.0	0.5	10%	4.2	5.8
1985	5.4	0.6	12%	4.5	6.5
1990	6.1	0.8	13%	5.0	7.4
1995	6.8	0.9	13%	5.5	8.4
2000	7.6	1.0	14%	6.0	9.4
2005	9.0	1.4	15%	7.1	11.5
2010	10.4	1.7	17%	7.9	13.5
2015	11.5	2.0	18%	8.6	15.3
2020	12.6	2.3	19%	9.3	17.0
2025	13.5	2.6	19%	9.9	18.4
2030	14.3	2.9	21%	10.3	19.9
2035	15.1	3.3	22%	10.7	21.2
2040	15.7	3.6	23%	10.9	22.5
2045	16.3	3.8	24%	11.1	23.5
2050	16.7	4.0	24%	11.3	24.4

uncertainty distribution of the CO_2 emission, is characterized in Table 16. The uncertainty in CO_2 emission increases over time, which is not surprising as the uncertainty accumulates for each new year in the calculation based on the preceding year. This increase in uncertainty is visualized in Figure 15, presenting the mean of the uncertainty distribution, the median, and the 5- and 95-percentiles as functions of time. For the final year, 2050, the mean value of the CO_2 emission was 16.7 Pg C with a coefficient of variation of 24%. The 90% confidence interval of emissions in 2050 is 11.3 to 24.4 Pg C. Figure 16 presents histograms of the uncertainty distribution for the years 1970, 2015 and 2100. These figures show that the uncertainty distribution changes over time from a normal distribution to a Poisson-like distribution. This can be explained from the exponential character of some input variables in the model.

Comparing our uncertainty analysis with the uncertainty analysis performed by Edmonds et. al. (1986), it appears that their coefficient of variation of the uncertainty in the 2050 CO_2 emissions is much larger than ours (120% versus 24%). However, the Edmonds-Reilly group included scenario variables (like population and GNP) in their analysis as we did not. The uncertainty of those variables does not show up in our output uncertainty, which probably is the reason for this discrepancy.

We also did a final regression analysis to rank the input variables according to their influence on the output uncertainty. For this analysis we did a new Monte Carlo sampling on the basis of a set of "overall" input variables and passed the samples through the energy model. We used a new set of input variables to be able to compare the variables at the same level of aggregation. If we would perform the rank regression on the basis of the

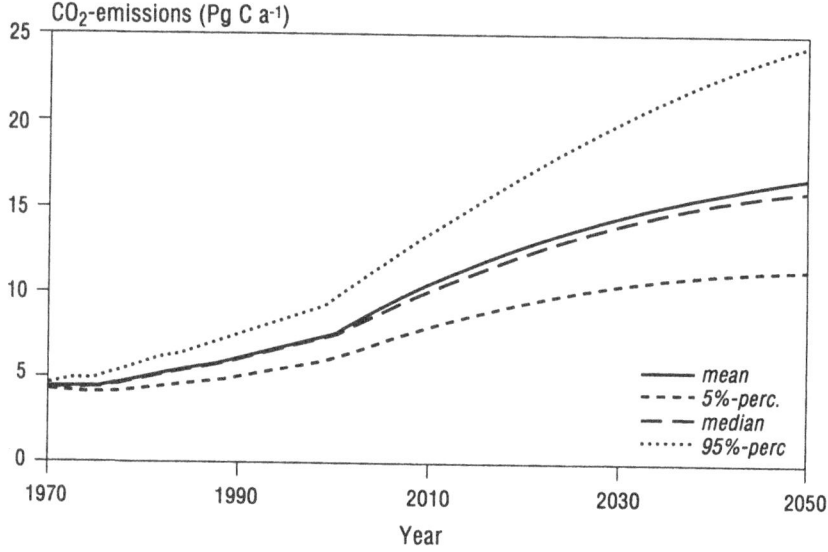

Figure 15: Percentiles for the uncertainty distribution of the global CO_2 emissions as a function of time.

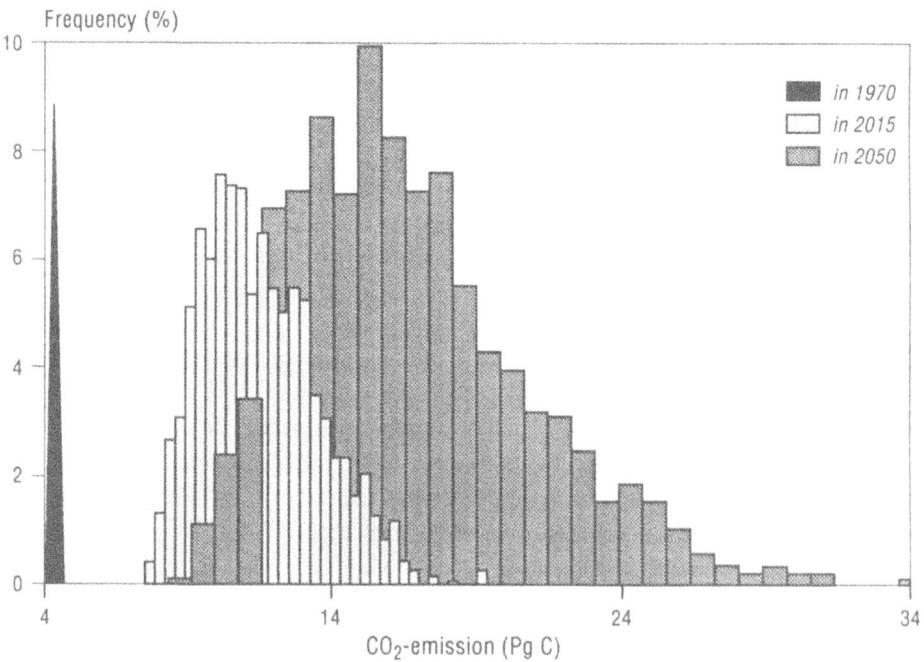

Figure 16: Histograms of the uncertainty distributions of the global CO2 emissions in 1970, 2015 and 2050.

TABLE 17

Input variables in the regression analysis

Variable	Uncertainty
sectoral energy consumption 1970	±30%
growth elasticities	±33%
conversion efficiency electr.generation	±10%
pay-back time	±50%
scaling factor c/s curves	±90%
autonomous decrease factor c/s curves	±50%
autonomous decrease factor electr.int.	+10%, -50%
indicator shares of GNP	±15%
CO_2 emission factors	±10%

elaborate set of variables, this would mean we would compare, for instance, the influence of the globally defined growth elasticities with influence of the regionally specified autonomous decrease factor of electricity intensity. The autonomous factor for one region will have much less influence than the growth elasticity path for all regions simultaneously. The set with "overall" input variables is listed in Table 17, along with the estimates of the uncertainty ranges we applied (distribution is still uniform).

Figure 17 shows the SRC for a selection of input variables as a function of time. The variables that act non-cumulatively in the energy model (CO_2 emission factor, energy consumption 1970) show a decline of influence in output uncertainty over time, while the influence of the exponential variables (growth-elasticity, autonomous decrease/electricity)

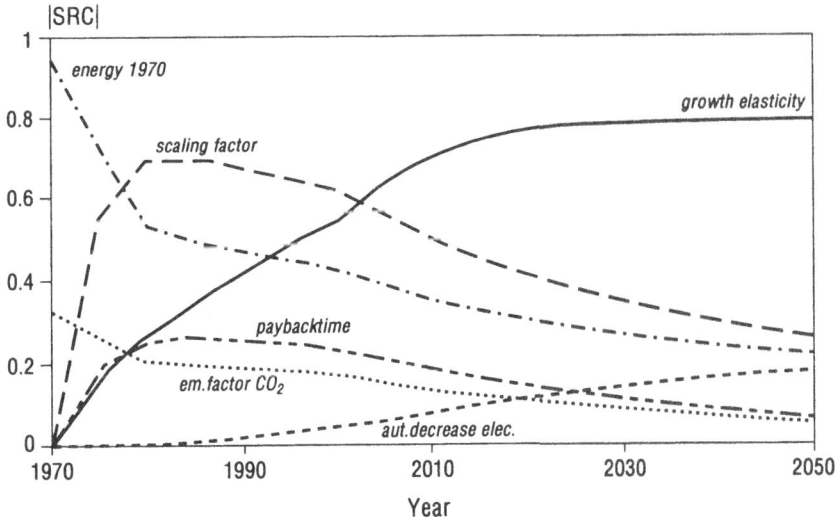

Figure 17: Relative importance of input variables according to their absolute SRC value.

increases. The variables that act via the costs/savings mechanism (pay-back time, scaling factor) show a maximum in 1980-1990. At the end of the simulation, in 2050, the growth elasticity is by far the most influential variable (SRC = 0.79), followed by the scaling factor for the costs/savings mechanism of heat (SRC = 0.25) and the energy consumption 1970 (SRC = 0.21). No large influence of autonomous efficiency improvement shows up, in contrast with the findings of Edmonds *et al.* (1986) and Nordhaus and Yohe (1983). This can be explained from the fact that we incorporated no autonomous decrease of total energy intensity in our energy model, but only autonomous decrease of electricity intensity.

7. Discussion of results and future extensions

Based on the preliminary analysis reported in Section 3.4 we have calibrated the Energy-Economy model to historical data for the period 1970-1990 and derived energy growth elasticities based upon cross-region data for the 13 world region for this period. It can be concluded that conservation cost curves as parametrized in Eq. (11) can at present only be applied for heat end-use; i.e. they do not apply to electricity end-use. In scenario calculations the use of the elasticities as derived from our analysis of historical data appear to lead to higher energy consumption and associated emissions as compared to mean IPCC estimates. The 1990 emissions estimate for commercial and non-commercial energy combustion, based on the set of emission factors as described in Appendix B, appears to be rather well in line with IPCC estimates for large regions. Comparison of individual IMAGE 2.0 regions differ up to 25%, which can be traced back to mainly differences in energy assumptions.

The Energy-Industry model in the present form appears to be quite capable of defining a broad scope of consistent scenarios, based on basic economic relations and specific assumption on the development of key economic indicators, technological and cost parameters and emission abatement assumptions. Of the different submodels making up the Energy-Industry system of IMAGE 2.0, the submodels that would most benefit from improvement is the Energy-Economy model. The Industrial Production and Consumption models will need improvement as knowledge on these type of emission categories develops over time. The following are some examples of model extensions under consideration:

*   Inclusion of activity levels based on physical output or capital stock and inclusion of more fuel types (e.g. coal and oil types, hydrogen - cf. paragraph 3.1);
*   Implementation of a dynamic mechanism to account for technology and fuel substitution as described in paragraph 3.2.1;
*   Improved treatment of combined heat-and-power schemes (cogeneration) since this is considered to be an important CO_2 control option;
*   Inclusion of an energy supply sector to generate fuel price paths endogenously, based on depletion dynamics of resources and a market clearing mechanism for (global) energy demand and supply;

* Inclusion of a "warning system" for major resource constraints, e.g. for biofuel and fuelwood production (availability of land to grow the energy crop), and for renewables and cogeneration (limited potential per region/sector);
* Improving the macro-economic realism of the model. At present, costs of fuel use and control measures do not affect sectoral activity levels in the model; some of these economic feedbacks may be included in the next version of the model. Also, work is underway to link the IMAGE 2.0 model with separate macro-economic models;
* Inclusion of a module to assess uncertainty estimates of key parameters. Including an uncertainty range in the output will clarify the state of knowledge regarding source strengths and relative importance of different reduction options;
* Including an halocarbon emissions module;
* Inclusion of feedbacks from climate change to energy demand.

Summarizing, a major advantage of this approach is that it focuses on the most important aspect of energy use in the context of sustainable development strategies: basic human needs for energy, i.e. need for energy functions, and their change over time as a function of growth in [physical] activity levels and change in energy costs and the state of technology. Secondly, it allows the explicit introduction of information on physical flows and engineering trends; this is important because most longer-term assessments, e.g. of material use or energy conservation techniques, rely on such information. Also, when data are available, or can be estimated, well founded emission calculations are possible. With a good user-interface the possible impact on emissions of various policy options can be examined. Moreover, since the described models are contained within IMAGE 2.0, these emissions can be related to a wide range of impacts on the atmosphere and biosphere.

Acknowledgements

This work is funded by the Dutch Ministry of Housing, Physical Planning and Environment, project no. 481507, and by the Dutch National Research Program on Climate Change and Global Air Pollution, project no. 851044. We thank Joseph Alcamo, Johannes Bollen, Rik Leemans and Rob Swart for their valuable comments on the manuscript.

Appendix A.   The Calculation of End-use Energy Costs

In the Energy-Economy model the cost of supplying useful energy for an energy function j are used to calculate the level of energy conservation measures taken at time step. The calculation of useful energy costs is based on the following [time-dependent] parameters for every conversion technique T_μ :
-   ε_μ the Conversion Efficiency (%)
-   I_μ specific Investment costs ($/kW$_{capacity}$)
-   OM_μ Operation & Maintenance costs ($/GJ$_{out}$)
-   L_t, L_e Technical and Economic Lifetime (year)
-   T_{in} year in which the technology enters the market.

In addition, for each technique providing a given energy function the following variables are specified :
-   LF_μ load factor [average capacity utilization factor] (fraction)
-   α_μ market share (fraction)

The costs of useful energy are based on annualized capital costs, fuel costs and operation & maintenance costs. To provide 1 unit of useful energy for an energy function j with techniques T_μ, the costs are :

$$c_j = \sum_\mu \alpha_\mu * a * I_\mu / (LF_\mu * 8760 * 0.0036 * \varepsilon_\mu) + p_{sf}/\varepsilon_\mu + OM_\mu \quad \$/GJ_{ue} \quad (A.1)$$

with the annuity $a = (r/100) / (1 - [1 + r/100]^{-L_e})$, based on an exogenously set interest rate r (in %), and p_{sf} the price of the secondary fuel under concern (in $/GJ$_{in}$).

The costs figures used for the simulations described in Section 6 are presented below in Table A.1. Generally only a distinction was made between techniques/fuel types, not between sectors. An exception was the transportation and residential sector.

Table A.1

Capital costs (US$/kW$_{in}$) and O&M costs (US$/GJ$_{out}$) used for end-use techniques (assumed to be equal for all regions and constant in time).

Type of cost[****]	Energy function (All sectors[*] [**])	End-use technique/energy carrier Coal	Oil	Gas	Fuelwood	Other biomass	Electricity
Investment	Heat	200[*]	150[**]	100	100	100	0
Investment	Electricity	0	0	0	0	0	100
O&M	Heat	0.5[***]	0.3[***]	0.2[***]	0	0	0
O&M	Electricity	0	0	0	0	0	100

[*]   For transportation: coal fired vehicles: 1300 US$/kW$_{in}$; oil and gas fired vehicles: 300 US$/kW$_{in}$.
[**]   For residential oil fired heating: 50 US$/kW$_{in}$.
[***] For residential 0 US$/GJ$_{out}$. For transportation: 0 for oil fired vehicles and 10 US$/GJ$_{out}$ for oil and gas fired vehicles.
[****] A technical lifetime of 15 year was assumed for all techniques and an interest rate of 10% was used for calculating the annuity a.

Appendix B.   Preliminary emission factors

With regard to emissions, every technique or activity is characterized by a [time-dependent] emission factor of greenhouse gases (g/GJ). For every technique i the data file contains an estimate of these factors. Estimation of emissions by this approach requires at least two and sometimes four distinct types of data (or assumptions when data are lacking): 1) fuel consumption (economic production/consumption) data per fuel type and per type of technology (process), respectively; 2) [default] emission factors (without abatement); and when applicable: 3) abatement factors, resulting in effective emission factors reduced by a certain fraction; and 4) the part of the activity level for which a specific abatement factor applies.

Emission factors for fuel combustion processes can be expressed either in units of fuel consumption (e.g. kg of coal or GJ of oil) or related to the overall activity level of the sector in consideration (e.g. tonnes of crude steel produced). For a specific compound the factor is related to fuel composition and/or to the energy function or device/technology used for the combustion of fuel. Emissions of CO_2 and SO_2, being directly dependent on the carbon and sulphur content of the fuel, are examples of the first category. The emission factors for other compounds may vary strongly between different energy functions or energy technologies, such as boilers, ovens, kilns, stoves and space heaters.

For the various gases we have applied emission factors based on average values per unit of fuel used for combustion and industrial production/consumption, respectively. As explained in the main text, heat is supplied by a single conversion technique with a single corresponding fuel for each sector. Consequently, emission factors EF_c are dependent on the region r, sector s and the input of fuel i. The preliminary values adopted for IMAGE 2.0 are presented in the Tables B.1 to B.3[12].

---

[12] In a next version of the model a more comprehensive approach will be used to feed the IMAGE model with more reliable data, derived from more detailed basic elements, based on more regional technology distributions and/or stock quality data, when the Emissions Database for Global Atmospheric Research (EDGAR) at RIVM has been established (Van der Maas et al., 1994). This is a separate project of RIVM and IMW-TNO, which is executed in collaboration with the Global Emissions Inventory Activity (GEIA), which is embedded in the International Global Atmospheric Chemistry (IGAC) Project, a core project of the International Geosphere-Biosphere Programme (IGBP).

TABLE B.1

Preliminary emission factors implemented in IMAGE 2.0 for energy consumption (combustion).

Fuel type	Sector[1]	Emission factor [2]					
		CO_2 (kg C/GJ)	CH_4 (g CH_4/GJ)	N_2O (g N_2O/GJ)	NO_x (g NO_2/GJ)	CO (g CO/GJ)	VOC (g/GJ)
Coal	ELECTR	25.8[5]	1	1.4	558[7]	50[9]	30
	IND/OFT	25.8[5]	2	1.4	251[7]	210[7]	30
	RES/COM/OTH	25.8[5]	300	1.4	183[7]	3580	400
	TRANS	25.8[5]	17	1.0	1037[7]	711	400
Oil	ELECTR	20.0	1	0.6	200	15	10
	IND/OFT	20.0	3	0.6	100	15	10
	RES/COM/OTH	20.0	2	0.6	50[7]	15	15
	TRANS	20.0	37	1.0[6]	750[7]	6650[8]	[10]
Gas	ELECTR	15.3	5	0.1	190	25	5
	IND/OFT	15.3	1	0.1	98	15	5
	RES/COM/OTH	15.3	1	0.1	50	10	10
	TRANS	15.3	37	1.0[6]	376	12589[7]	90
Fuelwood[3]	IND/TRANS	21.2	500	10.0	500	1500	800
	Other sectors	21.2	500	10.0	500	6600	800
Other biomass[4]	All sector	21.1	250	0.1	100	1500	80

Notes:
[1] Sectors: ELECTR = Electricity generation; IND/OFT = Industry/Other fuel transformation (non-electric); RES/COM/OTH = Residential/Commercial/Other; TRANS = Transportation
[2] Emission factors are the same for every region, except when indicated, and expressed in GJ input on LHV basis.
[3] Traditional biofuels such as fuelwood, charcoal, bagasse.
[4] Modern biofuels such as ethanol, methanol produced from energy crops e.g. sugar cane.

[5] For Eastern Europe: 26.3 g C/GJ (correction for high brown coal share)
[6] For Canada and USA: 10 g N_2O/GJ (correction for catalyst equipped cars)
[7] For OECD regions Canada, USA, Western Europe, Oceania and Japan:
- NO_x: for Coal in ELECTR, IND/OFT, RES/COM/OTH and TRANS: 457, 156, 142 and 831 g NO_2/GJ, respectively; for Oil in RES/COM/OTH and TRANS: 61 and 375 g NO_2/GJ, respectively.
- CO : for Coal in IND/OFT 80 g CO/GJ and for Gas in TRANS 2787 g CO/GJ.
[8] For OECD regions (in g CO/GJ): for Oil in TRANS: USA and Canada: 1680; Western Europe: 5750; Oceania and Japan: 1880 (an average emission factor is used which includes the fraction gasoline and diesel).
[9] For CIS and Eastern Europe: CO in Coal-ELECTR 210 g CO/GJ.
[10] For VOC in Oil-TRANS: OECD: 350; CIS and Eastern Europe: 390; China region: 100; Other regions: 450 (just calibrated to be in line with IPCC assumptions).

sources: CO_2: IEA (1991), p. 141
CH_4: IEA (1991), p. 179-181; 1985, except Coal in RES/COM/OTH, Coal in TRANS, Fuelwood, Other biomass: Berdowski et al. (1993)
N_2O: ELECTR, IND/OFT, RES/COM/OTH: De Soete (1993), p. 21; TRANS (non-catalyst): OECD (1991), p. 2-56; TRANS (catalyst cars) (USA and Canada): 10 g N_2O/GJ); data for fuelwood and other biomass are own preliminary estimates.

Appendix C.   Methodology of the Uncertainty Analysis

e basic approach of the Monte Carlo technique is to produce a large number of samples, ch sample being an ensemble of values of the input variables. The value of each dividual variable is selected from the uncertainty distribution of the variable. The ection is done in such a way that all selected values of a variable represent its certainty distribution. All selected values of a variable are combined with the values of other variables randomly, thus producing the samples.

The algorithm used for selecting the variable values and combining them to samples is led Latin Hypercube Sampling (Janssen *et al.*, 1992). With this efficient way of mpling, the number of samples required to represent all possible combinations of riables within their uncertainty distributions, is small. It is recommended to choose the mber of samples 2-5 times the number of input variables.

In the sampling as described above, the values for the variables are independently osen and combined. However, it is possible that with one value for a certain variable, other variable can only have a limited range of values; in other words those variables are rrelated. The UNCSAM package has a possibility to take correlation into account in the mpling of variables, however in a limited way. Correlation can be expressed between airs of variables only, as a number between -1 and 1.

When a set of samples is created, the model considered must be run for each sample as put, resulting in a set of values of the model output. This set represents the uncertainty istribution of the output and is fed back into UNCSAM, which then calculates the naracteristics of the uncertainty distribution. It also performs a regression analysis: the lative contribution of each input variable in the output uncertainty is calculated. The elative contribution can be expressed with several indicators (Janssen *et al.*, 1992). We hose the Standardized Regression coefficient (SRC):

$$\frac{\Delta y}{S_y} = SRC_i * \frac{\Delta x_i}{S_{x_i}} \qquad (C.1)$$

with x_i the ith input variable, y the output variable and S the standard deviation.

TABLE B.2

Preliminary emission factors implemented in IMAGE 2.0 for energy production (including tr:

Type of fuel production	Emission factor CH_4 (g CH_4/GJ)[1]			
	Region groups[2]			
	OECD	CIS/EE	CPA	Others
Coal	175	525	350	350
Oil	70	210	140	210
Gas	230	805	460	460

Note: [1]   Emission factors correspond to production, transport and distribution jointly and are
            GJ of fuel produced on LHV basis.
        [2]   Region groups: OECD: Canada, USA, OECD-Europe, Oceania, Japan;
            CIS/EE: CIS, Eastern Europe;
            CPA: China region;
            Others: Latin America, Africa, Middle East, India region, East Asia.

source: Pepper et al. (1992)

TABLE B.3

Global average emission factors for Industrial Production & Consumption.

Process source	Compound	Global emission factor[*]	Unit	Reference
Cement production	CO_2	499	kg CO2/ton cement	Marland et al. (
Adipic Acid production	N_2O	130	kg N2O-N/ton AA	Reimer et al. (1
Nitric Acid production	N_2O	17	kg N2O-N/ton HNO_3-N	Olivier (1993a)
Consumptive use of solvents	VOC	1000	kg VOC/ton solvent	Piccot et al. (19
Miscellaneous VOC sources	VOC	1000	kg VOC/ton	Piccot et al. (19
Copper smelting	SO_2	1060	kg S/ton	Spiro et al. (199

[*] Including global average abatement factors, if any.

References

Agrostat, 1990: *AGROSTAT.PC*. FAO, Rome. Data diskettes.

Alcamo, J. and Vries, B. de: 1992, Low Energy, Low Emissions : SO_2, NO_x and CO_2 in Western Europe, *Int. Env. Affairs* **4:3**: 155-185.

Alcamo J., Kreileman, G.J.J., Krol, M., Zuidema, G., 1994a: Modeling the Global Society-Biosphere-Climate System. Part 1: Model Descriptions and Testing. *Water, Air and Soil Poll.*, **76** (this volume).

Alcamo, J., Van den Born, G.J., Bouwman, A.F., De Haan, B., Klein Goldewijk, K., Klepper, O., Leemans, R., Olivier, J.G.J., Toet, A.M.C., De Vries, H.J.M., Van der Woerd, H.J.: 1994b, Modeling the Global Society-Biosphere-Climate System, Part 2: Computed Scenarios. *Water, Air and Soil Poll.*, **76** (this volume).

van Amstel, A.R. (ed.): 1993, *Proceedings of the International Workshop Methane and Nitrous Oxide: Methods in National Emission Inventories and Options for Control*, Amersfoort, The Netherlands, February 3-5, 1993. RIVM, Bilthoven

Ayres, R.U.: 1978, *Resources, environment and economics - Applications of the materials/energy balance principle*. Wiley Interscience, New York, 204 pp.

Beer, J.G. de, Worrell, E., Blok, K., Cuelenaere, R.F.A.: 1992, *ICARUS 2.1, The potential of energy saving for the Netherlands until 2000 and until 2010* (in Dutch). Database and Report 92-070. Department of Science, Technology and Society, Utrecht University, July 1993.

Benders, R.M.J., Biesiot, W. and Maurits, F.G.: 1993, *Electricity conservation and electrification in OECD Europe*. IVEM-report no. 57 (in Dutch). University of Groningen/IVEM, Groningen, 77 pp.

Berdowski, J.J.M., Olivier, J.G.J. and Veldt, C.: 1993, Methane emissions from fuel combustion and industrial processes. In: *Proceedings of the International Workshop Methane and Nitrous Oxide: Methods in National Emission Inventories and Options for Control*, Amersfoort, The Netherlands, February 3-5, 1993. RIVM, Bilthoven, pp. 131-141.

CEC, DG XI: 1992, *Development of a framework for the Evaluation of Policy Options to deal with the Greenhouse Effect. A scientific description of the ESCAPE model: Version 1.1*. CEC, Brussels, May 1992.

Corinair: 1991, *European Inventory of Emissions of Pollutants into the Atmosphere*. Information brochure. CITEPA No. 053-14/01/91.

EC (European Community): 1988, *Useful energy balance sheets 1985. Supplement to the Energy Statistics Yearbook*. Eurostat, Luxemburg.

Edmonds, J. and Reilly, J.: 1985, *Global energy - Assessing the future*. Oxford University Press, New York/Oxford.

Edmonds, J.A., Reilly, J.M., Gardner, R.H. and Brenken, A.: 1986, *Uncertainty in Future Global Energy Use and Fossil Fuel CO_2-emissions 1975 to 2075*. TR036, DOE3/Nbb-0081 Dist. Category UC-11, National Technical Information Service, U.S. Department of Commerce, Springfield, Virginia.

EPA (U.S. Environmental Protection Agency): 1986, *National Acidification Precipitation Assessment Programme, Version 5.2, 1980 Emissions Inventory*. US-EPA, Washington DC, Report No. EPA-600/7-86-0057a.

EPA (U.S. Environmental Protection Agency): 1990, *Policy Options for Stabilizing Global Climate*. Washington DC. Draft Report to Congress.

Fickett, A.P., Gellings, C.W. and Lovins, A.B.: 1990, Efficient Use of Electricity, *Scientific American*, **263:3**: 28-37.

Fisher, J. and Pry, R.: 1971, A simple substitution model of technical change , *Techn. Forecasting & Social Change* **3**: 75-88.

Grübler, A. and Nakicenovic, N.: 1991, *Evolution Transport Systems Past and Future*, IIASA Research Report 91-8. IIASA, Laxenburg.

Houghton, J., Callander, B. and Varkey, S. (ed.): 1992, *Climate Change 1992. The Supplementary Report to the IPCC Scientific Assessment*. Cambridge University Press, Cambridge.

IEA (International Energy Agency): 1991, *Greenhouse gas emissions: the energy dimension*. OECD/IEA, Paris. ISBN 92-64-13444-1.

IEA (International Energy Agency): 1992, *Energy balances of OECD countries 1989-1990*. OECD/IEA, Paris. Data diskettes dated 15-04-1992.

Janssen P.H.M., Heuberger P.S.C., Sanders R.: 1992, *UNCSAM 1.1: a Software Package for Sensitivity and Uncertainty Analysis; Manual*. RIVM, the Netherlands.

Johansson, T., Bodlund, B. and Williams, R. (ed.): 1989, *Electricity - Efficient end-use and new generation technologies and their planning implications*. Lund University Press. Lund

Johansson, T., Kelly, H., Reddy, A. and Williams, R.(eds.): 1993, *Renewable energy*. Island Press, Washington

Klein Goldewijk, K., Minnen, J.G. van, Kreileman, G.J.J., Vloedbeld, M. and Leemans, R.: 1994, Simulating the carbon flux between terrestrial environment and the atmosphere. *Water, Air and Soil Poll.*, **76** (this volume).

Kram, T.: 1993, *Greenhouse Gas Emissions and National Energy Options*. ETSAP-4 report. ECN, Petten. In prep.

Kreileman, G.J.J. and Bouwman, A.F.: 1994, Computing land use emissions of greenhouse gases. *Water, Air and Soil Poll.*, **76** (this volume).

Levine, M., Feng Liu and J. Sinton: 1992, *China's Energy System*. Annu. Rev. Energy Environ. 1992,17: 405-35

van der Maas, C.W.M., Berdowski, J.J.M., Olivier, J.G.J. and Bouwman, A.F.: 1994, *Information analysis of EDGAR: Atmospheric Database for Global Atmospheric Research* RIVM, Bilthoven, RIVM report no. 776001001 (in prep.)

Maddison, A.: 1991, *Dynamic forces in capitalist development - A long-run comparative view*. Oxford University Press, p. 109.

Marland G. and Rotty, R.M.: 1984, Carbon dioxide emissions from fossil fuels: a procedure for estimation and results for 1950-1982. *Tellus* **36B:4**: 232-261

Marland, G., Boden, T., Griffin, R., Huang, S., Kanciruk, P. and Nelson, T.: 1989, *Estimates of CO2 emissions from fossil fuel burning and cement manufacturing, based on the United Nations Energy Statistics and the U.S. Bureau of Mines cement manufacturing data*. Report no. ORNL/CDIAC-25; NDP-030.

Mount, T., Chapman, A. and Tyrrell: 1974, Electricity demand in the United States: an econometric analysis. In: Macrakis, S. (ed.), *Energy*. MIT Press, Boston

Nakicenovic, N.: 1989, *Technological progress, structural change and efficient energy use: trends worldwide and in Austria. International part*. Final report no. 700/76.7165/9. IIASA, Laxenburg, 304 pp.

Nordhaus, W.D. and Yohe, G.W., 1983: Future Carbon Dioxide Emissions from Fossil Fuels, in: *Changing Climate*. National Academy Press, Washington D.C.

Nørgård, S. and Viegand, J.: 1992, *Low electricity Europe. Sustainable options*. Technical University of Denmark, Lyngby, April

OECD: 1990, *Emission Inventory of Major Air Pollutants in OECD European Countries*. Environment Monographs. OECD, Paris, November 1990

OECD: 1991, *Estimation of greenhouse gas emissions and sinks. Final report from the OECD experts meeting, 18-21 February 1991*. Prepared for IPCC. Revised August 1991. OECD, Paris

OLADE (Latin American Energy Organization): 1993, *Energy-Economic Information System for Latin America and the Caribbean OLADE-SIEE* (Demonstration diskette) Quito (Ecuador)

Olivier, J.G.J.: 1993a, Working Group Report. Nitrous Oxide Emissions from Fuel Combustion and Industrial Processes. A Draft Methodology to Estimate National Inventories, in: R.A. van Amstel (ed.), *Proceedings of the International Workshop Methane and Nitrous Oxide: Methods in National Emission Inventories and Options for Control*, Amersfoort, The Netherlands, February 3-5, 1993, pp. 347-361

Olivier, J.G.J.: 1993b, Nitrous oxide emissions from industrial processes, in: R.A. van Amstel (ed.), *Proceedings of the International Workshop Methane and Nitrous Oxide: Methods in National Emission Inventories and Options for Control*, Amersfoort, The Netherlands, February 3-5, 1993. RIVM, Bilthoven, pp. 339-341

Olivier, J.G.J., Benders, R.M.J., Vries, H.J.M. de, Wijngaart, R.A. van den, and Swart, R.J.: 1994, *An end-use model to develop long-term emissions scenarios from energy related sources*. RIVM, Bilthoven. RIVM report no. 222901010. In prep.

Pepper, W., Leggett, J., Swart, R., Wasson, J., Edmonds, J. and Mintzer, I.: 1992, *Emission scenarios for the IPCC; an update. Assumptions, methodology, and results*. Prepared for IPCC Working Group 1. May 1992

Piccot, S.D., Buzun, J.A. and Frey, H.C.: 1990, *Emissions and cost estimates for globally significant anthropogenic combustion sources of* NO_x, H_2O, CH_4, *CO and* CO_2. Radian Corp., Research Triangle Park NC, Report prepared for EPA, no. EPA-600/7-90-010

Piccot, S., Watson, J. and Jones, J.: 1992, A global inventory of Volatile Organic Compounds emissions from anthropogenic sources, *J. Geophys. Res.* **97:D9**: 9897-9912

Reimer, R.A., Parrett, R.A., and Slaten, C.S.: 1992, Abatement of N_2O emission produced in adipic acid. In: *Proc. of 5th Int. Workshop on Nitrous Oxide emissions*, Tsukuba (JP), July 1-3, 1992

Schipper, L., Meyers, S., Howarth, R.B. and Steiner, R.: 1992, *Energy efficiency and human activity. Past trends, future prospects*. Cambridge Univ. Press, Cambridge

Slesser, M. and De Vries, B.: 1990, *The potential for economic growth in the EC in the context of greenhouse constraints.* IVEM, Groningen University. Report no. 44, 66 pp.

de Soete, G.: 1993, Nitrous oxide from combustion and industry: chemistry, emissions and control, in: *Proceedings of the International Workshop Methane and Nitrous Oxide: Methods in National Emission Inventories and Options for Control*, Amersfoort, The Netherlands, February 3-5, 1993. RIVM, Bilthoven, pp. 287-337.

Spiro, P., Jacob, D. and Logan, J.: 1992, Global inventory of sulfur emissions with 1°x1° resolution, *J. Geophys. Res.* **97:D5**, 6023-6036

Sterner, T.: 1985, Structural change and technology choice - Energy use in Mexican manufacturing industry, 1970-81. *Energy Economics*, **7:** 77-85

Subak, S., Raskin, P. and Hippel, D. von: 1992, *National greenhouse gas accounts: current anthropogenic sources and sinks.* SEI (Stockholm Environment Institute), Boston

Tol, R.S.J.: 1993, *The Climate Fund. Survey of Literature on Costs and Benefits.* Free University/Institute for Environmental Sciences, Amsterdam

UN: 1990, *Industrial Statistics Yearbook 1988. Volume II Commodity Production Statistics 1979-1988*, New York

UNCTAD: 1992, *Handbook of International Trade and Development Statistics 1991*, United Nations, New York

Von Hippel, D., Raskin, P., Subak, S. and Stavisky, D.: 1993, Estimating greenhouse gas emissions from fossil fuel consumption. To approaches compared, *Energy Policy* **21:6**: 691-702.

World Bank: 1979, *INDIA, Economic Issues in the Power Sector* Washington

World Bank: 1990, *World Tables 1989-90* (Data on Diskette) Washington

Yu, O.S. and Shu-Dong He: 1990, Determinants of energy technology transitions: a cost-benefit approach. In: Vasko, T., Ayres, R. and Fontveille, L. (eds.), *Life cycles and long waves.* Lecture notes in Economics and Mathematical Systems, no. 340. Springer Verlag, Berlin

WRI (World Resources Institute): 1990, *World Resources 1990-1991.* Oxford University Press, Oxford

DETERMINING THE POTENTIAL DISTRIBUTION OF VEGETATION, CROPS AND AGRICULTURAL PRODUCTIVITY.

R. LEEMANS AND G. J. VAN DEN BORN

Global Change Department, National Institute of Public Health and Environmental Protection, RIVM
P. O. Box 1, 3720 BA Bilthoven, the Netherlands.

Abstract. The terrestrial biosphere component of the Integrated Model to Assess the Greenhouse Effect (IMAGE 2.0) uses changes in land cover to compute dynamically the greenhouse gas fluxes between the terrestrial biosphere and the atmosphere. Potential land cover for both natural ecosystems and agrosystems, are determined with the Terrestrial Vegetation Model (TVM). TVM consists of separate submodels for the water–balance, global vegetation patterns, crop distribution and potential rain fed crop yield. All these submodels are based on local climatic, hydrological and soil characteristics and appropriate global data bases for those parameters are collected or compiled. The structure of all models, data bases and linkages between them and other modules of IMAGE 2.0 are described. Although computationally demanding, the models give an adequate description of the global vegetation and agricultural patterns. The only discrepancy occurs in regions where the vegetation and agricultural distribution depends on causes other than climatic, such as additional water storage and supply, anthropogenic influence and natural disturbance. Despite this discrepancy, we conclude that TVM simulates satisfactory global vegetation characteristics and that it can be adequately integrated with other models of IMAGE 2.0.

Keywords. biome, climate, crop yield, global data bases, land cover, potential vegetation, simulation models, soil.

1. Introduction

An adequate description of land cover is needed to determine the global fluxes of Greenhouse Gases (GHGs) between the terrestrial biosphere and the atmosphere. Such description is only adequate when it allows for dynamic changes through time, provoked by both anthropogenic and natural causes. Humans nowadays control large shifts in land use that modify or change the current land cover. Land use shifts involve processes such as changing forest cover to crop- and range lands; loss and degradation of productive crop and range lands through overgrazing, drought and other (natural and anthropogenic) factors; conversion of wetlands, urbanization, etc. Land–use shifts are also caused by natural processes such as vegetation dynamics, response to environmental change and disturbances, such as storms, fire and flooding.

Several land cover maps have been generated and used for land–cover studies. The earliest maps were all based on very coarse censuses of the world's climate and vegetation (e.g. von Humboldt, 1807). The perception displayed on these maps is that the vegetation of a given area is a manifestation of regional gradients in climate. This perception has been used to create large-scale climate classifications (e.g. Köppen, 1936; Holdridge, 1947; Budyko, 1986). These classifications all assume that vegetation and climatic patterns are in equilibrium with each other and that climate is the most conclusive factor for vegetation. Walter and Box (1976) made the last assumption explicit and asserted that it is valid for the global-scale distribution of biomes and vegetation types. Even in many

Water, Air, and Soil Pollution **76**: 133–161, 1994.

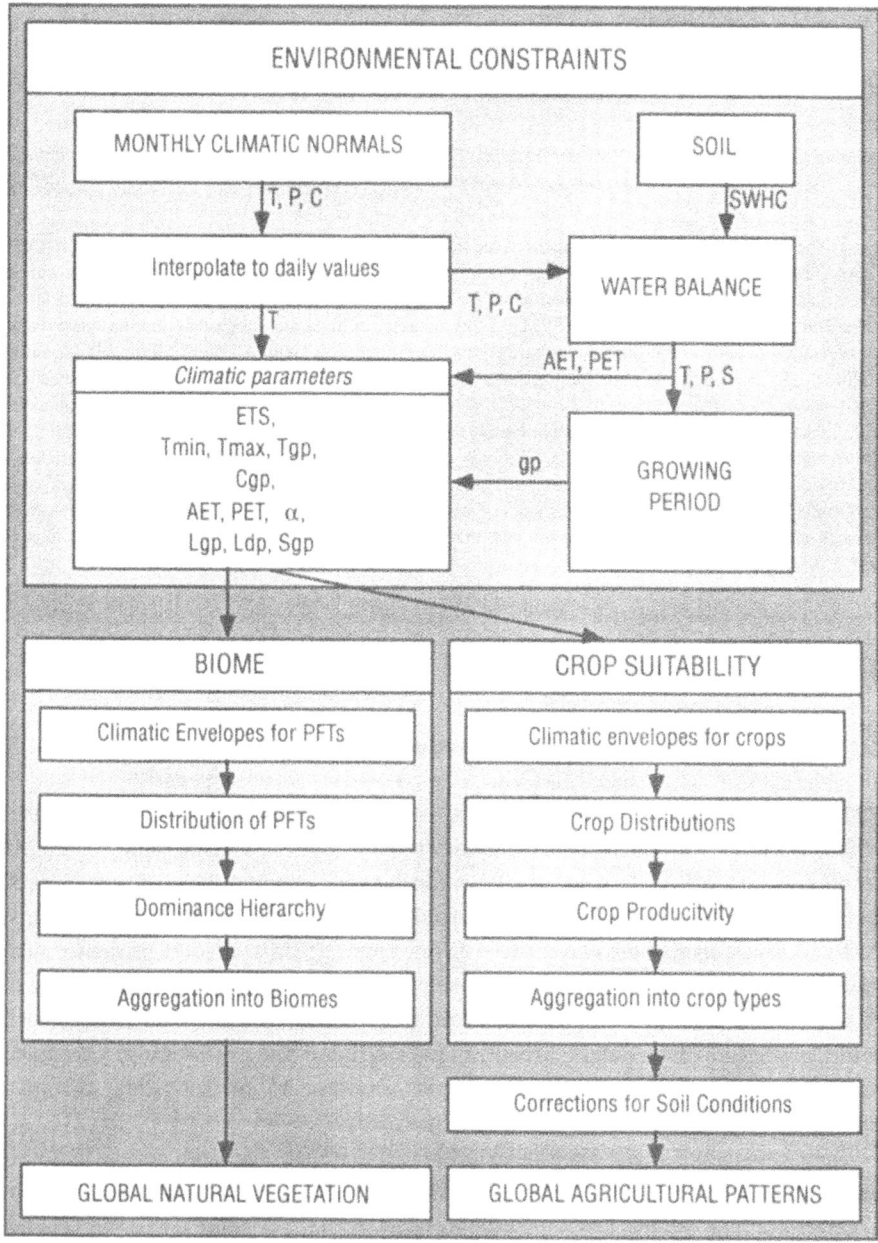

Figure 1: Flow diagram of the Terrestrial Vegetation Model. Input is provided by several global data bases. All climatic parameters and vegetation patterns are saved for use elsewhere in IMAGE or a separate analyses. (T: monthly mean temperature; P: monthly mean precipitation; C: monthly mean cloudiness; SWHC: soil water holding capacity; PET potential evapotranspiration; AET: actual evapotranspiration; S: soil moisture; gp: growing period; ETS: effective temperature sum; T_{min}: mean monthly minimum temperature; T_{max}: mean monthly maximum temperature; T_{gp}: mean temperature of the growing period; C_{gp}: mean cloudiness of the growing period; α: moisture index; L_{gp}: length of the growing period; L_{dp}: length of the dry period; S_{gp}: start of the growing period; PFTs: plant functional types).

recently published global vegetation maps the derivation of vegetation patterns from climate is obvious (e.g. vegetation map by Küchler (1947, in Espenshade and Morrison, 1991)).

Many climate–change impacts studies on natural ecosystems have adopted these climatic classifications (e.g. Emanuel *et al.*, 1985; Guetter and Kutzbach, 1990; Leemans, 1992) and they are often used for defining the vegetation patterns by climate (e.g. Henderson-Sellers, 1991) and C cycle modeling (Prentice and Fung, 1990; Smith *et al.*, 1992). Despite their frequent use, these simple climate classifications have several major disadvantages. First, they consist of an empirically derived set of climatic descriptors to delimit global patterns suggested as being biomes. Projections of these classifications into the future (or past) with fundamentally different climatic conditions could lead to misleading vegetation patterns because the empirical descriptors could be used outside their valid domain. Secondly, all these classifications use "biomes" as their basic unit. This could be sufficient if the primary objective is to describe current vegetation patterns, but is inadequate in the plausible prediction of shifts in future patterns (Leemans, 1994). Finally, the climatic descriptors used to delimit the different classes often do not describe important aspects of climate. For example, seasonality is not adequately considered in the Holdridge Life Zone Classification (Holdridge, 1947). These deficiencies limit the use of the climate classification approach for our purpose: a model that will simulate past, present and future (potential) land–cover under changing environmental conditions.

Here we present a Terrestrial Vegetation Model (TVM) that simulates the potential distribution of vegetation and major crops. We define potential distribution as that distribution that exists under equilibrium conditions, i.e. those areas where a specific crop or ecosystem can occur. For ecosystems the potential corresponds to a fully developed and undegraded system. For crops, it is defined as those regions where conditions are adequate for obtaining an economically feasible yield. The distribution here is only defined for completely rain fed conditions. TVM consists of a set of integrated models and data bases (Figure 1) and plays a central role in the simulation of the terrestrial biosphere within the IMAGE 2.0 Model (Alcamo *et al.*, 1994). The results are used by the Land Cover Change model (Zuidema *et al.*, 1994) to determine actual land cover, which in turn is used to determine GHG emissions. TVM uses climate, if necessary, combined with climatic change (de Haan *et al.*, 1994). It therefore provides a dynamic link between the impacts of global change and land cover.

The main assumption within TVM is that there is a strong linkage between climate, vegetation and crop distributions (Walter and Box, 1976; Leemans, 1992). However, we do not generate a specific climate classification, but use the limiting climatic factors to determine the distribution of 'species'. These factors should therefore describe plant responses more mechanistically and thus mimic important eco–physiological processes. Biomes or agricultural zones are therefore not simulated as such. Ideally, we should have chosen 'species' as the basic unit because it is relatively straightforward to determine the climate dependent, limiting processes for a species (e.g. Woodward, 1987). However, it is impossible to do this for all natural plant and crop species and varieties. The approach taken here is to aggregate the vast amount of species into Plant Functional Types (PFTs).

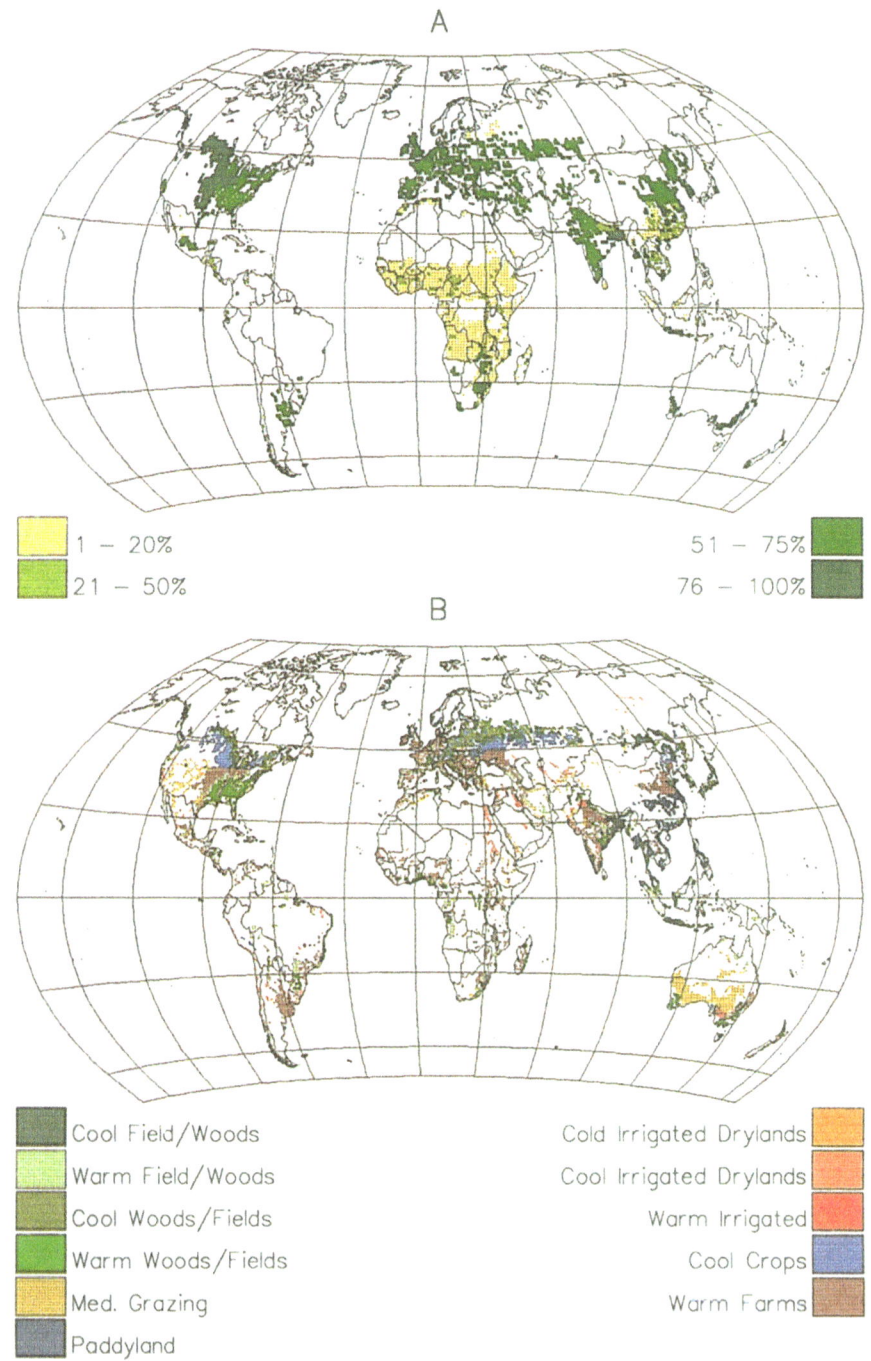

Figure 2: Landuse patterns described by different global land–cover data bases. (a) cultivation intensity by Matthews (1985) and (b) agricultural or mixed classes of Olson *et al.* (1985).

An adequately selected set of PFTs could mimic the wide range of species responses and are therefore a very suitable approach for realistic and comprehensive land cover modeling (Prentice *et al.*, 1989). A comparable approach has been adopted for crop species.

This paper first describes the global environmental data sets, necessary to develop a set of land cover models. Inferior key data sets could limit the development, testing and validation of the proposed set of models. We then continue to discuss the theory, implementation and testing of each model separately against the limitations of available global data sets. Validation runs and models simulations are presented. Finally we will discuss the linkages between this set of models and the other components of IMAGE 2.0.

2. Evaluation of relevant global data bases

Global maps on actual land cover are needed to develop, test and validate global land cover models. Few comprehensive global land cover data sets have been created, documented and published, although some potential vegetation data bases have been derived from climate classifications (e.g. Emanuel *et al.*, 1985; Guetter and Kutzbach, 1990). However, this approach does not result in an adequate land cover map. The most well known land cover data sets are those of Küchler (1947; in Espenshade and Morrison, 1991), Matthews (1985), Olson *et al.* (1985) and Melillo *et al.* (1993). The first two data bases are based on a rather inadequate classification and good documentation is lacking. This renders them less useful. The last data set covers only potential vegetation (Greenland and Antarctica have been left out), and is based on several other data bases (including Matthews, 1985) and several additional regional expert assessments. This has lead to a heterogeneous quality with large regional differences. The most acceptable map hitherto is the global data base compiled by Olson *et al.* (1985). Although there are still some flaws in the classification, it covers actual vegetation and its documentation is exceptionally good. Each item is clearly described by its vegetation structure and the composition of its major species is listed. However, several agricultural land cover classes are combined with natural land cover classes and it is impossible to separate the two satisfactorily. Grazing and pasture lands are not considered and completely hidden in the natural vegetation classes. Savanna and grassland ecosystems seems to be natural vegetation classes, but may be heavily influenced by anthropogenic activity. This becomes obvious when these regions are compared against the "Land Use Intensity" data base of Matthews (1985; Figure 2). The classes in the Olson *et al.* (1985) data base must, therefore, be evaluated carefully with respect to land use before they are used in any model exercise.

An alternative approach to obtaining a global data base on land cover is to use remote sensing techniques. Several countries (regions) are currently compiling these, but no comprehensive data base is available globally, although the advanced technology and methodology is promising and for a large part available (e.g. Tucker *et al.*, 1985; Matson and Ustin, 1991; Rock *et al.*, 1993;). The IGBP Data and Information Systems Programme (IGBP–DIS; Townshend *et al.*, 1991; Townshend, 1992) is setting up a research network for creating a global land–cover data base, but this project has just been initiated and no results are yet available. It is important, however, that our modeling approach be compatible with future results of these remote sensing projects. Remote sensing can provide an adequate and timely description of global land–cover patterns with a coverage that is not obtainable with current ground–based surveys. Further more, it can be used to determine trends objectively. Another, important advantage is that the data can easily be used to scale up from patch to region, allowing for an adequate parameterization and testing of processes at different temporal and spatial scales.

Despite its disadvantages, we have chosen to adopt the Olson *et al.* (1985) data base to describe the current land cover patterns of IMAGE 2.0. The data set is suitable for linking to initialize C models (for which it was originally intended for). The 51 classes of the Olson *et al.* (1985) data base can be linked to the proposed classification of the remotely sensed Land Cover data base (Townshend *et al.*, 1991), so that we can update to an improved data base, when it becomes available. The land–cover classes can further become congruous with the proposed land–cover classification of the FAO (Mücher *et al.*, 1993). This allows for maximum flexibility and compatibility with planned global data compilations worldwide.

For use in IMAGE 2.0, we have aggregated the original 51 classes into 17 major classes (Table 1). These classes represent the major differences in C characteristics between biomes (c.f. Klein-Goldewijk *et al.*, 1994). The land cover class 'wetlands' is important for C dynamics, but its distribution cannot (yet) be simulated directly as a function of large scale environmental factors, such as soil and climate. We assume that this class has been reduced considerably already by anthropogenic activities during the past centuries and its current distribution will remain stable throughout the time horizon of the IMAGE 2.0 simulations.

Another important data set for running the models concerns climate. Climate is difficult to characterize, because it continuously changes and no particular equilibrium climate can be defined (e.g. Gribbin and Lamb, 1978). Therefore climatologists characterize a "normal" climate of each station or region by using a long term average of weather records. The most commonly published climatic normal is for the 30–year period from 1931 to 1960, but recently normals for other periods have been released.

Climatic normals are crucial for describing the interactions between climate and other components of the biosphere, such as vegetation. Climatic parameters needed to describe such interactions should be capable of representing at least the seasonal, but preferable the monthly patterns of climate, and include the most important aspects of temperature (mean, minimum and maximum), precipitation (mean, range), radiation (cloudiness and/or sunshine duration) and, if possible, humidity. Several global sets of climatic normals for station have been compiled (e.g. Anonymous, 1987; Müller, 1982; Federal Climate Complex, 1992; Chadwyck–Healey, 1992), but these compilations do not provide a comprehensive global climate surface at the resolution required here.

Several studies compute the desired climatic parameter for each station and then interpolate it globally. The main disadvantage of this approach is that large deviations can occur from the actual patterns, because non–linearities between stations are neglected

TABLE 1

Aggregation of the Olson *et al.* (1985) data base into the 17 major land–cover types of the Terrestrial Vegetation model (TVM) in IMAGE 2.0. The column PFTs give the associated Plant Functional Types as a result of the BIOME model (cf. Table 3).

Nr.	Land–cover type in TVM	Olson *et al.* (1985) land–cover classes	PFTs in BIOME
1	Agricultural land	Cold irrigated drylands, cool crops, cool fields and woods, cool irrigated drylands, cool woods and fields, warm farms, warm fields and woods, warm irrigated, warm woods and fields, wetlands rice	
2	Ice	Antarctica, polar desert, ice	14
3	Cool Semidesert	Cool desert	13
4	Hot Desert	Sand desert, hot desert,	12
5	Tundra	Wooded tundra, tundra	4
6	Cool Grass and Shrubs	Siberian parks, Cool Grass and Shrubs, highland scrubs	4, 5
7	Warm Grass and Shrubs	Warm Grass and Shrubs	6
8	Xerophytic Woods	Succulent thorns, semiarid woods, mediterranean grazing, low shrubs	8
9	Taiga	Southern Taiga, eastern southern Taiga, northern taiga, main taiga	6, 7
10	Cool Conifer Forest	Cool conifer forests	5, 6
11	Cool Mixed Forest	Cool mixed forests	4, 5, 6, 7
12	Temperate Deciduous Forest	Warm deciduous forests	4, 5, 7
13	Warm Mixed Forest	warm mixed forests, warm conifer forests, tropical montane forests, broadleaved evergreen forests	3
14	Tropical Dry Woodlands and Savanna	Tropical dry forest, tropical savanna	2
15	Tropical Seasonal Forest	Tropical seasonal forests	1, 2
16	Tropical Rain Forest	Equatorial evergreen forests	1
17	Wetlands	Bogs and bog woods, coastal edges, moors and heaths, mangrove, swamp and marsh, water	

(Willmott *et al.*, 1985). Although the approach could sometimes be suitable for a solitary study, performing such interpolation for each parameter and time step would be computationally too inefficient for our IMAGE 2.0 purpose. We therefore need a high–resolution, gridded data base, which is interpolated separately for each monthly climatic normal. This data base should ideally be based on a large number of quality–checked stations, spread evenly throughout the world. Only three such data bases exists (Emanuel *et al.*, 1985; Legates and Willmott, 1990a,b; Leemans and Cramer, 1991).

The data base compiled by Emanuel *et al.* (1985) is mainly based on low–quality climatic records and was therefore not considered by us. Legates and Willmott (1990a,b) used an advanced interpolation scheme, but the included records do not refer to a specific climatic normal. The records span different time intervals between 1850 and 1980, sometimes varying more than a century. Important climatic trends are thus ignored or underestimated. The coverage is further strongly biased towards North America and

TABLE 2

The different reduction factors for each FAO soil type for the Zobler (1986) data base. (S_f =fertility, S_s=salinity, S_a=acidity, S_d=drainage and S_r=rooting depth, Q is the final soil reduction factor. For further explanation see text.)

FAO soil class	S_f	S_s	S_a	S_d	S_r	Q	FAO soil class	S_f	S_s	S_a	S_d	S_r	Q
Eutric Fluvisols	0.9	1.0	1.0	1.0	0.9	0.85	Haplic Xerosols	0.6	1.0	1.0	1.0	0.9	0.57
Calcaric Fluvisols	0.9	1.0	0.9	1.0	0.9	0.81	Calcic Xerosols	0.6	1.0	0.9	1.0	0.9	0.54
Dystric Fluvisols	0.7	1.0	1.0	1.0	0.9	0.66	Gypsic Xerosols	0.6	1.0	0.9	1.0	0.9	0.54
Thionic Fluvisols	0.5	1.0	0.5	1.0	0.5	0.25	Luvic Xerosols	0.6	1.0	1.0	1.0	1.0	0.60
Eutric Gleysols	0.7	1.0	1.0	1.0	0.5	0.42	Eutric Cambisols	0.9	1.0	1.0	1.0	0.9	0.85
Calcaric Gleysols	0.7	1.0	0.9	1.0	0.5	0.40	Dystric Cambisols	0.7	1.0	0.5	1.0	0.9	0.40
Dystric Gleysols	0.6	1.0	0.5	1.0	0.5	0.27	Humic Cambisols	0.9	1.0	1.0	1.0	0.9	0.85
Mollic Gleysols	0.7	1.0	1.0	1.0	0.5	0.42	Gleyic Cambisols	0.7	1.0	1.0	1.0	0.7	0.59
Humic Gleysols	0.7	1.0	1.0	1.0	0.5	0.42	Gelic Cambisols	0.7	1.0	1.0	1.0	0.5	0.42
Plinthic Gleysols	0.6	1.0	0.5	1.0	0.5	0.27	Calcic Cambisols	0.9	1.0	0.9	1.0	0.9	0.81
Gelic Gleysols	0.7	1.0	1.0	1.0	0.5	0.42	Chromic Cambisols	0.7	1.0	1.0	1.0	0.9	0.66
Eutric Regosols	0.7	1.0	1.0	1.0	0.9	0.66	Vertic Cambisols	0.9	1.0	1.0	1.0	1.0	0.90
Calcaric Regosols	0.6	1.0	0.9	1.0	0.9	0.54	Ferralic Cambisols	0.7	1.0	0.5	1.0	0.9	0.40
Dystric Regosols	0.5	1.0	0.5	1.0	0.9	0.35	Orthic Luvisols	0.9	1.0	1.0	1.0	1.0	0.90
Gelic Regosols	0.6	1.0	1.0	1.0	0.5	0.40	Chromic Luvisols	0.9	1.0	1.0	1.0	1.0	0.90
Lithosols	0.6	1.0	1.0	0.5	0.9	0.40	Calcic Luvisols	0.9	1.0	0.9	1.0	1.0	0.85
Haplic Kastanozems	1.0	1.0	1.0	1.0	0.9	0.90	Vertic Luvisols	0.9	1.0	1.0	1.0	0.9	0.85
Calcic Kastanozems	1.0	1.0	0.9	1.0	0.9	0.85	Ferric Luvisols	0.7	1.0	1.0	1.0	1.0	0.70
Luvic Kastanozems	1.0	1.0	1.0	1.0	1.0	1.00	Albic Luvisols	0.7	1.0	0.5	1.0	1.0	0.42
Haplic Chernozems	1.0	1.0	1.0	1.0	0.9	0.90	Plintic Luvisols	0.7	1.0	1.0	1.0	0.7	0.59
Calcic Chernozems	1.0	1.0	0.9	1.0	0.9	0.85	Gleyic Luvisols	0.9	1.0	1.0	1.0	0.7	0.66
Luvic Chernozems	1.0	1.0	1.0	1.0	1.0	1.00	Eutric Planosols	0.6	1.0	1.0	1.0	0.9	0.57
Glossic Chernozems	1.0	1.0	1.0	1.0	0.9	0.90	Dystric Planosols	0.5	1.0	0.5	1.0	0.9	0.35
Cambic Arenosols	0.6	1.0	1.0	1.0	0.9	0.57	Mollic Planosols	0.6	1.0	1.0	1.0	0.9	0.57
Luvic Arenosols	0.6	1.0	1.0	1.0	1.0	0.60	Humic Planosols	0.6	1.0	1.0	1.0	0.9	0.57
Ferralic Arenosols	0.5	1.0	1.0	1.0	0.9	0.47	Sollodic Planosols	0.5	1.0	1.0	1.0	0.9	0.47
Albic Arenosols	0.5	1.0	1.0	1.0	0.9	0.47	Gelic Planosols	0.6	1.0	1.0	1.0	0.5	0.40
Rendzinas	0.7	1.0	0.9	0.5	0.9	0.40	Eutric Histosols	0.6	1.0	1.0	1.0	0.5	0.40
Rankers	0.5	1.0	1.0	0.5	0.9	0.37	Dystric Histosols	0.5	1.0	0.5	1.0	0.5	0.25
Ochric Andosols	0.9	1.0	1.0	1.0	0.9	0.85	Gelic Histosols	0.6	1.0	1.0	1.0	0.5	0.40
Mollic Andosols	0.9	1.0	1.0	1.0	0.9	0.85	Eutric Podzoluvisols	0.7	1.0	1.0	1.0	1.0	0.70
Humic Andosols	0.9	1.0	1.0	1.0	0.9	0.85	Dystric Podzoluvisols	0.6	1.0	0.5	1.0	1.0	0.40
Vitric Andosols	0.9	1.0	1.0	1.0	0.9	0.85	Gleyic Podzoluvisols	0.7	1.0	1.0	1.0	0.7	0.59
Pellic Vertisols	0.7	1.0	1.0	1.0	0.9	0.66	Ortic Podsols	0.6	1.0	0.5	1.0	0.9	0.37
Chromic Vertisols	0.7	1.0	1.0	1.0	0.9	0.66	Leptic Podsols	0.6	1.0	0.5	1.0	0.9	0.37
Haplic Phaeozems	1.0	1.0	1.0	1.0	0.9	0.90	Ferric Podsols	0.6	1.0	0.5	1.0	0.9	0.37
Calcaric Phaeozems	1.0	1.0	0.9	1.0	0.9	0.85	Humic Podsols	0.6	1.0	0.5	1.0	0.9	0.37
Luvic Phaeozems	1.0	1.0	1.0	1.0	1.0	1.00	Placic Podsols	0.6	1.0	0.5	1.0	1.0	0.40
Gleyic Phaeozems	1.0	1.0	1.0	1.0	0.7	0.70	Gleyic Podsols	0.6	1.0	0.5	1.0	0.7	0.32
Ortic Greyzems	0.7	1.0	1.0	1.0	1.0	0.70	Orthic Ferrasols	0.6	1.0	0.5	1.0	0.9	0.37
Gleyic Greyzems	0.7	1.0	1.0	1.0	0.7	0.59	Xantic Ferralsols	0.6	1.0	0.5	1.0	0.9	0.37
Ortic Solonchaks	0.5	0.5	1.0	1.0	0.9	0.35	Rhodic Ferralsols	0.6	1.0	0.5	1.0	0.9	0.37
Mollic Solonchaks	0.6	0.5	1.0	1.0	0.9	0.37	Humic Ferralsols	0.6	1.0	0.5	1.0	0.9	0.37
Takyric Solonchaks	0.5	0.5	1.0	1.0	0.9	0.35	Acric Ferralsols	0.5	1.0	0.5	1.0	0.9	0.35
Gleyic Solonchaks	0.5	0.5	1.0	1.0	0.7	0.30	Plintic Ferralsols	0.5	1.0	0.5	1.0	0.7	0.30
Orthic Solonetz	0.5	0.5	1.0	1.0	0.9	0.35	Orthic Acrisols	0.7	1.0	0.5	1.0	1.0	0.42
Mollic Solonetz	0.6	0.5	1.0	1.0	0.9	0.37	Ferric Acrisols	0.7	1.0	0.5	1.0	1.0	0.42
Gleyic Solonetz	0.5	1.0	1.0	1.0	0.7	0.42	Humic Acrisols	0.7	1.0	0.5	1.0	1.0	0.42
Haplic Yermosols	0.6	1.0	1.0	1.0	0.9	0.57	Plinthic Acrisols	0.7	1.0	0.5	1.0	0.7	0.32
Calcic Yermosols	0.6	1.0	0.9	1.0	0.9	0.54	Gleyic Acrisols	0.6	1.0	0.5	1.0	0.7	0.32
Luvic Yermosols	0.6	1.0	0.9	1.0	0.9	0.54	Dystric Nitosols	0.7	1.0	1.0	1.0	0.9	0.66
Gypsic Yermosols	0.6	1.0	1.0	1.0	1.0	0.60	Eutric Nitosols	1.0	1.0	1.0	1.0	0.9	0.90
Takyric Yermosols	0.6	1.0	1.0	1.0	0.9	0.57	Humic Nitosols	1.0	1.0	1.0	1.0	0.9	0.90

Europe. These drawbacks are unacceptable, when the main objectives of a study concern global coverage and rapid climatic change. Leemans and Cramer (1991) have compiled their records mainly for the period 1931–1960 and have obtained a reasonable global coverage, although Africa and South America and Asia have somewhat lower station densities than the rest of the world. The data base was especially developed for ecological and agricultural climate–impact research by removing less relevant stations. They further used a simple correction scheme for topography and temperature. Mountainous regions are therefore better represented than in the other data bases and this is the main reason for inclusion of this data base into the IMAGE 2.0 model.

The third set of data needed to develop and run a set of land cover models contains soil characteristics. No high–quality soil data set exists, but several projects that lead to a globally comprehensive data set are underway (Bouwman and Leemans, 1994). However, several digitized implementations of the FAO soil–map–of–the–world (FAO/UNESCO, 1974) have been created (e.g. FAO/CSRC, 1974; Zobler, 1986). These data sets all include the major FAO–soil, texture and slope classes, but the resolution and digitization methodologies have been different. For the input of our models, we do not need soil type or texture class, but derivatives of this data, such as water holding capacity, soil fertility, salinity and soil depth. These parameters can only be derived indirectly from the existing data bases.

For soil water holding capacity we have used the approach of Prentice *et al.* (1992). This approach gives a very coarse resolution with only few classes, which is prone for further improvements. We have further created a soil properties data base, which is derived from the different soil classes of the Zobler (1986) data base. Each class is assigned a fertility, salinity, and acidity level, drainage and rooting conditions (Table 2). FAO is currently creating advanced data bases on these parameters (Brinkman, personal communication) and as soon as these are available we will include them.

As can be concluded from this section, availability and quality of data bases with a global coverage is still poor. Fortunately, several agencies have planned improvements by peer reviewing and disseminating existing data (e.g. Kineman, 1992) and this has already led to a very useful compilation of global data relevant for ecological studies (Kineman and Ohrenschall, 1992). With the initiatives of IGBP–DIS and START such data sets will become more easily available to the research community. This should be very beneficial for our type of modeling efforts. We are eager to evaluate such new data bases and to include them in our analyses as input, testing and validation data, hopefully increasing the reliability of our models and sub–models.

3. Determination of potential land cover

The above data sets were used as input, testing and validation of TVM. However, due to the limitations and regional differences in quality of these data sets, we cannot achieve an equal and comprehensive coverage of global vegetation and crop patterns. Simulations of some regions will be much better than others. A globally comprehensive calibration of

TVM is virtually impossible. Tuning the values of limiting factors empirically will improve the performance in some regions, but surely reduce that of other regions. An objective measure of global performance is difficult to obtain (Monserud and Leemans, 1992). This argues again for a more basic approach that represents our current understanding of the processes that determine vegetation and crop distribution.

3.1. ENVIRONMENTAL CONSTRAINTS

TVM calculates environmental constraints based on climate and soil characteristics. We first discuss the constraints caused by climate. Plant distributions are limited by mean and extreme temperatures and moisture conditions (Woodward, 1987). In addition, plants have to compete through their entire life cycle successfully to subsist. Each phase, such as establishment, growth, maturation and seed–setting, requires different environmental conditions including specific heat and moisture requirements. These requirements can be defined as different characteristics of climatic normals (e.g. summer or winter temperatures). Below we first present the model to determine moisture conditions, then a model to describe the growing season and finally some additional climatic indices.

3.1.1 The water balance model

The water balance model is based on the approach developed by Prentice *et al.* (1993). It is based on the assumption that all water is transmitted to plants through the soil. Roots seize soil water, which is transported through stems to the leaves, where it transpires. Soil characteristics determine the extent to which moisture can be carried through into dry seasons. Local, grid cell specific, water shortage can be releaved by irrigation (or drainage); excess water disappears as runoff. The model is a simple bucket model (Figure 3) that accounts for the hydrological budget of a single soil water store as driven by quasi daily precipitation, drainage, temperature, and radiation. Output is actual and potential

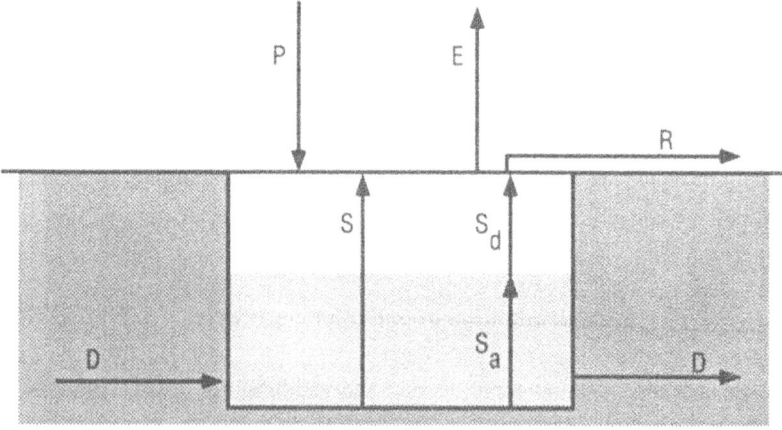

Figure 3: The different elements of the soil moisture model. (E=evapotranspiration; P=precipitation; D=drainage; R=runoff; S= soil moisture; S_a= actual soil moisture; S_d =soil moisture deficits).

evapotranspiration (AET and PET, respectively). We assume that direct evaporation from the soil, covered by vegetation, is small and can be neglected. Precipitation is assumed to be spread evenly throughout the month, allowing for adequate infiltration into the soil. Runoff caused by a too low infiltration rate is neglected, which can sometimes lead to an over estimation of actual soil moistures. The AET is based on a demand/supply function (Federer, 1982) between water availability and the potential evaporation (PET). The latter is based on an equilibrium transpiration of the landscape, which is mainly driven by incident radiation (proportion of possible hours of bright sunshine) and location. The full approach is described by Prentice *et al.* (1993) and is valid for many different environmental conditions.

The quasi daily input values are interpolated from the monthly climatic normals. For simplicity, we assume that the daily values are spread evenly through each month and that all precipitation comes as rain. Snow accumulation and drainage are not (yet) accounted for. These assumptions are valid for wet and moist conditions, such as the temperate forest regions, but are less so for dryer or colder conditions. Despite these shortcomings, we rely on these simplifications because use of more realistic interpolation schemes and models, such as weather generators (e.g. Hutchinson, 1987) or local water balances, have enormous data and/or computational demands, and are difficult to calibrate and test globally.

The main output of the soil moisture model is daily soil moisture availability (mm). This is the result of evaluating the daily P–AET and soil water capacity (difference between field capacity and wilting point). If daily soil moisture availability is larger than soil water capacity, the excess is removed as runoff (not tracked). Soil water capacity is derived from a soil data base (Prentice *et al.*, 1992). The initial soil water content at day 1 (January 1st) is set to the lowest value of soil water capacity or total annual precipitation. The model simulates soil water content for one year until the values of the last day converge to those of the first day. This conversion process could take a few iterations under some xeric or cold conditions, elsewhere it requires only one.

3.1.2 Additional climate constraints

The water balance model is also used in combination with the temperature regime to define the characteristics of the growing season. The growing season is defined as that period during the year when warmth and soil moisture are adequate for growth (including germination, growth and maturation).

The growing period for crops (Figure 4) is characterized by the annual pattern of daily evapotranspiration, precipitation and soil moisture values. Growth only occurs at temperatures above 5°C. The moisture season follows a more complex pattern (Anonymous, 1978). In regions with a distinct dry season (inadequate moisture availability), the growing period starts arbitrarily when precipitation equals half the PET. The rationale is that the first rains that fall on the surface of a soil with a large moisture deficit can be utilized immediately for seed germination and early growth. The soil moisture therefore does not need to be replenished completely (Anonymous, 1978). A successful crop life cycle requires further a period when the evapotranspirative demands

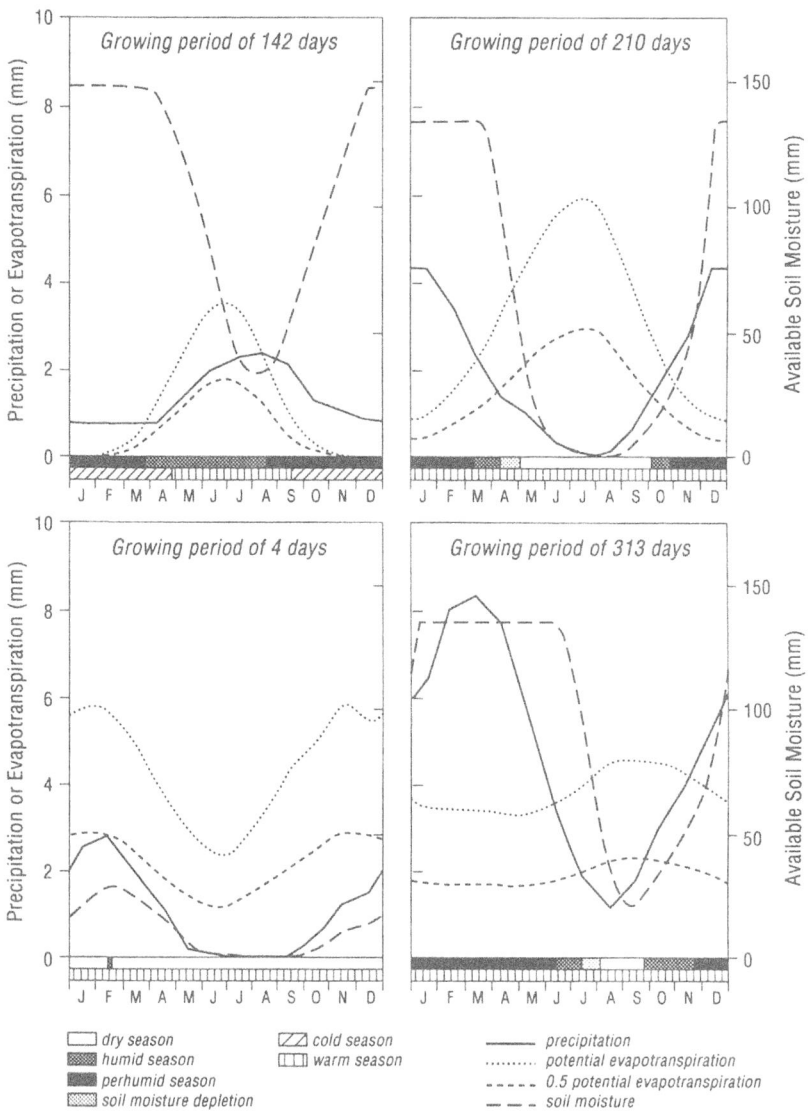

Figure 4: The growing period diagram for a boreal moist (Archangels, FSU), mediterranean (Creta, Greece), hot desert (Kalahari, Namibia) and tropical seasonal climate (Manaus, Brazil). The bars at the bottom of each graph indicate the moisture conditions (upper bar) and temperature conditions (lower bar) throughout a year.

of a crop under a full canopy cover can be met.

Besides moisture conditions and growing season characteristics, several other climate indices are computed. Aggregate seasonal and annual precipitation indices are determined by summing monthly mean precipitation (December–February, March–May, June–August and September–November). Temperatures of the coldest and warmest month are determined from the monthly mean temperature values. The next set of indices determines

TABLE 3

Climatic envelopes for each Plant Functional Types (PFTs) as defined in the BIOME model. The parameters for each PFT are (1) Effective Temperature Sum (ETS) base 0°C; (2) ETS, base 5°C; (3) mean temperature of the coldest month; (4) Mean temperature of the warmest month; and (5) annual α moisture index (AET:PET). The last column (6) gives the dominance class for each PFT. Only PFTs from the highest dominance class remain (cf. Prentice *et al.*, 1992).

	1	2	3	4	5	6
Trees:						
1. Tropical evergreen trees	none	none	≥ 15.5	none	≥ 0.80	1
2. Tropical deciduous trees	none	none	≥ 15.5	none	0.45 - 0.95	1
3. Warm–temperate evergreen trees	none	none	≥ 5.0	none	≥ 0.65	2
4. Temperate deciduous trees	none	≥ 1200	-15.0 - 15.5	none	≥ 0.65	3
5. Cool–temperate conifers	none	≥ 900	-19.0 - 5.0	none	≥ 0.65	3
6. Boreal evergreen conifers	none	≥ 350	-35.0 - -2.0	none	≥ 0.75	3
7. Boreal deciduous trees	none	≥ 350	< 5.0	none	≥ 0.65	3
Non-trees:						
8. Sclerophyll shrubs and Succulents	none	none	≥ 5.0	none	≥ 0.28	4
9. Warm grasses and shrubs	none	none	none	≥ 22.0	≥ 0.18	5
10. Cool grasses and shrubs	none	≥ 500	none	none	≥ 0.33	6
11. Cold grasses and shrubs	≥ 100	none	none	none	≥ 0.33	6
12. Hot desert plants	none	none	none	≥ 22.0	none	7
13. Cool desert plants	≥ 100	none	none	none	none	8
14. Polar desert	none	none	none	none	none	9

the amount of heat available for plant growth throughout the year. Effective Temperature Sums (ETS) are computed by using the interpolated daily temperature values:

$$ETS = \sum_{i=1}^{365} t_i - t_{th} \quad \text{for } t_i > t_{th}$$
$$ETS = 0 \quad \text{for } t_i \leq t_{th} \tag{1}$$

where t_i is the daily temperature and t_{th} a predefined threshold value. We used the values 0.0°C and 5.0°C for t_{th} (Table 3). The different threshold value are used for evergreens or deciduous species respectively (cf. Prentice *et al.*, 1992). A moisture index, α, is computed from the water balance model. α is defined as the ratio of annual AET and PET. The value ranges from 0 (very xeric conditions) to 1 (wet conditions). The indices described above are all used by both the BIOME and the crop distribution model.

The crop–yield model requires several additional climatic indices, averaged over the growing period. If a day is suitable for plant growth, then the climatic data for this day will be used to determine this average. These characteristics include the growing period average maximum and minimum temperatures, cloudiness and radiation. The maximum temperature mean refers to day temperatures (°C) and is used to determine photosynthetic rates. The minimum temperatures refers to a 24–hours temperature mean (°C) and is used to define soil and plant respiration rates. These processes continue even in the dry season, although at a slower rate. Therefore, average respiration rates were computed for a longer period than the actual growing period, namely for all days with temperatures >0°C and adequate soil moisture.

Some additional information on the growing period (length, continuous or discontinuous and the starting date) are stored as output. It is not required directly by the crop model, but they can be used to assess specific impacts, such as the timing of different ecological and phenological events, such as the timing of flowering or seed maturation. These data are particular important when conservation and species relations are of concern (Peters, 1992). All computed climatic indices determined are stored for off-line plotting and analyzing or to be used for advanced impact assessment, not necessary for the scope of IMAGE 2.0.

3.2   DISTRIBUTION OF VEGETATION PATTERNS

3.2.1   Estimating natural vegetation with the BIOME model

Several schemes have recently been developed to simulate global vegetation patterns (e.g Box, 1981; Neilson *et al.*, 1992; Prentice *et al.*, 1992). We have adopted the BIOME approach by Prentice *et al.* (1992). This model is conceptually less complex than many others, uses PFTs and a set of climatic parameters that can easily be derived from the climate data base. Previous applications have demonstrated that the BIOME model is capable of adequately simulating global natural vegetation patterns (Prentice *et al.*, 1992; Leemans, 1992).

PFTs are defined for distinct latitudinal zones (tropical, temperate and boreal or warm, cool and cold). The labels for these zones are only for convenience and do not imply a rigid boundary. Secondly, for each zone the model takes into account the major physiognomic adaptations of vegetation relating to (limiting) climatic factors such as evergreen *vs* deciduous, broad-leaf *vs* needle–leaved and woody *vs* herbaceous. These definitions show similarities to Küchler's (1949) physiognomic classification. The PFTs in BIOME (Table 3) represent an oversimplification of the variety of plant types that exist, but the aim was to construct a model that would only predict the distributions of major biomes and their transient zones rather than the distribution of many different types of species.

The BIOME model relates the distributions of each PFT to the earlier computed ETS (eq. 1), as well as mean temperatures of the coldest and warmest month, and the α index (Table 3). Since upper or lower limits (or both) were often considered meaningless to limit a specific PFT, these limits were not specified. For example, tropical evergreen trees are assumed to tolerate no frost and to have a high moisture requirement. Based on a worldwide regression of annual minimum temperature against mean coldest-month temperatures, 'no frost' implies a coldest month temperature of > 15.5 °C. Based on map comparisons, a 'high moisture requirement' means a α–index of at least 0.8. Relevant limits for all other plant types are based on similar reasoning. Finally, a 'dominance hierarchy' in which PFTs dominate over others (e.g. trees dominate over grasses) is

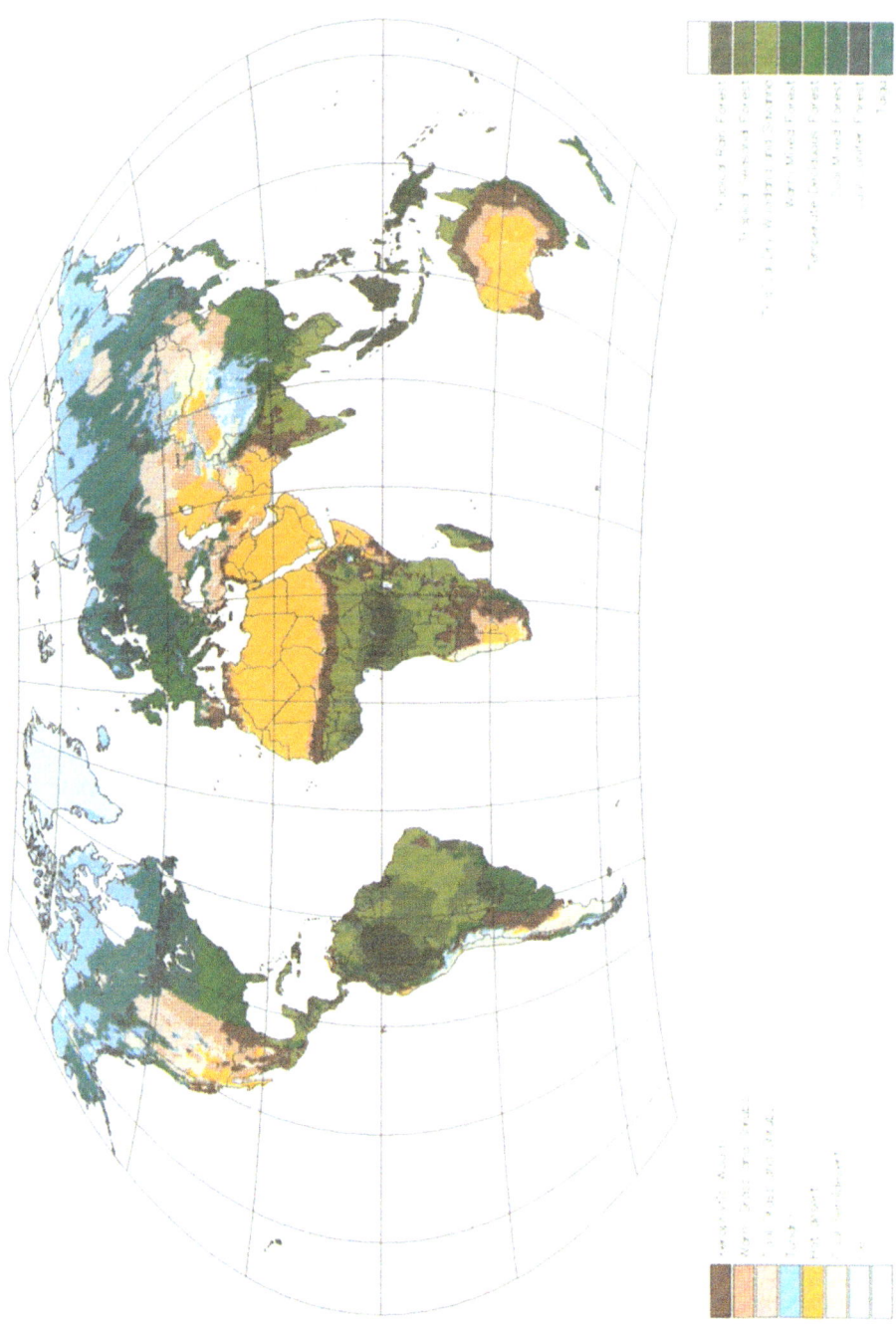

Figure 5: The potential vegetation cover for current climate as determined by the BIOME model.

applied to define the final potential vegetation (Table 3). To assign each location to a biome this hierarchy was strictly applied so that only PFTs from the highest level present were retained.

The total set of rules were used to assign each location a biome, which consisted of an array of unique PFTs that emerged from the rules. The combinations of PFTs were not described in advance, so the model should be quite capable of generating novel combinations under a changed climate. The model generates 15 different combinations for current climate (Figure 5 and Table 1).

3.2.2 Validation of the simulated patterns of the BIOME model

It is difficult to validate the simulated vegetation patterns (Figure 5). The calculations resemble other global maps, but large differences still occur in several regions. Using the Kappa statistic (Monserud and Leemans, 1992), we obtained a good correlation between the simulated patterns and Olson et al.'s (1985) land cover data base. However, such good pattern correlation can be misleading, because it does not actually represent a direct cell to cell comparison.

In this version of the IMAGE model, we want to calculate the climate–induced vegetation shifts with the BIOME model. In this case, cell to cell inaccuracies between the initial data base (Olson et al., 1985) and the simulated pattern will lead to large undesired shifts early during the simulation resulting in unrealistic changes with large consequences for the C cycle. The correlation of the cell to cell comparison, however, is very poor (more than 25% of the cells were different). When the difference pattern is investigated more closely, it become obvious that most discrepancies were along boundaries of specific vegetation zones. In addition, the model computed that some high altitude and latitude regions have too harsh conditions for plants to survive. This is surely due to flaws in the climate data base. Temperatures in these regions are based on some few, unrepresentative stations with extreme cold temperatures, which have too great an influence on data interpolations. Further more, the assumed lapse rate corrections could lead to colder than observed temperatures at high elevations. With the planned improved climate data bases (Cramer and Leemans, pers. comm.) these deficiencies can be reduced.

Apart from these data–related faults, there are still many cells with incorrectly computed vegetation, particularly grassland–forests or forest–grassland switches. These may be caused by other then climatic factors, such as complex soil conditions that are not considered by the one–layer soil water balance model, or from natural and anthropogenic disturbances, such as fires. As already stated above, it is not possible from the Olson et al. (1985) data base to distinguish between human–induced or natural grasslands.

All these factors lead to a poor cell to cell performance. Until better data bases become available, we cannot circumvent this serious problem. The only way to include reasonable climate–induced vegetation shifts is to initialize the simulations with a combination of the land cover patterns, resulting from BIOME and Olson et al. (1985). The first should give all natural vegetation classes, while the latter the agricultural land–cover classes. The location of human–induced grasslands must be considered carefully and compared with other land–use data bases (e.g. Matthews, 1985) to obtain a satisfactory initial land cover.

TABLE 4

Climatic crop requirements for 12 major crops varieties (GPL= length of the growing period (# days); MTR = temperature of the coldest month (°C); ETS= effective temperature sum (°C); MR: annual α moisture index (AET:PET); CT: crop type (cf. Figure 6); H_i: Harvest index).

	GPL	MTR	ETS	MR	CT	Hi	Commodity
Temperate maize	≥ 130	-20.0 – 15.0	≥ 1500	none	IV	0.40	Maize
Tropical maize	≥ 175	≥ 5.0	≥ 3000	none	III	0.30	Maize
Rice	≥ 135	≥ – 7.5	≥ 2250	≥ 0.95	II	0.30	Rice
Spring wheat	≥ 75	< 5.0	≥ 950	none	I	0.45	Cereals
Winter wheat	≥ 165	< 10.0	≥ 1250	none	I	0.35	Cereals
Millet	≥ 80	≥ – 25.0	≥ 1500	< 0.95	IV	0.25	Millet
Potatoes	≥ 90	< 15.0	≥ 750	none	I	0.55	Roots
Cassava	≥ 170	≥ 10.0	≥ 4500	none	II	0.60	Roots
Pulses	≥ 110	< 20.0	≥ 1000	none	I	0.35	Pulses
Sugar Beet	≥ 160	< 15.0	≥ 1000	none	I	0.65	Sugar
Sugar Cane	≥ 240	≥ 10.0	≥ 4500	none	III	0.95	Sugar
Soybeans	≥ 115	< 20.0	≥ 2000	none	II	0.40	Oil
Oilpalms	≥ 330	≥ 10.0	≥ 5500	none	II	0.35	Oil
Sunflower	≥ 330	< 10.0	≥ 1500	none	II	0.35	Oil
Rapeseed	≥ 330	< 10.0	≥ 900	none	II	0.35	Oil
Cottonseed	≥ 330	≥ – 5.0	≥ 2500	none	II	0.35	Oil

These initialization problems emphasize again the importance of comprehensive data collection projects with high quality standards.

3.2.3 Distribution and productivity of major crops

To speed up the computation of agricultural patterns, extreme cold and dry climates are omitted from the simulation of agricultural productivity. These climates are those where the temperature of the warmest month is less then 1.0 °C and/or where the α index is less then 0.05. For all other regions the potential productivity of a global set of important crops is determined.

As noted above, economic yield of each crop requires the successful completion of each life phases, including germination, growth, flowering, seed setting and maturation. The phenology of these events is also strongly influenced by climatic constraints and is uniquely timed for each phase.This time-dependent sequence is crop specific and, in the model, characterized by a required minimum length of the growing period (Table 4). This minimum growing period length is adjusted for the winter varieties of several cereal crops by splitting the growth period over two years. The final length of the growing season is adjusted accordingly.

We have also used a comprehensive set of climatic parameters to specify additional climatic crop requirements or constraints. Several crops require a cold period for their vernalization processes. This requirement is included by using a maximum temperature for the coldest month (Table 4). Since some crops cannot tolerate low temperature, we also specify a minimum value for the coldest month temperature. The crops further require adequate warmth during the growing period to attain adequate growth and size for harvest. These warmth requirements are simulated using ETS (eq. 1) above 5 °C with the minimum values given in Table 4. Finally, some crops need a dry period for seed maturation. This is simulated by defining a maximum value for the α index. Together, these climatic crop requirements delimit the distribution of crops.

If a crop can grow in a certain region, its productivity is determined using a simple photosynthetic model based on the crop models of de Wit (1965) and adapted from the specific approach by FAO (Anonymous, 1978). Here we present only a short summary, because the full implementation is given in Leemans and Solomon (1993).

The model simulates net dry–weight biomass production throughout the growing period. The original models were specifically developed to manage the annual path of crop growth, but we have adapted them to compute a mean growth rate during the growing period. This simplified model can capture the major patterns of crop areas, at the expense of lower spatial resolution. Total yield of a crop (B_y) is determined by both photosynthesis and respiration processes, given by:

$$B_y = H_i (B_g - R) \tag{2}$$

where B_g is the gross biomass production (t ha^{-1} dry weight), R is the respiration loss (t ha^{-1} dry weight) and H_i the harvest index, which converts biomass production into

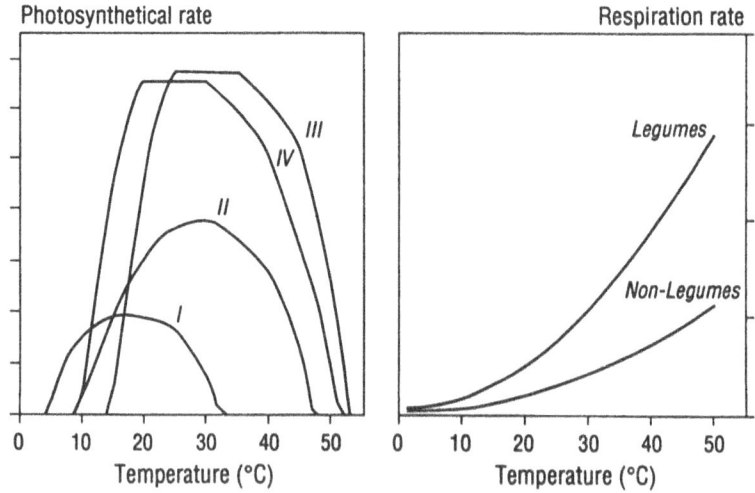

Figure 6: Photosynthetic (left) and respiration (right) responses for different crop types.

economically–useful yield (in t/ha dry weight; Table 4). Photosynthesis is governed by the total amount of irradiance, which is dependent on latitude and cloudiness fraction during the growing period and is also a function of temperature. This function is crop–specific and we account for four different crops types (Table 4 and Figure 6), each with a specific temperature range (temperate (type I and IV) *vs* tropical (type II and III) and photosynthetic response (Type I and II represent the C_3 photosynthetic pathway, while III and IV the C_4 pathway). Growth of C_3 plants is sensitive to atmospheric CO_2 concentrations and thier photosynthesis rates increase under increasing concentrations. C_4 plants are much less sensitive (Bazzaz and Fajer, 1992). A minimum, optimum and maximum photosynthetic rate can be observed for each crop type. Respiration rates increase exponentially with temperature. The response differs only between legume–crops and all others. The model is incorporated using the different indices computed from the climate data base (Leemans and Cramer, 1991).

The computation results in the potential distribution for each crop over the global grid, assigning each cell a potential rain fed productivity. To define the actual potential crop production, soil characteristics could also be of importance. The productivity is therefore adjusted for grid–specific soil conditions based upon the methodology defined by the Land Evaluation Computer System (LECS: Wood and Dent, 1983) and data from the global soil data base (Zobler, 1986). The method calculates the percentage of potential productivity for different climatic, soil and terrain conditions. Since topography and climatic conditions are already covered implicetely in the model and its underlying data bases, only soil characteristics are used here.

Reduction factors for fertility (S_f), salinity (S_s), acidity (S_a), rooting depth (S_r) and drainage conditions (S_d) were determined for each soil class of the data base. These soil reduction factors were aggregated into three major soil–quality reduction factors: fertility (Q_1), which is solely based on S_f; salinity and toxicity (Q_2), which are based on S_s and S_a; and rooting conditions (Q_3), which are based on S_r and S_d. The specific value for each quality is obtained from the lowest available score of the separate reduction factors:

$$Q_1 = S_f$$
$$Q_2 = \min(S_a, S_s) \tag{3}$$
$$Q_3 = \min(S_d, S_r)$$

The final soil reduction factor (Q) depicts the "unconstrained yield" reduction and the result of weighting the major soil–reduction qualities according the following equations:

$$Q_4 = \min(Q_1, Q_2, Q_3)$$
$$Q = Q_4 \frac{(Q_1 + Q_2 + Q_3 - Q_4)}{2} \tag{4}$$

Q is generally smaller than 1 (Table 2) and is multiplied with the potential rain fed yield to obtain a local potential yield. Together with the global soil data base (Zobler, 1986) the

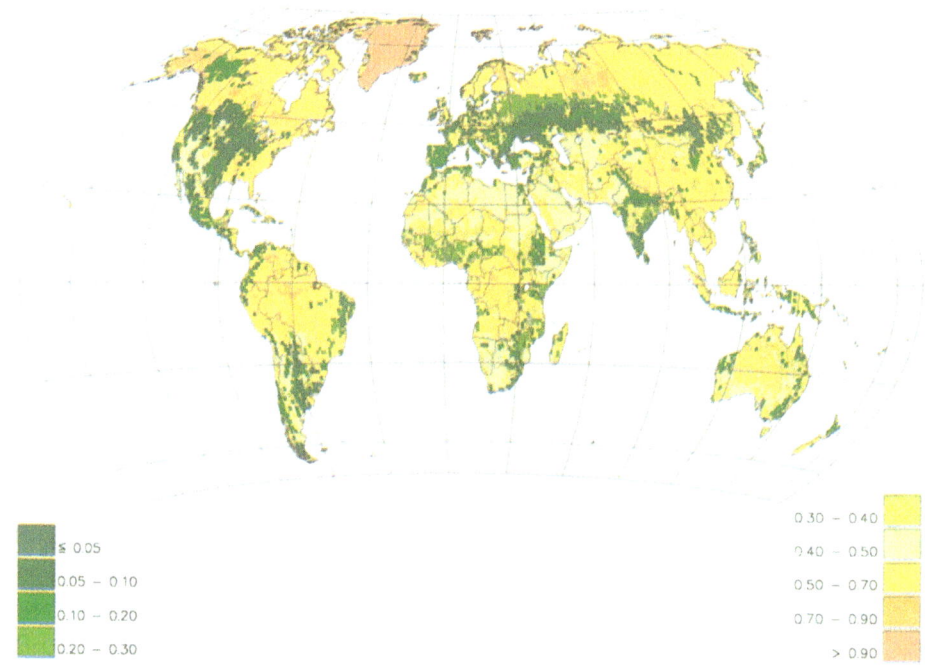

≤ 0.05	0.30 – 0.40
0.05 – 0.10	0.40 – 0.50
0.10 – 0.20	0.50 – 0.70
0.20 – 0.30	0.70 – 0.90
	> 0.90

Figure 7: The global distribution of the soil reduction factors (cf. Table 2) based on the Zobler (1986) soil data base.

final soil reduction factor for each cell of the global terrestrial grid can be determined (Figure 7). These reduction factors are also used in determining the response of soil properties on feedbacks in the C cycle (Vloedbeld and Leemans, 1993; Klein-Goldewijk *et al.*, 1994).

There are several disadvantages of this appraoch. First, only one generic set of yield reduction values is implemented. This implies that there is no discrimination between different crops, which are all adapted to different environmental conditions. Second, the initial computation of potential yield by climate can underestimate the important influence of soils. In the orignal LECS methodology climate, soil and terrain characteristics are included simulteneously, so that locally a more realistic yield can be determined.

The crop model will generate different distributions and productivity patterns for different climate change. The 16 crops listed in Table 4 contain the most important (tropical and temperate) varieties of major crops. The specific requirements of these varieties demand the simulation of each then by TVM. However, the IMAGE 2.0 Land Cover Model (Zuidema *et al.*, 1994) uses only major agricultural products or commodities. The linkage with this submodel is therefore achieved by aggregating the different crop varieties of a single crop type into commodities (Table 4).

Most crops are annuals or perennial and their migration is independent of natural processes. In contrast to the BIOME model, this crop model can therefore be dynamic with

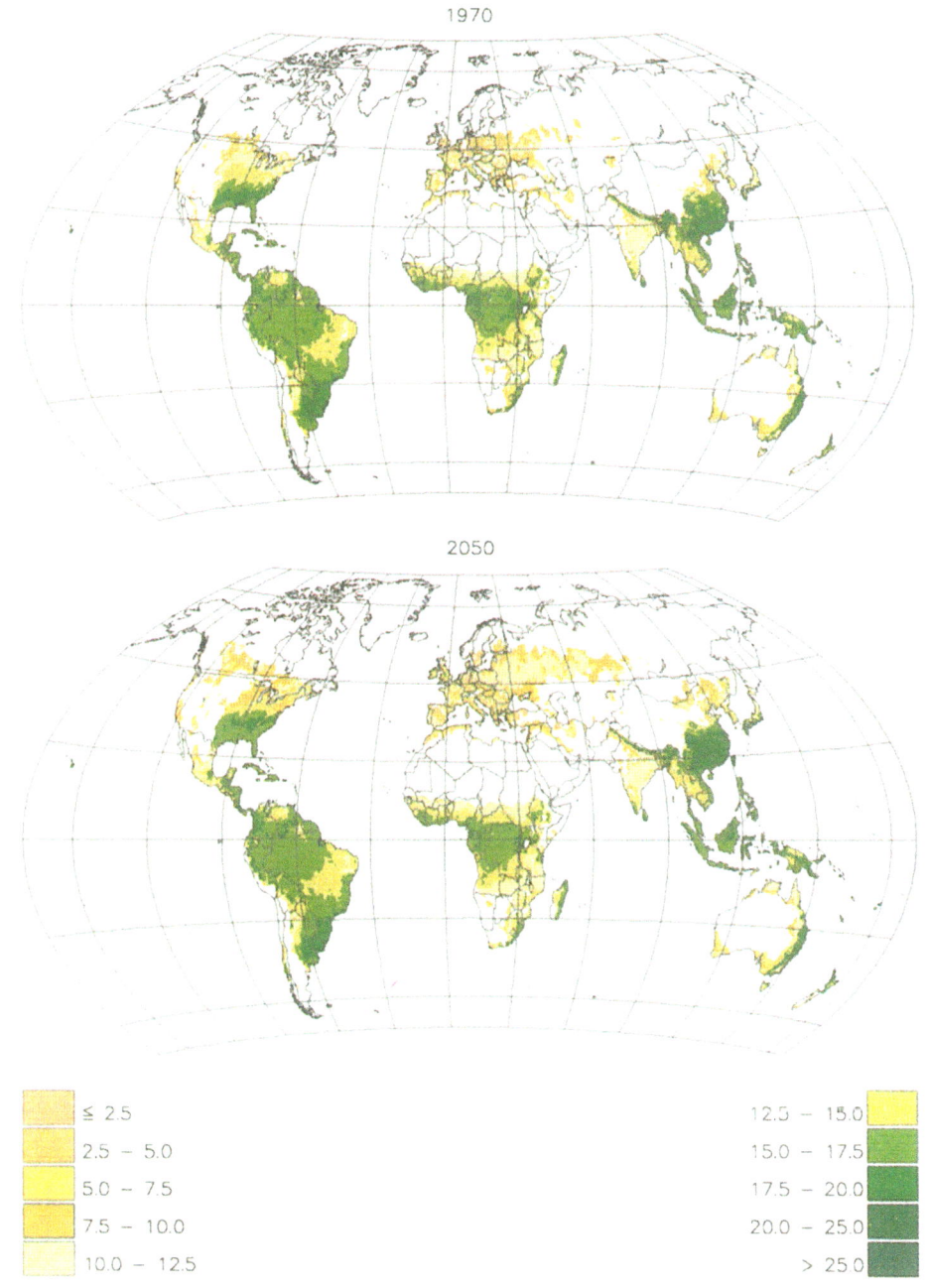

Figure 8: The distribution of maize as determined with the crop productivity model. The simulations show the current (1970) distribution and a plausible future distribution (2050) under changing environmental conditions. The latter is based on the IMAGE 2.0 "Conventional Wisdom" scenario (Alcamo *et al.*, 1994). Units: t dry weight ha^{-1} over the growing season

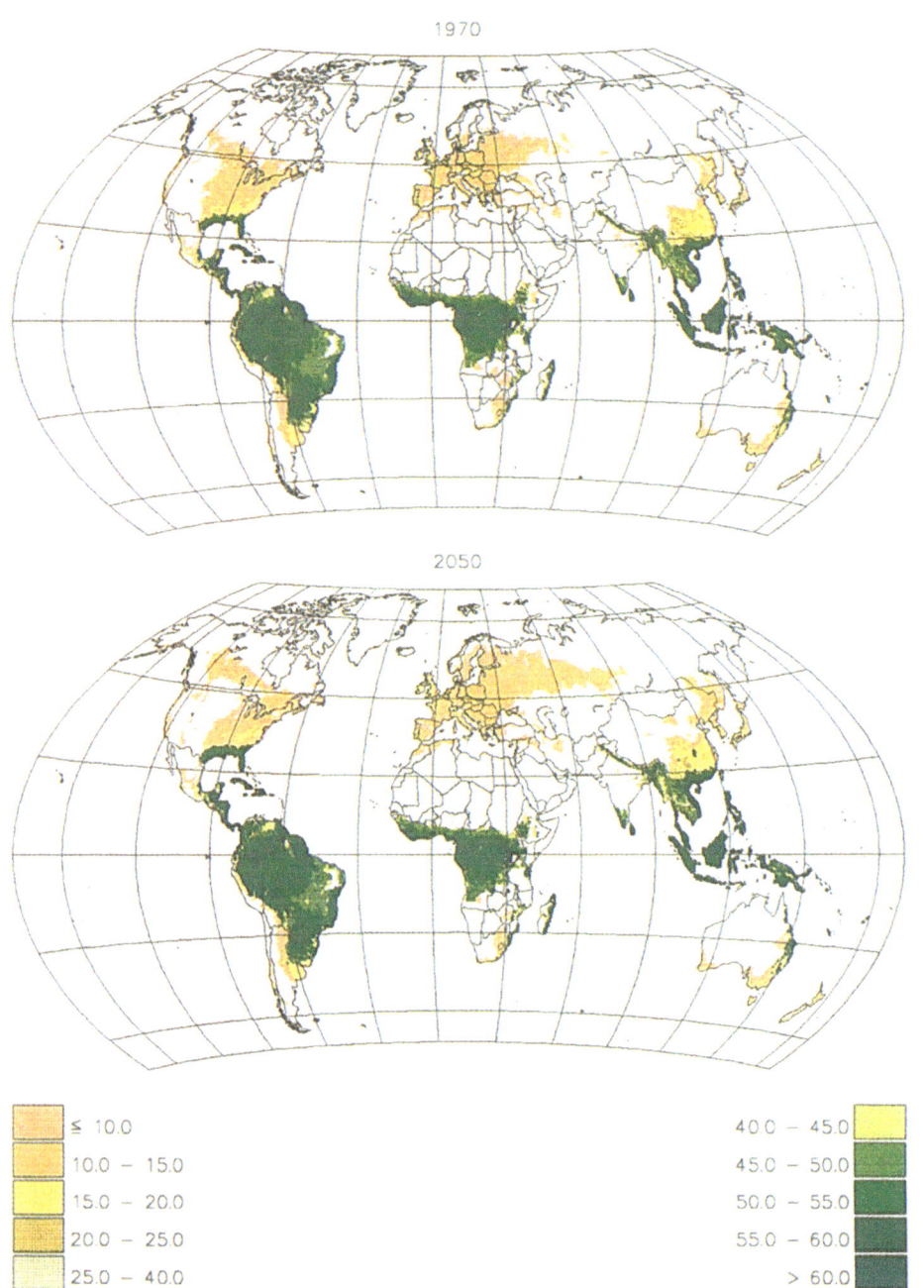

Figure 9: The distribution of sugar crops as determined with the crop productivity model. The simulations show the current (1970) distribution and a plausible future distribution (2050) under changing environmental conditions. The latter is based on the IMAGE 2.0 "Conventional Wisdom" scenario (Alcamo et al., 1994). Units: t dry weight ha⁻¹ over the growing season

short timesteps of a few years. However, to mimic the somewhat slower adaptation of regional farming systems, we have chosen to run the model with a similar timestep as the water balance and the BIOME model. This is computationally most efficient and allows for an adequate integration of the three models.

3.2.4 Validation of the simulated patterns of the CROP model

Figure 8 and 9 illustrates calculations of the crop model for maize and sugar. The distribution and yield of each crop coincides well with its current distribution (Leemans and Solomon, 1993). However, there are some problems closely related to the underlying data bases. As noted above, in higher altitude areas, the winter temperatures are probably too cold due to the lapse–rate correction scheme of the climate data base (Leemans and Cramer, 1991). This limits the distribution of some of the crops in an unrealistic way. Furthermore, some large agricultural areas are not adequately simulated because the underlying climate data base in these regions is too sparse. This is especially true for the Ganges valley in India and the Indus valley in Pakistan, where crop distribution maps indicate a large abundance of cereals, as compared to the low productivity simulated. Calculations will undoubtable improve when the underlying data bases are improved.

The other main source of error in the simulations involves the model's treatment of rain fed productivity only. Regions where crops depend mostly on irrigation water are not simulated by the model (Figure 10). The only solution to this problem is to improve the

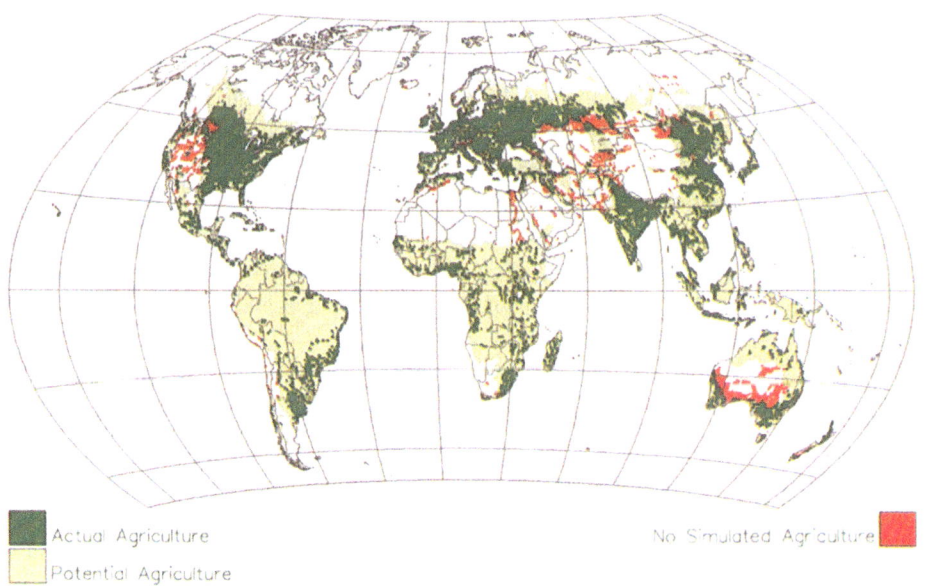

Figure 10: The fit of different crop distributions with the agricultural classes of Olson *et al.* (1985). The green regions are the current agricultural regions, while the light green regions can be used potentially. The red regions are those where agricultural use currently is listed, but where the crop model does not simulate any productivity.

hydrological model by including irrigation and drainage, which would allow for a realistic simulation of changing water availability under strongly changed climatic conditions.

4. Linkages and feedbacks between other IMAGE modules

TVM is linked in many ways to other IMAGE 2.0 submodels. The most obvious linkages are those concerning climate. The Atmosphere/Ocean System results in global patterns of climate change through time and these are linked with the current climate data base (de Haan *et al.*, 1994). Other linkages concern atmospheric concentrations of GHGs, albedo, and other environmental parameters. The simulated potential vegetation and crop patterns are immediately used by the Land Cover model (Zuidema *et al.*, 1994) to compute the human–induced shifts in land cover.

4.1   Shifts in Natural Vegetation Due to Climatic Change

The climate variables of TVM are updated at 10 years intervals based on the latitudinal average temperature and precipitation patterns computed by the Ocean/Atmosphere subsystem (de Haan *et al.*, 1994) of IMAGE 2.0. Different climatic patterns will lead to new potential vegetation patterns. The shifts are characterized as the equilibrium response, but used in a dynamic, transient mode. However, computing the potential vegetation patterns on an annual timestep is too computationally intensive, and leads to unrealistically fast responses and changes. Longer lived biomes, such as forests, are relatively resilient to gradual environmental changes (Prentice *et al.*, 1993). Rapid changes can only occur when extreme events and large scale disturbances occur (Overpeck *et al.*, 1990). We mimic such vegetation response by only determining the potential vegetation patterns once a decade. This results in smoother shifts.

   The shifts in vegetation can be used for a multitude of specific impacts assessments. Leemans and Halpin (1992), for example, have used location of major nature reserves and protected areas to determine the threats to the world's biodiversity. Such studies could broaden the scope of IMAGE 2.0 towards other global problems and their policies, such as the biodiversity crisis (Anonymous, 1992).

4.2   Shifts in Natural Vegetation Due to Changed CO_2 Concentrations

Gifford (1979) found that at extreme xeric conditions, plants could grow at high atmospheric CO_2-conditions were no plants grew at lower CO_2 levels. This effect is known as increased water–use efficiency (WUE). We have implemented it by assuming that WUE increases 30% with a doubling of atmospheric CO_2 concentrations (Körner, 1993). We further assumed a linear response between current and doubled CO_2 conditions and adjusted the lower α-limit for each PFT of the BIOME model accordingly (c.f. Vloedbeld and Leemans, 1993). This results in an change in the distribution of biomes and in an increase in crop productivity as atmospheric CO_2 concentrations increase. This could

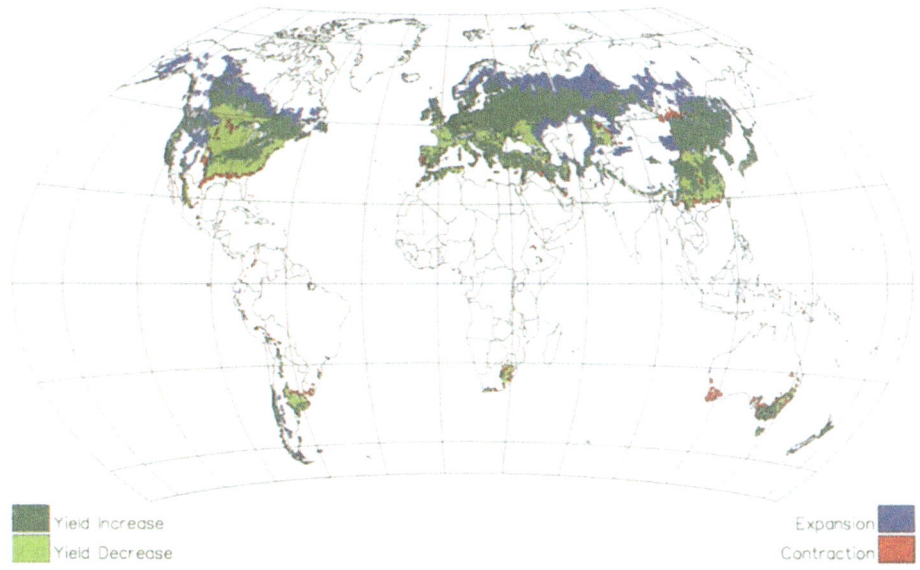

Yield Increase

Yield Decrease

Expansion

Contraction

Figure 11: A illustration of a risk assessment for global wheat productivity. The map displays the relative changes in productivity between current (1970) and future (2050) conditions according to the IMAGE 2.0 Conventional Wisdom scenario (Alcamo *et al.*, 1994). The blue regions (885 10^6 ha) are those where the crop could expand into, while the red regions (211 10^6 ha) are those where climate is probably not suitable anymore.

have large impacts on global agriculture, because it could offset some of the current land degradations and enhances agricultural productivity.

4.3 SHIFTS IN AGRICULTURAL PATTERNS

A very important linkage between the different submodels is the potential to simulate shifting agricultural zones (Figure 8, 9 and 11). The shifts depend on climatic changes and will have an immediate impact on land use and the subsequent land cover patterns. The changing land cover patterns are used to determine the C fluxes to the atmosphere and are linked to the other GHG emissions. These linkages allow for an assessment of the importance of the impact of climate change on agricultural systems for the final atmospheric composition. Declining productivities (c.f. Figure 11) could have particularly pronounced effects because a larger agricultural area would be needed to obtain a comparable yield.

The stored results on the climate indices and growing period characteristics provides the data to do several off–line analyses. A valid application would be to look at the change in individual crop productivity patterns, determine those regions where productivity declines and attach a risk factor to such decline (c.f. Figure 11). By closely analyzing the changes in the growing season patterns, one can actually define much more precisely such risk by linking it to the specific crop phenology. This would allow also for a more regional or even local impacts assessment.

5 Discussion and Conclusions

The different models that determine the potential natural and agricultural vegetation of the earth depend on an adequate resolution of seasonal temperature and the hydrological cycle. This has also been recognized by several other studies (e.g. Rind *et al.*, 1992; Anonymous, 1991). The simulation of the water balance and the resulting soil moisture content is computationally one of the most demanding submodels within the IMAGE 2.0 framework. However, the hydrological model is a simplified representation of the actual processes and it still does not account for drainage and/or irrigation, does not track runoff, and uses only a single–layer soil. Despite these limitations, the model still captures the main global features of the world's hydrological properties and patterns. In some areas the limitations of the model become obvious through the simulation of the potential vegetation and agricultural patterns. Here the integrated approach really assesses the capabilities of a model and shows the flaws and shortcomings in the underlying data.

The potential vegetation and crop models both simulate the main features of global vegetation fairly well. The models are strongly based on climate and several essential soil influences or vegetation feedbacks are neglected. Deviations from the actual patterns can therefore often be explained in terms of increased or decreased soil moisture storage, irrigation, fertility etc. However, improvements cannot be expected during the coming years, because better soil data bases with functional characteristics are not yet widely available on a global scale. For an adequate assessment of global change related–impacts, creating these data bases should get a very high research priority.

The integrated approach presented here demonstrates the potential value of a comprehensive set of models that determine alternative land cover patterns, such as cropland, range lands or natural vegetation. The models can be easily linked to different types of land use and land cover models (c.f. Alcamo *et al.*, 1994). This allows for a comprehensive simulation of the anthropogenic influences on global cycles, a field that is getting increasing attention by the global research community (Turner *et al.*, 1993).

The results of these models are further suited for a large array of regional and global impact assessments of global change. Examples of such applications involve: the vulnerability of protected areas under shifting vegetation zones, and the consequences for biodiversity and nature conservation (Leemans and Halpin, 1992), and the determination of risks associated with current productivity levels of specific crops with shifting agricultural patterns (Figure 11). These advanced analyses could well assist regional policy makers in assessing the seriousness of global change impacts. Finally, the relations between the direct CO_2 and indirect climatic effects can be handled in a single integrated framework and because different assumptions and/or scenarios can be analyzed in an objective way. This makes such integrative modeling approach a very powerful tool: it scrutinises theory and its underlying data sets and rejects the less valuable ones.

Acknowledgments

We thank Colin Prentice, Allen Solomon, Joseph Alcamo and his IMAGE team for fruitful discussions on the linkages of the different set of models. Eric Kreileman and Gé Zuidema improved the programming, so that it could easily be incorporated into the IMAGE 2.0 model. We appreciate the useful and critical comments of the IMAGE team, Oleg Anisimov, Wolfgang Cramer ands Sandra Brown on earlier drafts of this manuscript. The research was funded by the Dutch National Research Programme "Global Air Pollution and Climate Change" under agreement NOP 851042, 851045 and 852067, and The Dutch Ministry of Housing, Physical Planning and the Environment under agreement MAP 481507 to RIVM and contributes to IGBP–GCTE core project research.

References

Alcamo, J., G.J.J. Kreileman, M. Krol and G. Zuidema: 1994, Modeling the global society-biosphere-climate system, Part 1: model description and testing, *Wat. Air Soil Pollut.*, **76** (this volume).

Anonymous: 1978, *Report on the agro-ecological zones project. Vol 3. Methodology and results for South and Central America*, World Soil Resources Report 48/3, Food and Agriculture Organization of the United Nations, Rome, 251 pp.

Anonymous: 1987, *Agroclimatological data for Asia*, FAO Plant Production and Protection Series 25, Food and Agriculture Organization of the United Nations, Rome.

Anonymous: 1991, *Plant-Water Interactions in Large-Scale Hydrological Modelling*, IGBP-Report No.17, International Geosphere-Biosphere Programme, Stockholm, 44 pp.

Anonymous: 1992, Convention on Biological Diversity, *Biol. Int.*, **25**: 22-38.

Bazzaz, F.A. and E.D. Fajer: 1992, Plant life in a CO_2-rich world, *Sci. Am.*, **1992**: 18-21.

Bouwman, A.L. and R. Leemans: 1994, The role of forest soils in the global carbon cycle, *Soil Sci. Soc. Am. J.*, **(in press)**.

Box, E.O.: 1981, *Macroclimate and Plant Forms: an Introduction to Predictive Modeling in Phytogeography*, Dr. W. Junk Publishers, 258 pp.

Budyko, M.I.: 1986, *The Evolution of the Biosphere*, D.Reidel Publishing Company, 423 pp.

Chadwyck–Healey: 1992, *World climate disc: Global climate change data*, CD-ROM Chadwyck-Healey, Ltd., Cambridge.

Emanuel, W.R., H.H. Shugart and M.P. Stevenson: 1985, Climatic change and the broad-scale distribution of terrestrial ecosystems complexes, *Clim. Change*, **7**: 29-43.

Espenshade, E.B., Jr. and J.L. Morrison (eds): 1991, *Goode's World Atlas*, Rand McNally & Company, 368 pp.

FAO/CSRC: 1974, *Soil map of the world, 1.5M, UNESCO, Paris. 1/2 degree digitization*, University of Hew Hampshire, Durham, N.C.

FAO/UNESCO: 1974, *Soil Map of the World, 1:5,000,000*, Food and Agriculture Organisation.

Federal Climate Complex: 1992, *International Station Meteorological Climate Summary Version 2.0*, CD-ROM Naval Oceanography Command Detachment Asheville, USAFETA OL-A and National Climate Data Center, Asheville.

Federer, C.A.: 1982, Transpirational supply and demand: Plant, soil and atmospheric effects evaluated by simulation, *Water Resour. Res.*, **18**: 355-362.

Gifford, R.M.: 1979, Growth and yield of CO_2-enriched wheat under water-limited conditions, *Austr. J. Plant. Physiol.*, **6**: 367-378.

Gribbin, J. and H.H. Lamb: 1978, Climatic change in historical times, in: J. Gribben (ed), *Climatic Change*, Cambridge University Press, pp. 68-82.

Guetter, P.J. and J.E. Kutzbach: 1990, A modified Köppen classification applied to model simulations of glacial and interglacial climates, *Climatic Change*, **16**: 193-215.

de Haan, B.J., M. Jonas, O. Klepper, J. Krabec, M.S. Krol and K. Olendrzynski: 1994, An atmosphere-ocean model for evaluation of climate scenarios, *Wat. Air Soil Pollut.*, **76** (this volume).

Henderson-Sellers, A.: 1991, Developing an interactive biosphere for global climate models, *Vegetatio*, **91**: 149-166.

Holdridge, L.R.: 1947, Determination of world plant formations from simple climatic data, *Science*, **105**: 367-368.

Hutchinson, M.F.: 1987, Methods of generation of weather sequences, in: A.H. Bunting (ed), *Agricultural Environments. Characterization, Classification and Mapping*, C.A.B. International, pp. 149-157.

Kineman, J.J.: 1992, *Global Ecosystems database Version 1.0 (on CDROM) User's guide*, Key to Geophysical Records Documentation No. 26, USDOC/NOAA National Oceanic and Atmospheric Administration, Boulder, Colorado, 121 pp.

Kineman, J.J. and M.A. Ohrenschall: 1992, *Global Ecosystems database Version 1.0 (on CDROM) Disc-A, Documentation manual*, Key to Geophysical Records Documentation No. 27, USDOC/NOAA National Oceanic and Atmospheric Administration, Boulder, Colorado, 240 pp.

Klein-Goldewijk, K., J.G. van Minnen, G.J.J. Kreileman, M. Vloedbeld and R. Leemans: 1994, Simulating the carbon flux between the terrestrial environment and the atmosphere, *Wat. Air Soil Pollut.*, **76** (this volume).

Köppen, W.: 1936, Das geographische System der Klimate, in: W. Köppen and R. Geiger (eds), *Handbuch der Klimatologie*, Gebrüder Borntraeger, pp. 1-46.

Körner, C.: 1993, CO_2 fertilization: The great uncertainty in future vegetation development, in: A.M. Solomon and H.H. Shugart (eds), *Vegetation Dynamics and Global Change*, Chapman and Hall, pp. 53-70.

Küchler, A.W.: 1947, A geographic system of vegetation, *Geogr. Rev.*, **37**: 233-240.

Küchler, A.W.: 1949, A physiognomic classification of vegetation, *Ann. Ass. Amer. Geog.*, **39**: 201-210.

Leemans, R.: 1992, Modelling ecological and agricultural impacts of global change on a global scale, *J. Sci. Ind. Res.*, **51**: 709-724.

Leemans, R.: 1994, The use of plant functional type classifications to model the global land cover and simulate the interactions between the terrestrial biosphere and the atmosphere, in: T.M. Smith, H.H. Shugart and F.I. Woodward (eds), *Plant Functional Types Classifications*, Cambridge University Press, (in press).

Leemans, R. and W. Cramer: 1991, *The IIASA database for mean monthly values of temperature, precipitation and cloudiness on a global terrestrial grid*, Research Report RR-91-18, International Institute of Applied Systems Analyses, Laxenburg, 61 pp.

Leemans, R. and P. Halpin: 1992, Global change and biodiversity, in: B. Groombridge (ed), *Biodiversity 1992: Status of the Earth's Living Resources*, Chapman and Hall, pp. 254-255.

Leemans, R. and A.M. Solomon: 1993, The potential response and redistribution of crops under a doubled CO_2 climate, *Clim. Res.*, **3**: 79-96.

Legates, D.R. and C.J. Willmott: 1990a, Mean seasonal and spatial variability in gauge corrected, global precipitation, *Int. J. Climatol.*, **10**: 111-127.

Legates, D.R. and C.J. Willmott: 1990b, Mean seasonal and spatial variability in global surface air temperature, *Theor. Appl. Climatol.*, **41**: 11-21.

Matson, P.A. and S.L. Ustin: 1991, The future of remote sensing in ecological sciences, *Ecology*, **72**: 1917-1945.

Matthews, E.: 1985, *Atlas of archived vegetation, land-use and seasonal albedo data sets*, Technical Memorandum 86199, NASA, New York, 23 pp.

Melillo, J.M., A.D. McGuire, D.W. Kicklighter, B. Moore III, C.J. Vorosmarty and A.L. Schloss: 1993, Global climate change and terrestrial net primary production, *Nature*, **363**: 234-239.

Monserud, R.A. and R. Leemans: 1992, The comparison of global vegetation maps, *Ecol. Modelling*, **62**: 275-293.

Mücher, C.A., T.J. Stomph and L.O. Fresco: 1993, *Proposal for a global land use classification*, Final Report LUIS FAO, ITC and WAU, Rome, Enschede, and Wageningen, 37 pp.

Müller, M.J.: 1982, *Selected Climatic Data for a Global Set of Standard Stations for Vegetation Science*, Dr. W. Junk Publishers, 306 pp.

Neilson, R.P., G.A. King and G. Koerper: 1992, Toward a rule-based biome model, *Landscape Ecol.*, **7**: 27-43.

Olson, J., J.A. Watts and L.J. Allison: 1985, *Major World Ecosystem Complexes Ranked by Carbon in Live Vegetation: A Database*, Report NDP-017, Oak Ridge National Laboratory, Oak Ridge, Tennessee, 164 pp.

Overpeck, J.T., D. Rind and R. Goldberg: 1990, Climate-induced changes in forest disturbance and vegetation, *Nature*, **343**: 51-53.

Peters, R.L.: 1992, Conservation of biological diversity in the face of climate change, in: R.L. Peters and T.E. Lovejoy (eds), *Global Warming and Biological Diversity*, Yale University Press, pp. 15-30.

Prentice, I.C., W. Cramer, S.P. Harrison, R. Leemans, R.A. Monserud and A.M. Solomon: 1992, A global biome model based on plant physiology and dominance, soil properties and climate, *J. Biogeogr.*, **19**: 117-134.

Prentice, I.C., M.T. Sykes and W. Cramer: 1993, A simulation model for the transient effects of climate change on forest landscapes, *Ecol. Model.*, **65**: 51-70.

Prentice, I.C., R.S. Webb, M.T. Ter-Mikhaelian, A.M. Solomon, T.M. Smith, S.E. Pitovranov, N.T. Nikolov, A.A. Minin, R. Leemans, S. Lavorel, M.D. Korzukhin, H.O. Helmisaari, J.P. Hrabovszky, S.P. Harrison, R.W. Emanuel and G.B. Bonan: 1989, *Developing a Global Vegetation Dynamics Model: Results of an IIASA Summer Workshop*, IIASA research report RR-89-7, International Institute of Applied Systems Analysis, Laxenburg, Austria, 48 pp.

Prentice, K.C. and I.Y. Fung: 1990, The sensitivity of terrestrial carbon storage to climate change, *Nature*, **346**: 48-51.

Rind, D., C. Rosenzweig and R. Goldberg: 1992, Modelling the hydrological cycle in assessments of climate change, *Nature*, **358**: 119-122.

Rock, B.N., D.L. Skole and B.J. Choudhury: 1993, Monitoring vegetation change using satellite data, in: A.M. Solomon and H.H. Shugart (eds), *Vegetation Dynamics and Global Change*, Chapman and Hall, pp. 153-167.

Smith, T.M., R. Leemans and H.H. Shugart: 1992, Sensitivity of terrestrial carbon storage to CO_2 induced climate change: Comparison of four scenarios based on general circulation models, *Clim. Change*, **21**: 367-384.

Townshend, J., J. Cihlar, C. Justice, J.-P. Malingreau, S. Ruttenberg, F. Sadowski, D. Skole and P. Teillet: 1991, *A new high resolution global dataset for land applications. IGBP-DIS's pilot land cover project working group.*, Universite de Paris, Paris.

Townshend, J.R.G.: 1992, *Improved Global data for land Application: A proposal for a New High Resolution Data Set*, IGBP-Report No.20, International Geoshere-Biosphere Programme, Stockholm, 87 pp.

Tucker, C.J., J.R. Townshend and T.E. Goff: 1985, African land-cover classification using satellite data, *Science*, **227**: 369-375.

Turner, B.L., R.H. Moss and D.L. Skole: 1993, *Relating Land Use and Global Change: A Proposal for an IGBP-HDP Core Project*, IGBP Report No.24 and HDP Report No. 5, International Geosphere-Biosphere Programme and the Human Dimensions of Global Environmental Change Programme, Stockholm, 65 pp.

Vloedbeld, M. and R. Leemans: 1993, Quantifying feedback processes in the response of the terrestrial carbon cycle to global change - the modeling approach of image-2, *Water Air Soil Pollut.*, **70**: 615-628.

von Humboldt, F.H.A.: 1807, *Ideen zu einer Geographie der Pflanzen neben einem naturgemalde der Tropenländer.*

Walter, H. and E. Box: 1976, Global classification of natural terrestrial ecosystems, *Vegetatio*, **32**: 75-81.

Willmott, C.J., C.M. Rowe and W.D. Philpot: 1985, Small scale climate maps: a sensitivity analysis of some common assumptions associated with grid point interpolation and contouring, *Am. Cart.*, **12**: 5-16.

de Wit, C.T.: 1965, *Photosynthsis of leaf canopies*, Agricultural Research Report 663, Centre for Agricultural Publicaton and Documentation, Wageningen.

Wood, S.R. and F.J. Dent: 1983, *LECS: A Land Evaluation Computer System*, Manual AGOF/IN S/78/006 Manual 5 and 6, Ministry of Agriculture, Government of Indonesia, United Nations Development Programme, and Food and Agriculture Organization, Rome, 221 and 157 pp.

Woodward, F.I.: 1987, *Climate and Plant Distribution*, Cambridge University press, 174 pp.

Zobler, L.: 1986, *A World Soil File for Global Climate Modeling*, Technical Memorandum NASA, New York, 32 pp.

Zuidema, G., G.J. van den Born, J. Alcamo and G.J.J. Kreileman: 1994, Simulating changes in global land cover as affected by economic and climatic factors, *Wat. Air Soil Pollut.*, **76** (this volume).

SIMULATING CHANGES IN GLOBAL LAND COVER AS AFFECTED BY ECONOMIC AND CLIMATIC FACTORS

G. ZUIDEMA, G.J. VAN DEN BORN, J. ALCAMO, G.J.J. KREILEMAN

National Institute of Public Health and Environmental Protection
P.O. Box 1, 3720 BA, Bilthoven, the Netherlands

Abstract. This paper describes two global models: (1) an Agricultural Demand Model which is used to compute the consumption and demand for commodities that define land use in 13 world regions; and, (2) a Land Cover Model, which simulates changes in land cover on a global terrestrial grid (0.5⁰ latitude by 0.5⁰ longitude) resulting from economic and climatic factors. Both are part of the IMAGE 2.0 model of global climate change. The models have been calibrated and tested with regional data from 1970-1990. The Agricultural Demand Model can approximate the observed trend in commodity consumption and the Land Cover Model simulates the total amount of land converted within 13 world regions during this period. Some degree of the spatial variability of deforestation has also been captured by the simulation. Applying the model to a "Conventional Wisdom" scenario showed that future trends of land conversions could be strikingly different on different continents even though a consistent scenario (IS92a from the IPCC) was used for assumptions about economic growth and population. Sensitivity analysis indicated that future land cover patterns are especially sensitive to assumed technological improvements in crop yield and computed changes in agricultural demand.

Keywords: land cover, land use, agricultural demand, climate change, global change.

1. Introduction

Large scale transformation of land cover can lead to large scale "global changes" such as reduction of biodiversity, land degradation and desertification. As land cover is transformed on the large scale, it can also affect regional and global patterns of climate. It is well known, for example, that changes in vegetation cover alter the surface fluxes of heat and moisture and can affect the rate of emissions of some key greenhouse gases such as N_2O and CH_4 and reduce C storage in global vegetation (e.g., IPCC, 1990). In turn, a change in climate can influence the potential growing boundaries of agricultural crops (e.g. Parry *et al.*, 1988; Leemans and Solomon, 1993) and natural vegetation (IPCC, 1990; Leemans, 1992). When combined with changing demand for food or feed, these changing boundaries may lead to large scale land cover transformations.

In this paper we present an "Agricultural Demand Model" (ADM) used to estimate the economic factors leading to global land use and land cover transformations. We also present a "Land Cover Model" (LCM) for studying global land cover transformations as they are affected by both economic and climatic factors. We describe the premises and implementation of these models, their calibration and testing with historical data, and results from an example scenario.

These two models are part of the IMAGE 2.0 model whose goal is to provide a scientifically based overview of climate change issues to support climate policy analysis (Alcamo *et al.*, 1994a and b). This model consists of three main sub-systems, the Energy-

Water, Air, and Soil Pollution **76**: 163–198, 1994.

Industry sub-system, the Atmosphere-Ocean sub-system and the Terrestrial Environment sub-system. The added value of the integrated model is that it allows consistent computations of all greenhouse gas emissions and fluxes and the change in climate which feeds back to the land cover simulation. Specifically, the economic and population driving forces used to calculate regional crop demands in the ADM are also used to compute regional energy/industrial demand and consumption and the resulting greenhouse gas emissions (de Vries *et al.*, 1994). At the same time, the land cover changes computed by the LCM are used to calculate the flux of C and other greenhouse gases from the terrestrial environment to the atmosphere (Klein Goldewijk *et al.*, 1994; Kreileman and Bouwman, 1994). These emissions and fluxes then drive the Atmosphere-Ocean sub-system of IMAGE 2.0 which calculates new spatial patterns of temperature and precipitation for each model time step (de Haan *et al.*, 1994). These new patterns are used by the IMAGE 2.0 Terrestrial Vegetation model (Leemans and van den Born, 1994) to update estimates of potential crop productivity which are then used by the LCM for simulating land conversions in the next model time step. This integration of agricultural and land cover calculations with other climate change calculations is a first attempt to simulate the coupled dynamics of land cover and climate in geographic detail for the whole world.

The two models described in this paper belong to the Terrestrial Environment sub-system of IMAGE 2.0. Other submodels in this sub-system are: the Terrestrial Vegetation Model (Leemans and van den Born, 1994), the Terrestrial Carbon Model (Klein Goldewijk *et al.*, 1994), and the Land Use Emissions Model (Kreileman and Bouwman, 1994). Taken together, these five models provide a framework for simulating global land cover transformation on a grid basis, together with its effect on greenhouse gas emissions and C flux from the biosphere to the atmosphere.

2. Agricultural Demand Model

2.1 OBJECTIVES AND APPROACH

The demands for both crop and animal commodities are the driving forces in the ongoing expansion of agriculture and in turn, are driven by population and income growth. The growth of income influences demand by increasing per capita food consumption and the composition of diet (Sanderson, 1988). A common trend seems to begin with starchy foods, then a more diversified vegetarian diet, and finally a diet containing a substantial amount of meat. Although this trend seems to be almost universal, it depends on factors such as local availability of food, religion, culture, and urbanization (Sanderson, 1988).

The objective of the Agricultural Demand Model (ADM) is to estimate the demand for commodities that lead to large-scale land cover transformations, and to link the global agricultural system with the global climate system. In this respect, the objectives of the model differ from macro-economic type global food models, e.g. the IIASA Basic Linked System (Fischer *et al.*, 1988). Agricultural commodities are divided into "crops", "animal products" and "wood products". These commodities lead to changes in the three most

TABLE 1

Commodity classes in the Agricultural Demand Model and assumed upper limits of caloric intake for crops and animal products.

CLASSES OF CROPS	MAXIMUM INTAKE kcal capita^{-1} day^{-1}	CLASSES OF ANIMAL PRODUCTS	MAXIMUM INTAKE kcal capita^{-1} day^{-1}
Temperate cereals	1500	Meat, cattle	400
(wheat, barley, rye, oats)		Milk, cattle (excl. butter)	400
Rice	1500	Meat, Pigs	350
Maize	500	Meat, poultry (incl. eggs)	250
Tropical cereals (millet, sorghum)	350	Meat, sheep and goats	250
Pulses	150		
Roots and tubers	400		
Oil crops	700	Total for all commodities	
Sugar crops	700	(crops and animal products)	3500

CLASSES OF WOOD PRODUCTS

Industrial wood (timber)
Fuelwood[1]

[1] Fuelwood demand is computed in the IMAGE 2.0 Energy Demand model (de Vries, et al., 1994).

important types of global land cover (on the basis of total spatial coverage): cropland, range land, and forest (FAO, 1992a).

Crops and animal products are further broken down into 13 classes (Table 1). Crops are grouped into eight classes according to their photosynthetic pathway and climate/soil requirements (Leemans and van den Born, 1994). According to AGROSTAT (FAO, 1992a) these crops occupy about 90% of the world's total harvested area (year 1970, see Table 2) and 90% of total vegetable caloric intake by humans (1970). A ninth category "Other Crops" takes into account remaining commodities such as fibre crops, coffee, vegetables and fruits. The "Other Crops" are not simulated as such, but are assumed to take up an area proportional to the area simulated for the eight selected crops. Animal products are grouped into five classes which account for virtually all of the consumption of meat and milk products and 70% of the total caloric intake of animal products. The remaining intake of calories is related to animal fats, a by-product indirectly taken into account. The consumption of fish is not taken into account at all, since this has no direct influence on land use.

Wood products are broken down into two groups: fuelwood and timber. We do not compute demand for timber in this version of the model. The demand for fuelwood is computed as part of the regional energy economy in the IMAGE 2.0 Energy Economy model (de Vries et al., 1994).

Calculating agricultural demand is done as follows. First, the regional per capita consumption of different commodities (crops and animal products) is computed from economic data. Next, the per capita consumption in each region is converted into a total regional demand for crops and animals by taking into account the region's population and

TABLE 2

Extent of agricultural land for each region in 1970

IMAGE CROP GROUPS	Code	Canada	USA	Latin America	Africa	OECD Europe	Eastern Europe	CIS	Middle East	India+ S. Asia	China+ C.P. countries	East Asia	Oceania	Japan	WORLD
Area Harv.[1]	A	16,090	85,275	82,130	108,465	52,261	36,154	138,937	34,070	187,282	140,108	34,060	11,480	4,348	930,661
Est. Area[2]	B	215	10,192	12,343	11,683	19,291	3,288	10,691	5,734	16,065	14,598	13,487	1,262	2,065	120,914
Tot. Area Harv.	A+B	16,305	95,467	94,474	120,148	71,552	39,442	149,628	39,804	203,347	154,706	47,547	12,742	6,412	1,051,575
Area Olson[3]	C	84,891	368,816	287,640	322,811	178,036	90,245	377,925	84,058	252,962	229,154	111,768	100,221	21,100	2,509,595
Tot. Area Harv \(A+B)\ Area Olson	C	0.19	0.26	0.33	0.37	0.40	0.44	0.40	0.47	0.80	0.68	0.43	0.13	0.30	0.42
Perc. Total Area Harvested of each crop group															
Temperate cereals	75%	31%	11%	11%	54%	48%	71%	67%	13%	20%	2%	80%	8%	31%	
Rice	0%	1%	7%	3%	0%	0%	0%	2%	27%	27%	42%	0%	46%	13%	
Maize	3%	24%	27%	15%	5%	19%	2%	3%	3%	11%	13%	1%	0%	11%	
Tropical cereals	0%	6%	4%	24%	0%	0%	2%	3%	19%	8%	0%	3%	0%	9%	
Pulses	0%	1%	8%	10%	3%	6%	3%	3%	12%	5%	1%	1%	3%	6%	
Roots and tubers	1%	1%	5%	9%	5%	10%	5%	1%	1%	8%	6%	1%	5%	5%	
Oil crops	19%	29%	18%	16%	6%	5%	10%	9%	17%	14%	22%	8%	3%	15%	
Sugar crops	0%	1%	6%	1%	3%	2%	2%	1%	2%	0%	1%	2%	1%	2%	
Other crops	2%	7%	16%	12%	23%	9%	4%	11%	5%	7%	12%	4%	34%	9%	

[1] Area harvested based on FAO AGROSTAT, x 1000 ha (FAO, 1992a)

[2] Estimated area for crops with only production statistics available, based on FAO AGROSTAT, x 1000 ha (FAO, 1992a)

[3] Total of 10 classes (cool crops, warm farms, paddyland, warm irrigated, cold irrigated drylands, cool irrigated drylands, cool field/woods, warm woods/fields, cool woods/fields, warm field/woods) given by Olson et al. (1985).

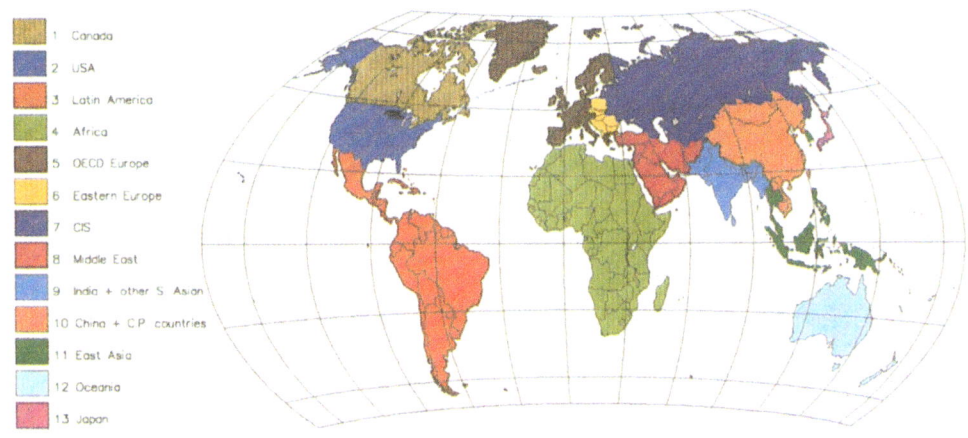

Figure 1: The 13 world regions of IMAGE 2.0.

the additional amount of crops needed as feed for animals. The total regional demand is then converted into a land use demand for each region by including interregional trade of crops and animal products. Therefore land may be used in one region to satisfy commodity demand in another. These steps are described in detail below. In the LCM the computed regional land use demand, together with suitability of land for agriculture, drive land cover conversions. All calculations in the ADM are performed separately for the 13 world regions as shown in Figure 1. Because of the differences between countries it would be desirable to take into account every country individually, but this is not considered feasible because of the enormous data requirements and the difficulty in prescribing future economic driving forces for every country (including commodity trade between nations).

2.2 CALCULATION OF COMMODITIES

2.2.1 Per Capita Commodity Consumption
We compute the consumption of commodities by estimating the change in per capita consumption of various crops and animal products relative to the change in income, called the income elasticity coefficient. We use a conventional semi-log function to describe the decreasing marginal consumption of food as income increases (e.g., Blakeslee *et al.*, 1973; FAO, 1971; Bouwman *et al.*, 1992).

TABLE 3

Income elasticities for crops and animal products for 13 regions

REGION / PRODUCT NAME	Canada	USA	Latin America	Africa	OECD Europe	Eastern Europe	CIS	Middle East	India + S. Asia	China + C.P. countries	East Asia	Oceania	Japan
Meat, cattle	-0.200 *	-0.200 *	-0.069	0.803 *	-0.037	0.465	0.557	0.503	-0.194	4.781 *	1.009 *	-0.119 *	3.082
Milk, cattle	-0.150 *	0.098	0.220 *	1.059 *	0.292	0.099	-0.218 *	0.208	1.525	2.598 *	1.782 *	-0.090	0.622
Meat, pigs	0.193	0.202	0.995	5.000 *	1.135	0.849	0.505	0.542	-0.131	0.702 *	0.759	1.395	1.475
Meat, poultry	0.440	1.181	0.810 *	2.909 *	0.685	1.171	1.802	2.591	5.000 *	1.799	1.404	1.694	0.780
Meat, sheep	-1.158	-1.174	-1.459	0.393	0.233	-0.159	-0.161	-0.023	1.689 *	3.284 *	0.643	-0.770 *	-0.863
Temperate cereals	0.127	0.618	0.141	0.093 *	-0.076 *	-0.161 *	-0.180 *-	0.180 *	0.160 *	0.381 *	0.126 *	-0.300 *	0.042 *
Rice	2.437	2.836	0.224 *	0.736 *	0.980	-0.056	0.912	0.761	0.044 *	-0.022 *	-0.056 *	2.000 *	-0.374
Maize	0.169	3.472	0.007	-0.060 *	1.613	-0.382	-0.208	-0.137	-0.076	-0.006 *	0.438	-1.421	2.957 *
Tropical cereals	-5.000	5.000	-1.283	-0.361 *	-2.462	0.000	-0.069	-1.303	-0.635	-0.624	-0.911	0.000	-0.944
Pulses	3.157	0.130	-0.335 *	-0.108	0.065	-0.173	-0.239 *	0.212	-0.204 *	-0.129 *	0.679	2.063	-0.334
Roots and tubers	-0.151	0.027	-0.191 *	-0.138 *	-0.245	-0.377	-0.294 *	1.270	0.384 *	-0.134 *	-0.099 *	0.245	-0.083
Oil crops	1.241	1.083	0.807 *	0.180 *	0.873	0.722	0.991	1.221	1.172 *	0.832 *	1.452	3.670	0.624
Sugar crops	-0.195	0.140	0.174 *	0.649 *	-0.129	0.282	0.268	0.491	0.378 *	1.378 *	0.503	-0.377	0.053

* Unlike the other values these elasticity values are not derived from 1970-1990 data. See section 2.2.1.

$$CNS_{t,i,r} = CNS_{t_0,i,r} \left(1 + x_{i,r} \ln \frac{GNP_{t,r}}{GNP_{t_0,r}} \right) \tag{1}$$

Where:

CNS = consumption (kcal capita^{-1} day^{-1})
GNP = Gross National Product (US \$ capita^{-1} year^{-1})
x = income elasticity coefficient
t = index for time
i = index for commodity
r = index for region.

Income elasticity coefficients (x in this Equation) were estimated by assuming a log-linear relationship between actual changes in commodity consumption and income in each of 13 world regions during the period 1970-1990 (Table 3). The following data sources were used: consumption from the FAO food balance sheets (FAO, 1992a); GNP data from World Bank (1989; 1992) and WRI (1992); and deflation factors from UNCTAD (1991). Although most (60%) of the computed elasticity coefficients showed a reasonable correlation with GNP, some of the coefficients were replaced as follows:

- For Latin America, Africa, and India plus South Asia, the irregular trend in GNP between 1970 and 1990 makes it difficult to estimate a reasonable income elasticity coefficient. For these regions, elasticity coefficients were chosen so that their future consumption patterns would resemble the present consumption of industrialized regions when the future GNP per capita in Latin America, Africa, etc. is similar to current GNP per capita in industrialized regions.
- A consumption maximum was set for all commodities (see below; Table 1). Where necessary, elasticities were replaced so that this maximum was not reached at a rate faster than the fastest regional rate of increase between 1970 and 1990.
- Cultural and religious differences between regions are also taken into account by replacing elasticity coefficients of some commodities to avoid that an increase in income would lead to an unreasonable increase in consumption.

In addition to the preceding replacements of elasticity coefficients, we also set an upper limit of 3500 kcal capita^{-1} day^{-1} for caloric intake. If we recall that the model takes into account about 80% of the total caloric intake of humans, this would amount to an upper limit of about 4250 total kcal capita^{-1} day^{-1} as a regional average; this is a 25% increase over current average intake in industrialized regions (3400 kcal capita^{-1} day^{-1}). If this limit is exceeded, the consumption of all individual crops and animal products is scaled down until total consumption is within this limit. A consumption maximum was also set for individual commodities, equal to the highest regional consumption level of each commodity in 1990 plus 10% (see Table 1). Elasticity coefficients are summarized in Table 3.

2.2.2 Regional Demand for Crops and Animals

After computing the per capita commodity consumption for each region with Equation (1),

TABLE 4

Historical productivity data and assumptions of the Conventional Wisdom Scenario for 5 classes of animals products in OECD Europe.

	GNP per capita	Cattle	Cattle	Pigs	Poultry	Sheep and goats
	US $ yr^{-1}	(meat) kg animal^{-1}	(milk) kg animal^{-1}	(meat) kg animal^{-1}	(meat) kg animal^{-1}	(meat) kg animal^{-1}
1970	7,600	204	3,242	107	10.1	8.2
1980	11,700	238	3,707	121	11.9	11.3
1990	14,500	260	4,344	129	12.1	10.1
2100	63,300	260	4,344	129	12.1	10.1

we multiply these figures by population figures for each region to obtain total regional commodity consumption by humans (accounting also for losses and non-dietary uses such as seed). Next we calculate the number of animals needed to satisfy meat demand in a region by multiplying the regional consumption of meat and dairy products by the animal productivity, i.e. the production of meat or dairy products per animal. Animal productivity varies widely in the world; as an example, the productivity of cattle (1990) ranged from 80 kg of meat per slaughtered animal in India to 255 kg/animal in USA. Data for OECD Europe are given in Table 4. Later, we discuss assumptions about future animal productivity. We corrected for "non-productive" cattle, i.e. cattle not used within a year for either meat or dairy products, to get the total number of cattle in a region. The ratio of non-productive to total cattle in some developing regions is rather high (0.70 - 0.80), and tends to decrease in higher income regions because of improved management of livestock.

Once the total number of animals has been estimated from data on meat consumption, animal productivity, and fraction of productive cattle, we estimate the feed required for animals by multiplying the number of animals by a feed requirement in total digestible nutrients (TDN) per animal (Gohl, 1981). The feed requirement per animal is based on animal productivity, type of production (e.g. milk or meat), and body weight (cattle only).

The total TDN feed requirement per animal class in each region must now be apportioned between "concentrates" from feed crops, and "roughage" from range land, pasture, and crop residues. This is an important calculation because the crop areas to provide these feed concentrates can be quite large. For the period 1970-1990, data are available for consumption of concentrates but not for roughage. Consequently, we assumed that the amount of roughage was equal to the difference between the computed total feed and concentrates. For computing future crop demands, the ratio of roughage to concentrates is a scenario variable; we also assume that concentrates will be provided by the same mix of crops as in 1990.

2.2.3 Regional Land Use Demands

The last factor to consider in estimating land use demands is that a significant amount of regional demands can be met by imports. Imports allow the consumption of a crop (or

TABLE 5

Summary of key coefficients used for each region.

General key coefficients:
- Population
- Gross National Product (US $ capita^{-1} yr^{-1})

Agricultural key coefficients:
- Management factor for crops (Indexed to technological development)
- Fertilizer use (Gg N yr^{-1})
- Fraction of non-productive cattle
- Fraction of total feed coming from a type of crop
- Ratio of feed from roughage *vs.* concentrate
- Production of meat per animal (kg animal^{-1})
- Net export of crop products (t)

TABLE 6

Summary of input data for the Agricultural Demand Model.

Caloric value of animal and crop products (region dependent)
- Energy of animal products (kcal kg^{-1})
- Energy of crop products (kcal kg^{-1})

Human consumption (region dependent)
- Consumption of animal products 1990 (kcal capita^{-1} day^{-1})
- Consumption of crops products 1990 (Mcal capita^{-1} yr^{-1})

Conversion figures (region independent)
- Grain weight to Total Digestible Nutrients (TDN kg kg^{-1})
- Harvested weight to dry weight (kg kg^{-1})
- Rice conversion of paddy to husked (kg kg^{-1})
- Oil conversion of processed crop to crop (kg kg^{-1})
- Sugar conversion of processed crop to crop (kg kg^{-1})
- Fraction harvested of a crop

Feed intake figures (region dependent)
- Amount of feed necessary for animals (TDN kg animal^{-1} day^{-1})

Consumable fraction of total production (region dependent)
- Fraction of total consumption of animal products used as food
- Fraction of total consumption of crops used as food or feed

Elasticities (region dependent)
- Income elasticity coefficient for animal products
- Income elasticity coefficient for crops

Caloric consumption (region independent)
- Maximum caloric intake (kcal day^{-1})
- Maximum consumption per animal product (kcal day^{-1})
- Maximum consumption per crop product (kcal day^{-1})

animal product) in a region where it is hardly produced, while exports lead to extra demand for land in the exporting region. Trade of commodities for the period 1970-1990 is taken from the AGROSTAT trade data base (FAO, 1992a), while future import/exports are prescribed.

Summing up to this point, the total number of animals per region are computed from meat consumption, animal productivity, and fraction of productive cattle, while the total crop demand per region is computed by adding the total consumption of crop products by humans to the total demand for feed concentrates by animals. Imports and exports of commodities are added to these totals to obtain the final land use demand for agriculture in each region. Table 5 summarizes the key coefficients needed for these calculations, while table 6 summarizes the input data for the ADM.

2.3   TESTING OF THE AGRICULTURAL DEMAND MODEL

Based on income and population from 1970 to 1990 and the previously described methodology, we can compute the resulting per capita demands for eight crops and five animal products. Representative results are given in Figure 2 for the USA, Latin America, and India plus South Asia. In some cases computed demands agree well with the 1970-1990 trend, e.g. cereals in the USA and cattle meat in Latin America. But in some circumstances elasticity coefficients were replaced to avoid unreasonable future trends, as in the case of cereals and rice in the region of India plus South Asia. Of course, replacement of the elasticity coefficients lead to poorer agreement with current data and give some indication of the uncertainty of estimating future per capita intake of food. We

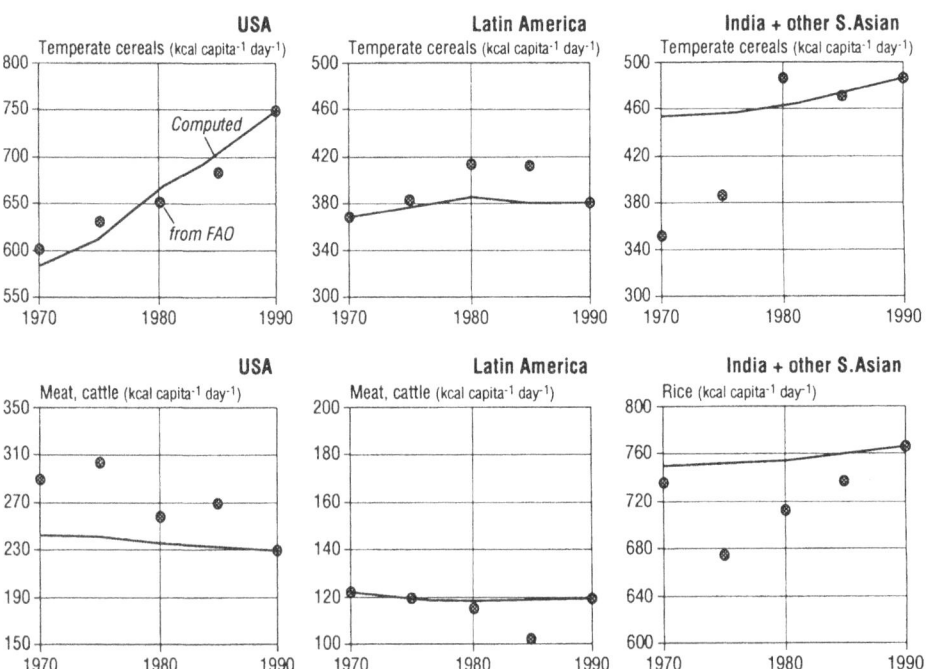

Figure 2. Computed per capita consumption of selected commodities for USA, Latin America, and India plus South Asia, 1970 - 1990 versus FAO (1992a) statistics.

are aware that the above is not a true test of the ADM since the data which were used to make the model are the same as those used for testing. Global data sets of these kind are however sparse, so we didn't have any choice. Some additional testing of the ADM is reported in Section 5.1 of this paper.

3. Land Cover Model

3.1   OBJECTIVES AND APPROACH

The objective of the Land Cover Model (LCM) is to simulate grid-scale changes in global land cover by reconciling *regional demands* for land with the *local potential* of land. Land cover is simulated on a grid of 0.5^0 latitude and 0.5^0 longitude. The regional demands for land are calculated by the Agricultural Demand Model (ADM) described above, and the local potential of land comes from the Terrestrial Vegetation Model (TVM) (Leemans and van den Born, 1994). The TVM takes into account the change in temperature and precipitation, and soil characteristics at each location following climate change, and calculates potential productivity of crops and potential vegetation (climate: Leemans, 1992; moisture: Prentice *et al.*, 1992; crops: Leemans and Solomon, 1993; vegetation: Prentice *et al.*, 1993). Changes in land cover lead to changes in land use related emissions and albedo, which are then used by the Ocean Atmosphere sub-system of IMAGE 2.0 to recalculate climate.

   Our first working hypothesis is that land cover transformations occurring on the local scale within world regions can be related to the sum of demographic and economic changes taking place in world regions (As noted above, IMAGE 2.0 takes into account 13 regions, Figure 1). Or put another way, we assume that regional-scale changes in agricultural demand are satisfied by changing land cover anywhere within the region, depending on the most suitable location for a particular land use. This is an obvious simplification because land cover is also affected by local and national economic factors, such as price subsidies.[1]   However, under certain circumstances this is a reasonable assumption e.g., in Africa, workers travel across borders in large numbers to meet the labor demands of agriculture in other African countries (Udo, 1987). Also, the vigorous trade that exists within a world region sometimes indicates that a particular country satisfics at least part of its demands for crops from the more suitable soils of a neighbouring country. Moreover, total regional demands for agricultural products are known to influence land cover changes in other regions as in the case of import of meat to North America in the 1970s which has been linked to the expansion of range land in various parts of Latin America (see, e.g., Williams, 1992, Turner *et al.*, 1993). These

---

[1]   Three of the world regions of the IMAGE 2.0 model correspond to countries (Canada, Japan, and the USA). For these regions, the model actually relates agricultural demands on a country-level to land cover changes within the country.

TABLE 7

Land use rules and assumptions used in the Land Cover Model

- Hierarchy of satisfying land use demands: (1) agricultural land (2) range land (3) exploited forest.

- Agricultural land expands only when current land is unable to satisfy demands.

- New agricultural land is allocated adjacent to current agricultural land

- New agricultural land is allocated to land with highest potential crop productivity

- Grassland expands only if it is replaced elsewhere by agricultural land, or if the number of animals in a region increases.

- New grassland is allocated adjacent to current agricultural land, grassland or savanna.

- Urban fuelwood demand (from residences and businesses) in the regions of Africa, India plus S. Asia, and East Asia is satisfied by clearing existing forests.

- Agricultural land taken out of production (because of disuse or decrease of potential productivity below a certain level) will revert to its climate-potential land cover.

---

examples suggest that regional driving forces can at least partly explain local land cover changes.

This assumption is also adopted out of pragmatism, because it is impossible at this time to track all the regional, national and local factors that will influence land cover changes over the long simulation period of the model. Indeed, the Land Cover model of IMAGE 2.0 is intended to be only a first step in incorporating driving forces on all these scales.

Our second working hypothesis is that socio-economic forces that drive land cover conversion can be represented as heuristic *land use rules* which represent many poorly understood and complex forces. These rules are defined to match actual land conversions occurring in each world region from 1970 to 1990 (FAO, 1992a), and are simple and straightforward, i.e. new agricultural land is located adjacent to current agricultural land in order to take advantage of existing agricultural infrastructure and labor. The complete set of land use rules used in this version of the model are given in Table 7.

The advantages of this approach are:
- Quantitative calculations of land cover can be made on the basis of qualitative knowledge.
- This qualitative knowledge is expressed as easy-to-understand rules which make model assumptions transparent.
- Several types of socio-economic driving forces can be taken into account simultaneously, which is almost always the case in the real-world (see, e.g. Stern *et al.*, 1992).
- New rules can be implemented fairly easily to reflect other assumptions about driving forces.

3.2 LAND COVER SIMULATION AND DERIVED LAND USE RULES

Starting with the above two working hypotheses, our basic approach is as follows:

- Compute the demand for three types of land use that lead to large land cover conversions: agricultural land, range land, and exploited forest. The demands for the first two are computed by the ADM described above as the demand for various crops and animals in each region. The only wood product currently taken into account by the model is fuelwood consumption, and this is computed by the IMAGE 2.0 Energy-Economy model (de Vries *et al.*, 1994).
- Compute potential crop productivity and potential natural vegetation based on climate and soil for each grid cell using the TVM.
- Reconcile land use demand and potential on a grid cell by grid cell basis by using land use rules (table 7). Demands for land use are satisfied in the following sequence: agricultural land, range land, and forest land[2].

As already noted, land use rules are tested by simulating known land use transformations between 1970-1990, focusing especially on changes in the extent of agricultural land (based on statistics). Spatial data of the location of cultivated crops are not yet available for the entire world to check our 1990 calculations, so we use regional total data from FAO's AGROSTAT data base (FAO, 1992a). However, later in this paper we compare some calculations with country-scale data.

For our initial global land cover we use the ecosystems data base of Olson *et al.*, (1985) available on a global grid of 0.5^0 latitude and 0.5^0 longitude. Each grid cell of this data base is assigned to the dominant ecosystem in that grid cell. Since Olson's data were compiled mostly in the 1960s and 1970s, we use these data to initialize the Land Cover model in 1970. In order to simplify calculations we modified the aggregation of Olson's 51 ecosystems into 17 biomes given by Prentice *et al.* (1992) to 17 land cover types (Figure 3A) that are particularly relevant to assessing ecosystem impacts, terrestrial C content, albedo and other factors related to global change (see also Leemans and van den Born 1994; Klein Goldewijk *et al.*, 1994). Olson's data base does not include a category for built-up land (settlements, infrastructure, or industrial areas). However, even in densely populated Europe, 10% or less of its total territory is built-up. Consequently, very few grid cells would have built-up land as the dominant type. Nevertheless, because of the important competition between built-up and agricultural land in various parts of the world, we may include this category in future versions of the model.

3.2.1 Simulation of Agricultural Land
In order to correctly simulate the amount of arable land in a region, we must first estimate the percentage of each agricultural grid cell that will actually be used for crop production.

---

[2]   We use, agricultural land for all land used for crowing crops for a period of five years or more, rangeland for all types of grassland from Olson's database including savanna, forest land for all types of forest from Olson's database (Olson *et al.*, 1985).

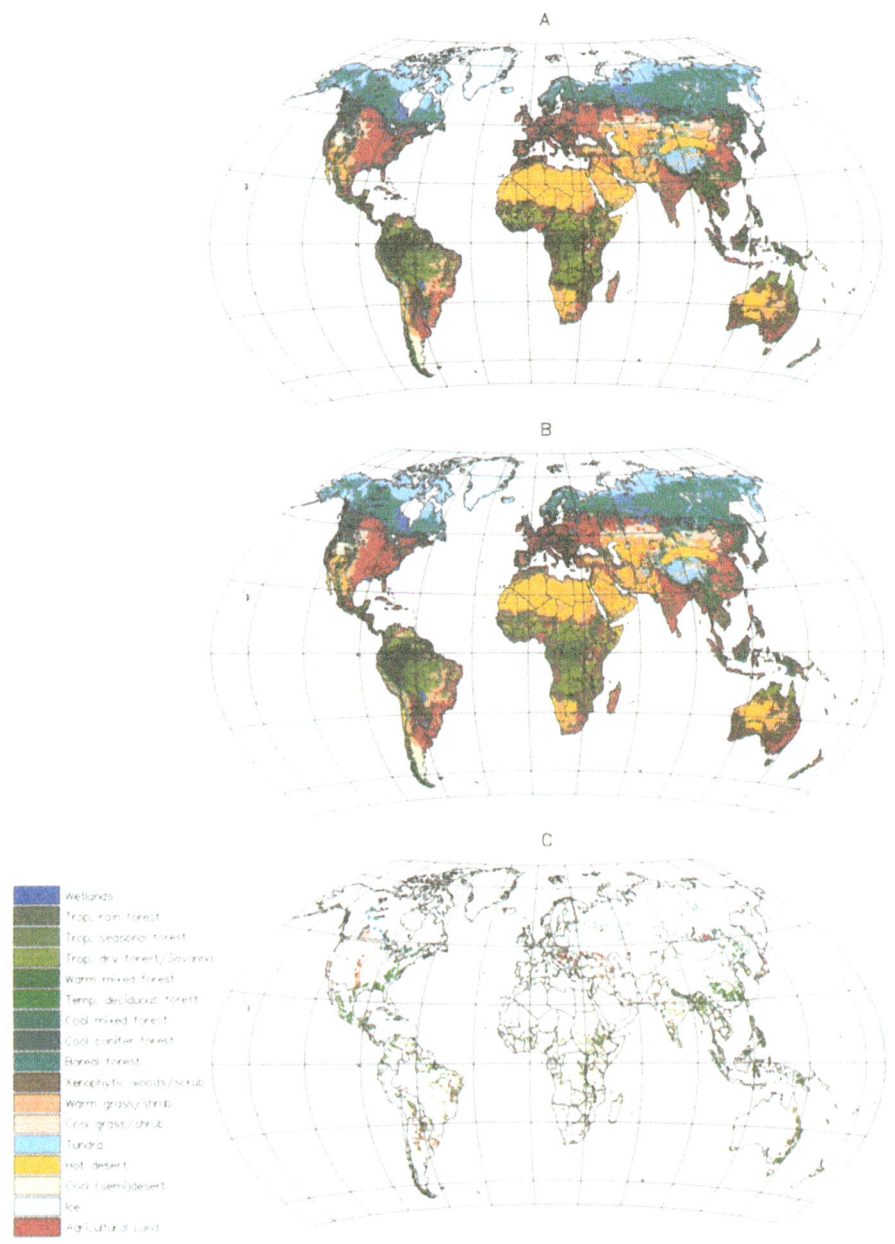

Figure 3A-C: Land cover types 1970 - 1990; (a) Initial land cover types 1970; (b) Computed land cover types 1990; (c) Converted land cover types 1970 to 1990

This is because part of an agricultural grid cell is fallow land, built up land, or is otherwise unsuitable for agricultural purposes. In addition, parts of the grid cell are also used for permanent pastures. On the other hand, grid cells having agriculture, but not as the dominant type are not taken into account. However, the agricultural area calculated using the agricultural grid cells in a region given by Olson's data base overestimate the area as compared to estimates given by FAO (1992a) for all regions. The fraction available for crop production is called the "usable area factor" and is computed as the ratio of two estimates of total agricultural area in a region: the area given by AGROSTAT for 1970 (FAO, 1992a) versus the area calculated by using Olson's agricultural grid cells (Table 2). Each grid cell in a region is assigned the same usable area factor.

The agricultural simulation begins by allocating the usable area in each existing agricultural grid cell, using the "usable area factor", to the eight types of crops until their regional demand (calculated by the ADM) is satisfied. All existing agricultural grid cells must be allocated before new land is used for agriculture. The allocation begins with the grid cell having the highest "composite productivity".[3] Equation (2) is used to calculate the relative area of the grid cell to be used for each of the crops. This Equation takes into account that crops with a higher regional demand (Dem_j) or a lower maximum potential productivity (MPP_j) require a greater amount of land than crops with lower demands or higher productivity. The Equation also assigns more of the grid cell's land to the most suitable crops for that grid cell, that is, to those crops with higher potential productivity class (PP_{ij}).

$$F_{ij} = \frac{\dfrac{Dem_j}{MPP_j} PP_{ij}}{\sum\limits_{j=1}^{i} \dfrac{Dem_j}{MPP_j} PP_{ij}}$$

(2)

---

[3] Composite productivity is defined as the sum of the productivity classes of each of the eight crops. Each crop in each grid cell is assigned a productivity class based on its potential productivity calculated by the TVM. Example productivity classes:

Crop A		Crop B	
potential productivity (ton ha^{-1})	productivity class (-)	potential productivity (ton ha^{-1})	productivity class (-)
< 0.1	0	< 0.1	0
0.1 - 2.0	1	0.1 - 4.0	1
2.1 - 4.0	2	4.1 - 8.0	2
> 4	3	> 8.0	3

As an example, to compute the composite productivity of grid cell x having crops A and B:
If the productivity of crop A in grid cell x is 2.6 ton ha^{-1}, then its productivity class is 2
If productivity of crop B = 3.1 ton ha^{-1}, then its productivity class is 1
Therefore, the composite productivity class in grid cell x is 2 + 1 = 3

Where:

F_{ij} = fraction of grid cell i allocated to crop j
Dem_j = regional demand for crop j (t)
PP_{ij} = potential productivity class of crop j in grid cell i (t ha^{-1})
MPP_j = maximum potential productivity of crop j (t ha^{-1})
i = index for grid cell
j = index for crop class.

After all usable area in this grid cell is allocated to crops, the same procedure is followed for the grid cell with the next highest composite productivity class, and so on. If the demand for the eight classes of crops cannot be met on the existing agricultural land, additional land is allocated. The land use rule applied here is that new agricultural land will tend to be located near to existing infrastructure and labour for agriculture, that is, close to existing agricultural land. After the demands for the eight main classes of crops are satisfied, we take "Other Crops" into account. Since the potential productivity and demand are not known for such a diverse class, we take a simpler approach and use the 1990 ratio of "Other Crops" area to total harvested area, and we assume that this ratio is constant in time, to calculate the needed area for "Other Crops". We allocate "Other Crops" to land where at least one of the other eight classes of crops can grow, using grid cells with the highest composite productivity class first[4].

The result of the preceding allocation scheme is that it somewhat optimizes the use of agricultural land, taking into account demand and location of favorable land. Since the calculations of the potential productivity do not yet account for irrigation (Leemans and van den Born, 1994), agricultural areas that depend on irrigation are not yet simulated. Irrigation will be included, however, in the next version of the TVM.

We should note here that Esser (1989) uses a somewhat similar rule-based approach to compute the rate of clearing of forests. In his model, a forest grid cell is selected for clearing over another grid cell if any of the following rules apply to it: (i) more agricultural land is in the vicinity of the grid cell, (ii) it has a higher natural primary productivity, (iii) it has a higher soil fertility, or (iv) it had a faster clearing rate between 1950 and 1980. Esser did not attempt, however, to explicitly include demand for land into his model.

3.2.2 Simulation of Range land

FAO defines range land as "land used permanently (five years or more) for herbaceous forage crops, either cultivated or growing wild" (FAO, 1992a). There are no consistent assessments of the extent of global range lands, but data regarding permanent pastures (FAO, 1992a) indicates that this type of range land area only slightly increased in most regions during the 1970s and 1980s. Since livestock numbers increased at a greater rate throughout the world during the 1970s and 1980s, the density of livestock per hectare also

---

[4]   If during allocation of existing agricultural land grid cells are found which have only a potential productivity for crops for which there is no demand, then this grid cell is allocated to "other crops" as well.

increased in many countries, bringing with it degrading soil conditions where carrying capacity is exceeded. Indeed, large-scale overgrazing is well documented around the world, leading to severe land degradation in Africa (UNEP, 1992), India (WRI, 1988) and elsewhere. In some cases this increase in livestock density has been paced by attempts to increase productivity of range land, as in China (WRI, 1988). Range land in the LCM is taken to include the following land cover types: warm grass/shrub, cool grass/shrub, and the portion of agricultural land not used for crops (see above). Tropical dry/savanna is also used as range land, but is assumed to sustain a lower density of animals than the other land cover types, as noted below.

Demand for range land is satisfied after the demand for agricultural land has been met. The Agricultural Demand model described above is used to compute the regional demand for cattle, sheep, pigs, and poultry, although only the first two have major land requirements. Since sheep require much less land than cattle, we follow Webster and Wilson (1980) and assume that nine sheep need the same amount of land as one cow.

Since we do not know the dynamics of changing livestock densities around the world, we make simple assumptions to simulate changes in range land. First, we assume that each range land grid cell has the average regional animal density. This is estimated by dividing the total number of animals that require range land in 1970 by the total area of range land in 1970 for each region. This means that in each region every range land grid cell (these are all grid cells having grass as the dominant land cover type) has the same average animal density. Savanna lands however are an exception, and have a constant low density of 10 animals km^{-2}. The average animal density is recalculated each time step.

Second, we adopt the land use rule that range land only expands if it is replaced elsewhere by expanding agricultural land, or if the number of animals in a region increase. An exception is made for regions where livestock densities are now relatively low (below 12.5 animals km^2). If new range land is needed it is added as either "warm grass/shrub", or "cool grass/shrub", depending on local climate. Since new range land is always expanded at the cost of forests (all grass land is already used as range land and other land cover types other than forest have a climate unsuitable for either cool or warm grass/shrub) we assume the animal density on new range land to be 100 animals km^{-2}, which is comparable to new range land carved out of forests in Latin America (see, e.g., WRI, 1990). New range land is located according to the land use rule that it will be adjacent to existing range land or agricultural land; this assumes that the infrastructure and labor necessary for livestock raising will be available there. Locations are selected randomly around already existing areas, but only where there is a potential for range land (cool or warm grass) according to the TVM. If the number of animals decreases, the amount of range land nevertheless remains the same (if the local climate is suitable at that time). This implies that some grassland may not be grazed.

3.2.3 Simulation of Wood Exploitation
This version of the Land Cover model takes a very simple approach towards wood exploitation, and only includes fuelwood. Harvesting of timber and other wood products will be included in the next version of the model.

Fuelwood used in rural areas usually comes from scavenging branches or thinning trees in and among agricultural areas, and typically does not lead to clearing of existing forests (Prasad, 1987; Smil, 1980; WRI, 1986) By comparison, the fuelwood (and charcoal derived from fuelwood) used in households and businesses in the cities of developing countries sometimes leads to large-scale forest clearing around settlements (e.g. Hall and de Groot, 1987; Williams, 1992; UNEP, 1992). This occurs partly because commercial enterprises find it profitable to clear-cut forests and transport wood long distances to urban consumers. With this situation in mind, we adopt the land use rule that urban fuelwood demand (from residences and businesses) in the regions Africa, India plus South Asia, and East Asia, is satisfied by clear-cutting existing forested areas. It is assumed that fuelwood use does not lead to large-scale forest clearing in other regions (although it may lead to forest thinning). According to another land use rule, the locations of these clear-cut grid cells are at the boundaries of forest and either grassland or agricultural land. The amount of area required is computed from the C content of forest grid cells computed by the IMAGE 2.0 Terrestrial Carbon Model (Klein Goldewijk *et al.*, 1994). We recognize that this procedure may overestimate the land cleared for fuelwood, and therefore this approach will be reassessed for the next version of the model.

3.3 CALIBRATION AND VALIDATION OF THE LAND COVER MODEL

The LCM was calibrated by adjusting a "management factor" for each region and each crop class until computed cultivated areas were close to actual cultivated areas between 1970 and 1990 based on FAO estimates from AGROSTAT (FAO,1992a). This "management factor" represents the gap between the theoretically-feasible production of crops based on climate and soil conditions, and the actual production which is limited by less-than-perfect technology and know-how. As a result of calibration, the simulated crop areas for most world regions and crop classes are close to actual areas between 1970 and 1990. The only exception was temperate cereals in the regions of Africa, India plus S. Asia, and Oceania where inaccurate climate data led to very low estimates of the potential productivity of this crop.

The amount of forest area required for fuelwood was computed by the method described above. The model estimates that about 31,000 km^2 of African forest area were cleared for fuelwood during the 1980s, or about 10% of the computed total deforested area of Africa. In the regions of India plus S.Asia and East Asia, a total of about 48,000 km^2 was computed, also about 10% of the total deforested area in these regions. Although we cannot compare these figures to authoritative estimates, we suspect that the model overestimates the Asian area required for fuelwood.

After the Land Cover model has been calibrated, we obtain the land cover patterns for 1990 presented in Figure 3B. Figure 3C depicts the computed differences in land cover between 1970 and 1990. We have not yet checked our calculations on a grid-by-grid basis because detailed spatial data are only now becoming available. However, in the following paragraphs we make a qualitative comparison of these calculations with other information.

For the United States, agricultural land replaces forested area in the Delta States and

TABLE 8

Average annual deforestation rate 1981-1990 (thousands ha year^{-1}). Only the 10 countries with the highest deforestation rates as given by FAO (1992b) are listed.

Latin America	FAO	IMAGE 2	Africa	FAO	IMAGE 2	Asia	FAO	IMAGE 2
Brazil	3671	2527	Zaire	732	371	Indonesia	1212	774
Mexico	678	833	Sudan	482	459	Thailand	515	334
Bolivia	625	590	Tanzania	438	214	Myanmar	401	741
Venezuela	599	461	Zambia	368	212	Malaysia	396	587
Paraguay	403	284	Angola	174	92	Philippines	316	153
Colombia	367	247	Ghana	137	92	Bangladesh	174	29
Ecuador	238	186	Mozambique	135	89	Vietnam	137	180
Peru	217	151	Madagascar	135	60	Kampuchea	131	90
Nicaragua	124	60	Centr.Afr.Rep.	129	31	Laos	129	59
Honduras	112	150	Cameroon	122	123	New Guinea	113	277

grassland in the Northern Plains and Mountain regions. This is consistent with known trends (Daugherty, 1987). However, for the East Coast, the model computes a conversion of forested to agricultural land, which did not occur between 1970 and 1990. On the other hand, the overall decrease in grassland computed by the model actually occurred in these decades (Daugherty, 1987).

In Latin America, the model computes that large forested areas of Brazil are replaced by grassland and agricultural land, although the amount of deforestation in Rondonia and other interior areas of Amazonia is apparently underestimated (see, e.g. Skole and Tucker, 1993). Large areas of Bolivia, Mexico, and Venezuela, are also estimated to be replaced by grassland and agricultural land. These trends are consistent with FAO statistics (FAO, 1992a). Also consistent with these data is the increase in agricultural land and grassland in India plus South Asia and the simultaneous decrease in savanna and forest land.

Since deforestation rates are indirectly computed by the model (based on expansion of grassland and agricultural land and clearance for fuelwood), they serve as a useful independent check of model calculations. The rate of deforestation is also of particular interest from the perspective of global climate change because conversion of tropical forests is one of the main sources of greenhouse gases from the biosphere to the atmosphere (e.g. IPCC, 1990). On a regional average basis, the computed rates are comparable but on the low side of FAO estimates for the 1980s (Table 8). We should note that these FAO figures have a large uncertainty and are regularly updated and improved. Moreover, the land cover model uses a land cover classification different from FAO's, which makes data comparisons difficult. Nevertheless, our figures for deforestation are too low for Africa probably because soil degradation is not yet included; soil degradation causes range land and agricultural land to be taken out of production, and leads to additional demands for forest land elsewhere on the continent.

As noted above, it would be desirable to compare grid-cell calculations of deforestation with observations, but these observations are not yet available for the whole world. However, we have taken a step in that direction by comparing our country-scale calculations with FAO estimates (FAO, 1992b) for the ten countries in Africa, Asia, and Latin America with the highest deforestation rates in the 1980s (Table 8). Most computed rates are within a factor of two of the estimates. Calculations for Brazil are also below FAO estimates, but Skole and Tucker (1993) have suggested that FAO estimates for Amazonian deforestation are too high.

The good agreement of model calculations with country-scale deforestation data is partly due to the tendency of the model to compute larger deforestation rates in countries with large standing forests (e.g. Brazil and Zaire). Nevertheless, this cannot account for the reasonable model agreement with FAO estimates for so many different sized countries. One conclusion is that the simple hypotheses of the model largely explains the magnitude and spatial variability of deforestation between 1970 and 1990. Nevertheless, some obvious improvements of the LCM (described in Section 5.2) should greatly improve the spatial details of the simulation.

4. Scenario Analysis

4.1   INPUT ASSUMPTIONS

In this section we apply the Agricultural Demand and Land Cover models to a "Conventional Wisdom" scenario for 1990-2100. This is of course only one example of an endless choice of possible scenario's ranging from business as usual to sustainable development to destabilization of regions to liberalisation of trade etc. The basic driving forces of this scenario, including population and economic growth, are taken from the IPCC intermediate scenario "IS92a" (Pepper *et al.*, 1992). Global population figures range from 5.2 billion in 1990, 9.8 billion in 2050 to 11.5 billion in 2100. The complete scenario as it was applied to all submodels of IMAGE 2.0 is described in Alcamo *et al.* (1994b).

The following variables are used to create a scenario:
• changes in crop yield due to fertilizer and other technological inputs;
• changes in animal productivity due to technology;
• feed ratio of roughage to concentrates
• import and export of food
• fraction of non-productive cattle

4.1.1   Crop Yield
Crop yield has improved throughout the world in the last decades owing to increased input of fertilizer and the application of technology in the form of tractors, irrigation schemes, and management know-how (Evenson, 1988). Average yield per hectare has increased for almost all commodities and regions, independent of soil and climate conditions. For developing regions the productivity for wheat increased slowly from about 0.6 t ha^{-1} in 1900 up to about 0.9 t ha^{-1} in 1960 (Scandizzo and Diakosawas, 1987) and increased more rapidly up to about 2.3 t ha^{-1} in 1990 (FAO, 1992a). In the USA the productivity increased

from about 1.6 t ha^{-1} up to 2.7 t ha^{-1} from 1960 to 1990 (FAO, 1992a). For scenario analysis we distinguish between the effect of fertilizer inputs (which can be related to fertilizer application scenarios as well as other environmental problems such as N_2O emissions) and other technological inputs.

To gauge the effect of increased fertilizer inputs on crop yield, we use a typical response curve of cereal yield to fertilizer from Addiscot et al. (1991). This curve is representative of a variety of crops including grasses and different legume-cereal rotations over a range of environmental conditions (Bock and Hugert, 1991; Hesterman, 1988; Martinez, 1990; Tisdall et al., 1993). We recognize that using one curve for all crops under a great variety of environmental conditions is far from satisfactory, although responses appear to be very similar. Future versions should specify response curves for all crops or crop groups distinguished in this model. Since it only depicts the effects of N fertilizer (rather than phosphate or other fertilizer), we implicitly use N fertilizer as a surrogate for total fertilizer inputs. For the Conventional Wisdom scenario we compute the change in crop yield per hectare due to fertilizer inputs to each region by using the N fertilizer assumptions for each region from IPCC's IS92a scenario (Pepper et al., 1992). Because some regions are already near the maximum yield given by the response curve, the addition of fertilizer will lead to only a small increase in yield over the period 1990-2100. It should be noted, however, that we do not take into account the likely environmental problems that ensue from high fertilizer application rates such as NO_3 contamination of ground water or eutrophication of surface waters.

To estimate the effect of technological inputs other than fertilizer on future crop yields we analyzed the increase in actual crop yield data from 1970 to 1990. We first subtracted the component of this increase that can be attributed to increased fertilizer inputs by using actual fertilizer application rates during this period together with the response curve mentioned above. We assume that the remaining increase in yield can be attributed to non-fertilizer inputs to agriculture. There might also be a small influence of CO_2 fertilization on crop yield, but the extend of this on a global level is unknown and is therefore ignored in this version of the model. Based on this analysis, crop yields in industrialized regions increased by about 40% due to non-fertilizer inputs. For the Conventional Wisdom scenario we assume that an increase in technological (non-fertilizer) inputs in the developing regions could increase yield to the same extent experienced historically in industrial regions. It is assumed that technological inputs will improve yield by 1.5% per year over the period 1990-2000, and 1% per year over the period 2000 to 2025 under the condition of 2% per year economic growth (Table 9). If economic growth assumptions are lower, the increase in yield is proportionately lower. The resulting index of crop yield is given in Table 9.

With regards to industrialized regions, this scenario assumes that yield continues to increase through conventional and bio-technical developments. However, yield improvements between 1990 and 2100 are assumed to be only half their level between 1970 to 1990 (Table 9). As a general rule we assume conservatively that the rate of technological improvements will decrease over the simulation period.

The result of the preceding assumptions is that yield per hectare increase in

TABLE 9

Index* for improvement in crop yield due to N fertilizer and other technological inputs. Conventional Wisdom scenario.

Agricultural improvement indices	YEAR	Canada	USA	Latin America	Africa	OECD Europe	Eastern Europe	CIS	Middle Eas	India + S. Asia	China + C.P. countries	East Asia	Oceania	Japan
Yield increase due to application of N fertilizer [1]	2000	1.09	1.06	1.10	1.05	1.07	1.14	1.15	1.11	1.14	1.06	1.13	1.04	0.87
	2025	1.49	1.32	1.61	1.13	1.18	1.26	1.58	1.17	1.67	1.07	1.43	1.17	0.83
	2050	1.52	1.32	1.85	1.09	1.20	1.30	1.62	1.06	1.74	1.09	1.59	1.18	0.81
	2100	1.47	1.30	1.96	1.08	1.21	1.33	1.62	1.13	1.78	1.15	1.68	1.17	0.78
Yield improvement due to technology [2]	2000	1.10	1.10	1.08	1.08	1.10	1.00	1.00	1.08	1.15	1.15	1.15	1.15	1.10
	2025	1.23	1.23	1.33	1.33	1.23	1.13	1.25	1.21	1.40	1.40	1.40	1.40	1.23
	2050	1.30	1.28	1.46	1.46	1.30	1.18	1.33	1.33	1.53	1.53	1.53	1.48	1.28
	2100	1.30	1.28	1.51	1.61	1.30	1.18	1.43	1.48	1.68	1.68	1.68	1.58	1.28
Total yield increase [3]	2000	1.20	1.17	1.19	1.13	1.18	1.14	1.15	1.20	1.31	1.22	1.30	1.20	0.96
	2025	1.83	1.62	2.14	1.50	1.45	1.42	1.98	1.41	2.34	1.50	2.00	1.64	1.02
	2050	1.98	1.68	2.69	1.59	1.56	1.53	2.15	1.41	2.65	1.66	2.42	1.74	1.03
	2100	1.91	1.66	2.95	1.73	1.57	1.56	2.31	1.67	2.98	1.93	2.81	1.84	0.99

*   Index 1990 = 1.00

1)   Future N application figures are from IPCC: IS92a,b,e scenario. The index figures are derived after a conversion from absolute supply figures to application per hectare. The relative yield impact is calculated using the yield response curve developed by Addiscot et al., 1991. The agricultural land area figures are calculated in the IMAGE model.

2)   Yield improvement due to technology is based on historical analysis of the period 1970-1990. Technological improvement includes: management, plant treatment, storage improvement, mechanization etc. Application of fertilizer is excluded.

3)   Final result is the multiplication of fertility index and technology index.

industrialized regions in the year 2100 is smaller than the increase in developing regions over the same period. In the USA the increase over this period was 66%, compared to 195% in Latin America and 198% in India plus other South Asian countries (Table 9). As already noted, the environmental impact of fertilizer application and technology use in the future is likely to be large, but is not accounted for in the current version of the model.

4.1.2 Animal Productivity and Related Input Variables
The global average productivity per animal from 1970-1990 increased for most animals. Poultry had the highest increase (30%), meat from cattle and pigs increased by about 15%, and milk production per cow by 10%. The increase in productivity of most animals is higher in developing than industrialized regions for most animals.

As for future productivity, we assume that animal productivity in a developing region reaches the current level of OECD Europe (Table 4) when the average GNP/cap of the developing region approaches the current GNP/cap in OECD Europe. For the Conventional Wisdom scenario, this amounts to a 20% increase in productivity of beef and milk cattle, pigs and poultry between 1990 and 2025. The increase in industrial regions is somewhat lower, i.e. only 10% over the same period. In the period 2025-2050 the rate of increase is assumed to be half that of the preceding period, and after 2050 is assumed to be negligible for all regions.

A related scenario variable is the assumed fraction of total cattle in a region that are unproductive. For the Conventional Wisdom scenario, we assume that the fraction of non-productive cattle in a region reaches the current OECD Europe figure (30 %) when the income per capita of this region reaches the current income of OECD Europe.

Another associated scenario variable, the ratio of roughage to concentrate feed, is assumed to remain constant at its 1990 value.

Since each of these variables are related to technological developments in agriculture they have a very high uncertainty, and should receive more research attention. The sensitivity of one of these variables, animal productivity, is discussed later in the paper.

4.1.3 Import/Export scenario
For the Conventional Wisdom scenario we assume that current exports from the developed world increase by 50% from 1990 to 2050 and level off afterwards; net exports from the developing regions are assumed to double their 1990 level by the year 2100. Export of animal products is assumed to stay constant at its 1990 value, while the export of sugar is assumed to be zero. Imports in various regions for various commodities are assumed to grow in proportion to their 1990 values.

4.2 SCENARIO RESULTS

The results of the Conventional Wisdom scenario are discussed for the regions USA, Latin America, and India plus South Asia because of their contrasting economic and demographic development. Results are also presented for the world as a whole.

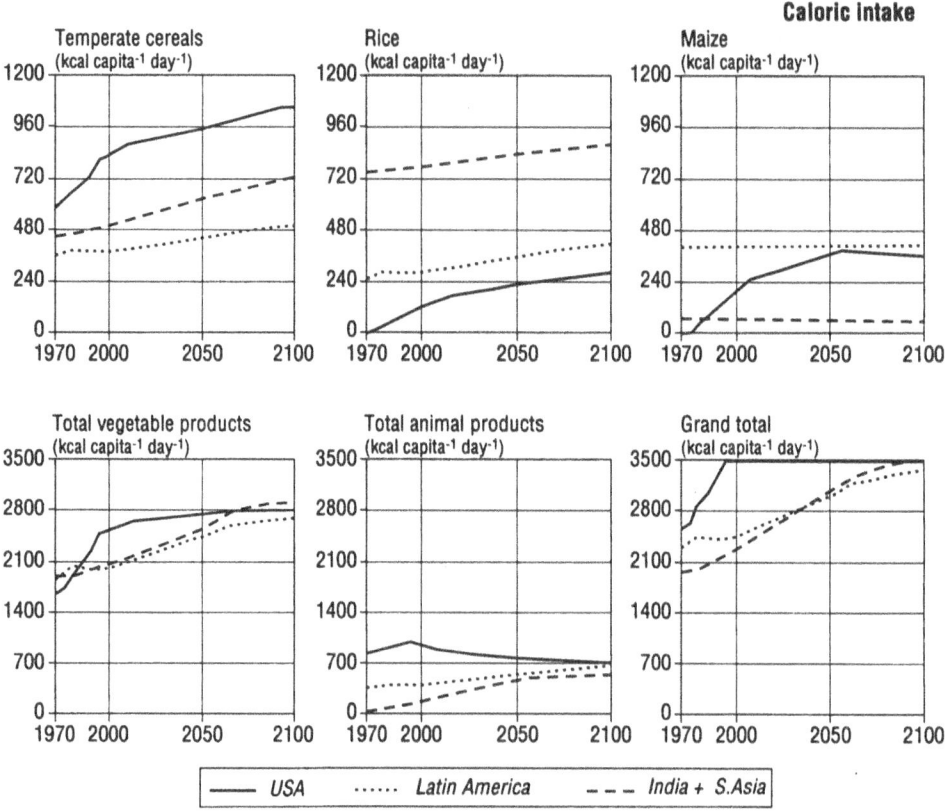

Figure 4: Computed per capita caloric intake of selected commodities.

4.2.1 USA

Two main factors influence the trend in caloric intake of agricultural commodities in the U.S. (Figure 4). First, because of the economic growth assumptions in the Conventional Wisdom scenario, total caloric intake of all agricultural products reaches its (assumed) maximum before year 2000, and afterwards remains constant. Second, although total intake of commodities remains constant, the individual commodities have distinct trends because their elasticity coefficients are different. However, these trends are slowed because the sum of the intake of products cannot exceed the assumed maximum. Hence, the increase in consumption of one commodity is compensated by a decrease in another as shown by the consumption of animal products which decreases slightly after 1990 (from about 950 kcal capita^{-1} day^{-1} to 690 kcal capita^{-1} day^{-1}), and is compensated by a slight increase in intake of crop products (from 2250 to about 2800 kcal capita^{-1} day^{-1}) (Figure 4).

The decrease in total intake of animal products is mostly because of a decrease in cattle meat and milk consumption; the consumption of pigs and poultry is more or less stable.

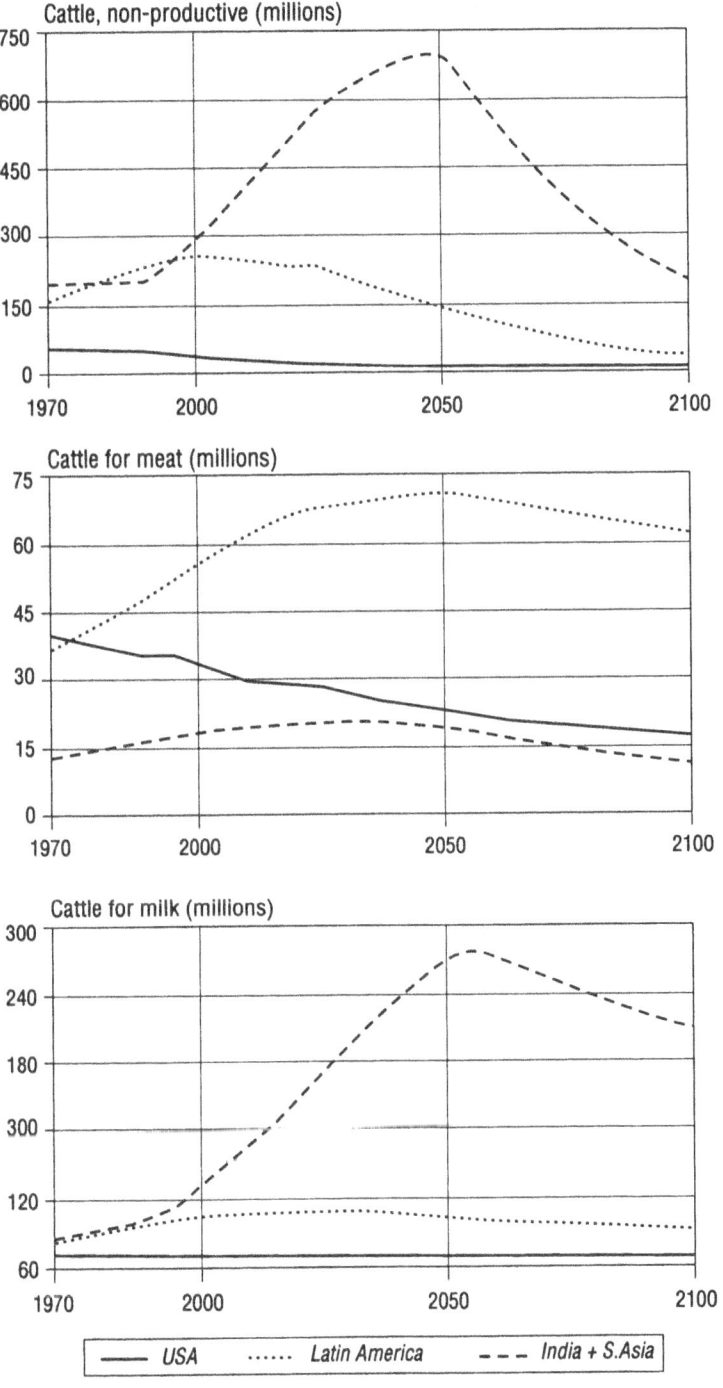

Figure 5: Computed number of different types of cattle.

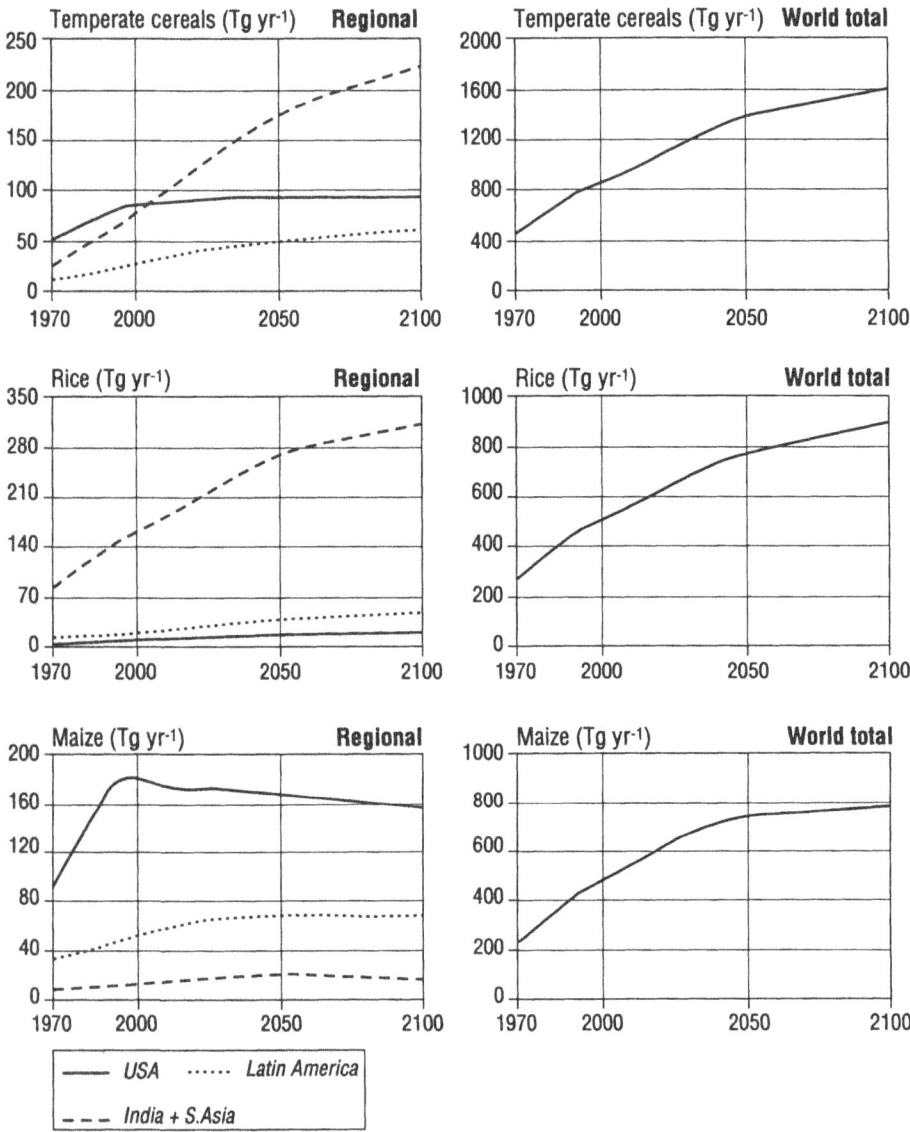

Figure 6: Computed total demands for selected crops.

The decreasing per capita demand for meat, together with the assumed decrease of non-productive cattle, leads to a 25% decrease in the total number of cattle over the period 1970 to 2000 (Figure 5). Fewer cattle require less feed and smaller crop areas to produce this feed. At the same time the slight increase in human consumption of crops is more than balanced by (assumed) improvements in crop yield. The trend in the total demand for crops (Figure 6) reflects a slight increase in population and a slight increase in intake of crops and a decrease in demand for animal feed. The net result is a decrease in agricultural

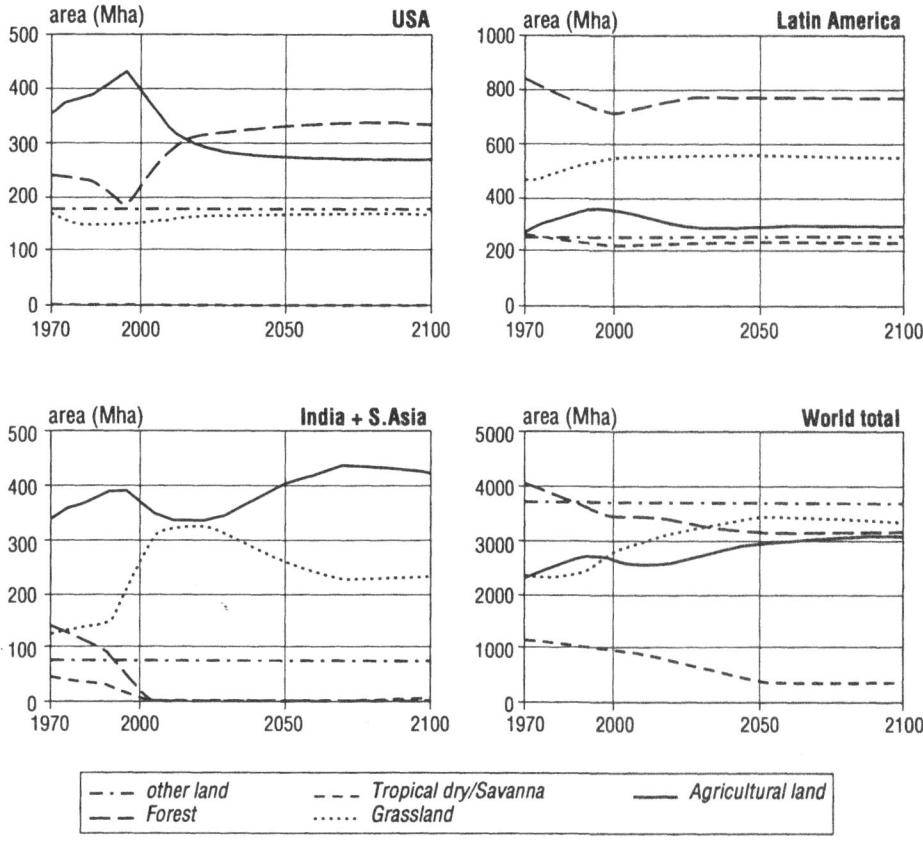

Figure 7: Computed areas of different land cover types.

area after year 2000 (Figure 7). Abandoned agricultural land is replaced by land cover suitable to local climate, most commonly, temperate deciduous forest. The temporal trend of land conversions (Figure 7) shows a peak in agricultural land area in the coming few years, a stabilization of grassland area, and a steady increase of forest.

4.2.2 Latin America

Increasing population and economic growth in Latin America lead to large increases in consumption of many commodities during the simulation period although the maximum per capita intake of food is not reached (Figure 4). In the period 1990-2100, total intake of crop products increases from about 2000 to almost 2700 kcal capita^{-1} day^{-1}. This is caused by increasing demand for oil crops, temperate cereals, rice and sugar. The intake of other crops remains at the same level (maize) or decreases (roots, pulses). In the same period the demand for animal products increases from about 380 to about 650 kcal capita^{-1} day^{-1} (Figure 4) owing to increased demand in particular for milk, and meat from pigs and poultry.

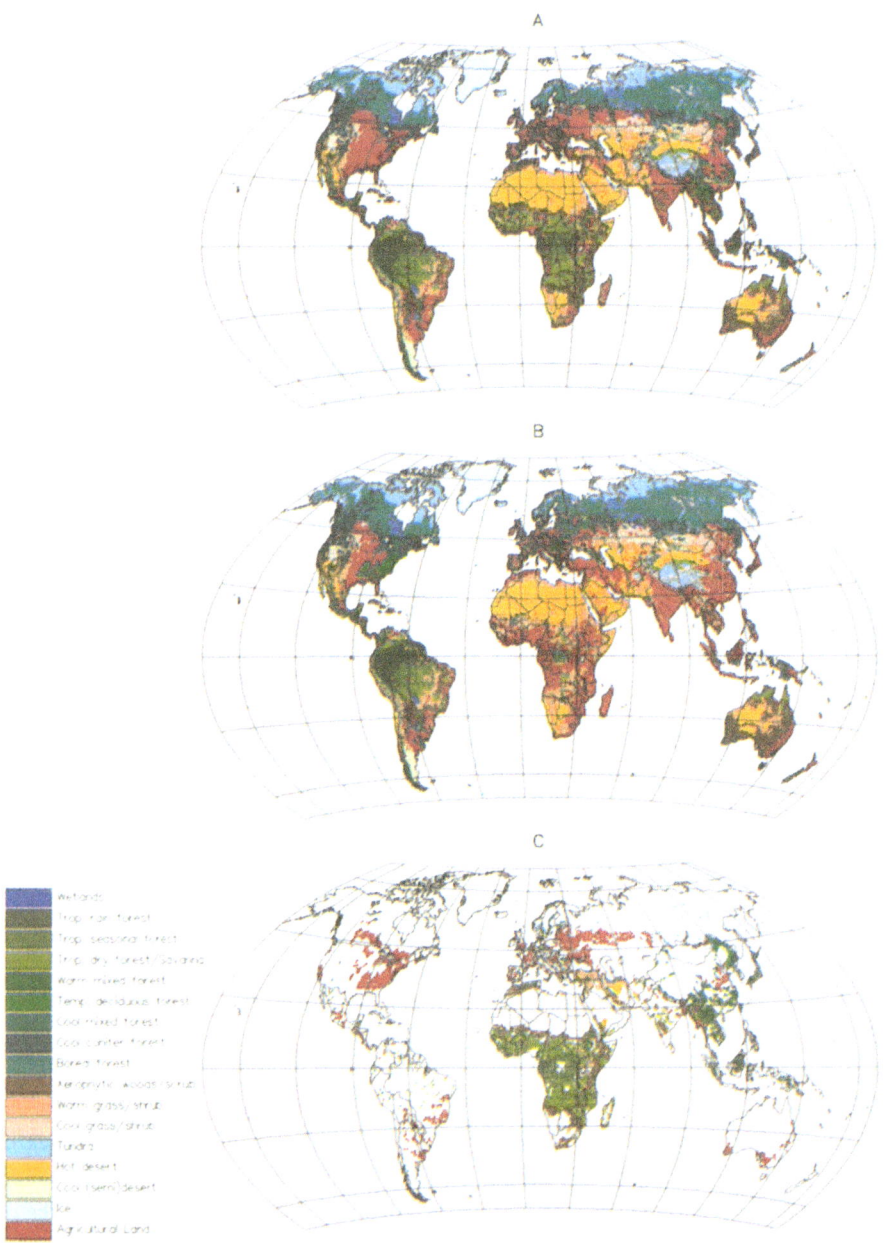

Figure 8A-C: Land cover types 1990 - 2050. (a) Computed land cover types 1990; (b) Computed land cover types for "Conventional Wisdom" scenario; (c) Converted land cover types from 1990 to 2050.

Although the intake of animal products increases throughout the simulation period, the total number of most types of animals peaks and then decreases (Figure 5). This is because of the increase in animal productivity (assumed to increase along with economic growth, see section 4.1.2). The trend in the total demand for crops (Figure 6) reflects a steady increase in per capita intake in crops due to an increasing population, but decreasing numbers of animals (leading to reduced demand for feed crops). The net result is a moderate increase in demand for most crops (Figure 6). However this trend in the demand for crops is more than balanced by the assumed improvements in crop yield per hectare in Latin America which leads to a net decrease in agricultural land from year 2000 to 2025. Afterwards the amount of agricultural land remains almost constant (Figure 7). This has important implications on the rate of land cover conversions. Up to year 2000, agricultural area expands, mainly at the expense of tropical forest areas in Brazil, Venezuela and Central America (Trend for 1970 to 1990 shown in Figure 3C). After year 2000, decreasing demand for crop land together with a peak in the numbers of animals leads to abandonment of some agricultural land along the coast of Argentina and Brazil, and in the interior of Argentina and Mexico (Figures 8A, B, C). This land reverts to the land cover appropriate for its local climate. In the case of the Brazilian coast this is mostly tropical seasonal forest; for the Argentinean coast, broad leaved/warm mixed forest; and for the interior of Mexico and Argentina, xerophytic woods/shrub (Figures 8A, B, C). We must emphasize at this point that we are describing the results of only one of many feasible scenarios.

4.2.3 India plus South Asia
In the India Region (India plus South Asia, see Figure 1), the caloric intake of crop products increases from about 2000 kcal capita^{-1} day^{-1} to almost 2900 kcal capita^{-1} day^{-1} over the period 1990-2100. This is caused by the assumed increase in income which leads to per capita increases in consumption of temperate cereals, rice, oil crops, and sugar (Figure 4). In the same period the demand for animal products increases from about 120 kcal capita^{-1} day^{-1} to about 550 kcal capita^{-1} day^{-1}. The increase in animal products reflects higher intake of milk and poultry meat. Following assumed increases in income, total caloric intake almost reaches the assumed maximum of 3500 kcal capita^{-1} day^{-1} around the year 2100. The increasing per capita intake of food coupled with rising population leads to steady increases in the demand for many important crops (Figure 6). This, together with the rise in the number of animals till 2050 (Figure 5) rapidly increases the amount of grassland and agricultural land at the expense of savanna and forest which disappear by the year 2005 (Figure 7). Afterwards the demand for agricultural land subsides because of (assumed) improvements in crop yield related to fertilizer use, technology improvements and a decreasing number of cattle. The land requirement for livestock is also not satisfied, and implicitly animal densities exceed their sustainable density on grassland which presumably accelerates soil deterioration. On top of that agricultural land continues to expand onto former grassland (Figures 8A, B, C). After 2050, the situation improves somewhat as improved crop yield and animal productivity allow both crop and livestock demands to be met (although the density of animals might be unrealistically high).

4.2.4 World and Other Regions

Results summed up for the whole world shows a tripling in the demand for some major food crops (wheat, rice, maize), a doubling for tubers, roots and pulses, a very strong increase for oil crops and sugar products, and a decrease in demand for the tropical cereals. The total number of animals in the world increases by a factor of three by 2050. After 2050 the number of all animals except meat cattle and poultry levels off or decreases. Both agricultural and grassland areas continue to increase until 2050 at the expense of forest and savanna (Figure 7).

The rate of deforestation reaches its maximum in the decade 1990-2000, and afterwards decreases. The global rate computed for year 2050 is between 4.4 and 13.1 M ha yr[1] (depending on whether the land cover class "tropical dry forest/savanna" is included or not). By comparison, the IPCC intermediate scenarios (IS92a,b,c) have a rate of 15.2 M ha yr[1] in year 2050 Pepper et al., 1992). The main factors leading to deforestation in this Conventional Wisdom scenario are the expansion and later stabilization of agricultural land and grassland, rather than the demand for fuelwood (demand for forest products other than fuelwood is however not included).

In relative terms, the amount of global savanna is reduced more drastically than forests; three quarters of its 1970 area disappears by the year 2050. After 2060, the areas of all major land types stabilize because of improvements in crop yield and animal productivity, and slowing of population and economic growth.

5. Discussion and Conclusions

At this point we remind the reader that the ADM and LCM described in this paper are components of the integrated IMAGE 2.0 model which allows for consistent computations of all greenhouse gas emissions and fluxes resulting from land use and the change in climate which then feeds back to the land cover simulation. This integration of agricultural and land cover calculations with other climate change calculations is a first attempt to simulate the coupled dynamics of land cover and climate in geographic detail for the whole world. Since this is only our first attempt to model agricultural demand and land cover change, there are great possibilities to improve both the ADM and LCM, as described in the following paragraphs.

5.1   IMPROVEMENTS OF THE AGRICULTURAL DEMAND MODEL

The goal of the ADM is to provide very long term estimates of agricultural demand related to changes in land cover, rather than to answer detailed questions about the global food system. Altough the results are reasonable improvement is expected of the following modifications to the model:

• *Price mechanism.* Since the ADM makes long term estimates of demand, it does not require the macro-economic detail of models such as the IIASA Basic Linked System model (Fischer *et al.*, 1988) and FAO's Agriculture: Towards 2010 (FAO, 1993).

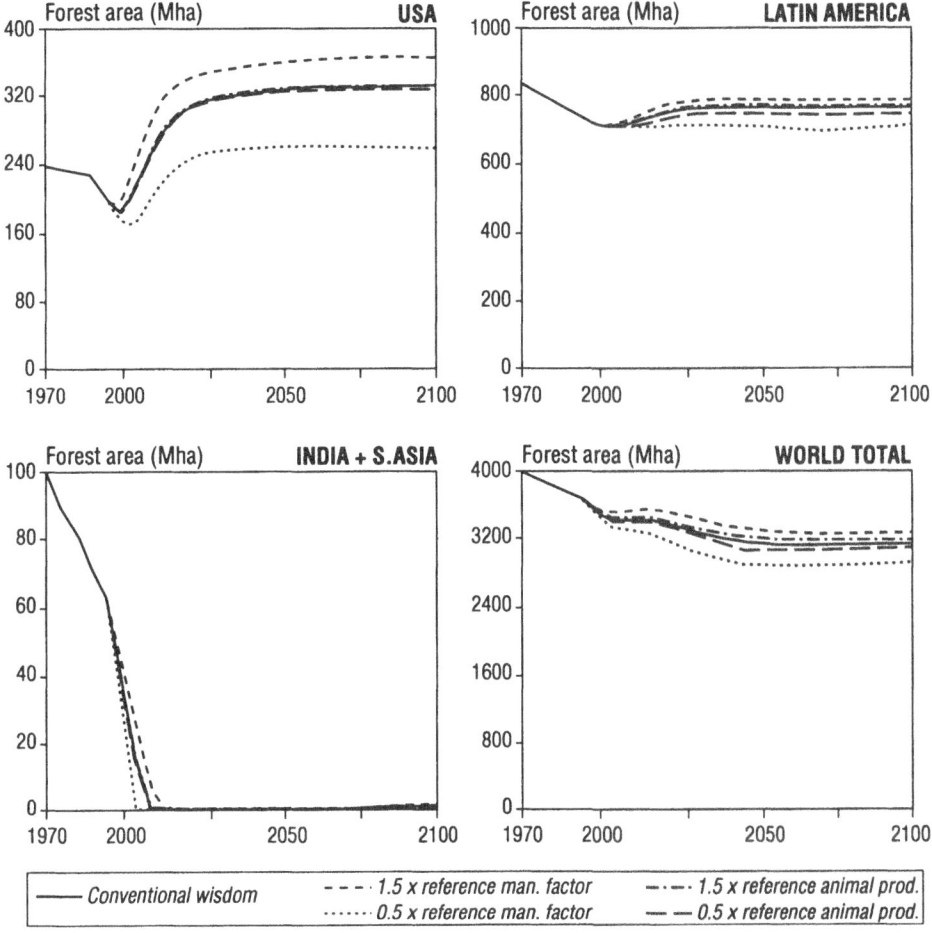

Figure 9: Sensitivity of calculated forest area to assumed animal productivity and crop yield (due to technological improvements).

Nevertheless, the ADM currently makes the unreasonable assumption that all demand will be satisfied regardless of price. In addition, import/export assumptions, such as increasing export in developed countries and increasing import in developing countries, do not directly take into account world food prices. Although detailed price calculations are unnecessary because of the model's long time horizon (till 2100), it would be worthwhile to constrain demand by at least a simple price mechanism, as given for example, by the SWOPSIM model (Roningen *et al.*, 1991).

• *Technological improvements.* Among the most uncertain of model variables are those that are affected by technological improvements in agriculture: animal productivity, fraction of non-productive cattle, and ratio of roughage to concentrate feed. These variables have an influence on the number and feeding habits of livestock in a region, and consequently on range land demand and crop feed requirements. However, it is unclear whether the uncertainty of these variables is important to model calculations. For example, a preliminary sensitivity analysis of future animal productivity (Figure 9) indicates that model calculations may not be particularly sensitive to this variable. Additional sensitivity analyses must be conducted before a large effort is invested in reducing their uncertainty.

• *Demand for timber*. Although the demand for timber products worldwide is not a major cause of deforestation (Stern *et al.*, 1992), timber extraction in parts of Asia sometimes leads to complete forest destruction (see, e.g. Myers, 1991). These demands are currently not included in the model, but should be in future versions in order to more completely account for all main driving forces of deforestation.

• *Sensitivity analysis*. As part of this paper we have only presented some preliminary and incomplete results from testing and sensitivity analysis of the model. Much more testing and sensitivity analysis is necessary for identifying how the model can be improved.

5.2   IMPROVEMENTS OF THE LAND COVER MODEL

While the agreement of calculated and FAO estimates of deforested areas on a country-scale was quite encouraging, special attention is needed to improve estimates of total area and spatial detail of land cover conversions. These might be improved by the following modifications of the model:

• *Expert Knowledge*. Analysis by and discussion with knowledgable Geographers or regional agricultural experts can help with identifying and fixing invalid assumptions.

• *Transportation Corridors*. Historically, society has first settled the coastlines of continents and islands, and then corridors to the interior such as large rivers, and nowadays, roads. These transportation corridors partly explain the current pattern of deforestation in the tropics (see, e.g. Skole and Tucker, 1993; Turner *et al.*, 1993; Williams, 1992). Adding new land use rules to the LCM which take into account these corridors are likely to improve the spatial definition of calculations.

• *Soil Degradation*. Soil erosion and other processes in Africa and elsewhere ("desertification") take large areas of range land and agricultural land out of production and displaces these areas elsewhere within a region. As noted previously, the accuracy of land cover simulations may be improved by including these processes.

• *Crop Yields*. The amount of land necessary for crops depends on assumptions about future improvement in crop yields due to fertilizer and other technological inputs (independent of climate or soils). A sensitivity analysis (Figure 9) showed that model calculations are quite sensitive to these assumptions. Hence, special effort should be given to improving the scenarios of future crop yield.

• *Range land Requirements*. The LCM currently makes rather crude assumptions about the amount of range land required per head of livestock. Model calculations may be improved by linking the amount of grassland needed to support livestock with the productivity of grasslands computed in the Terrestrial Carbon Model of IMAGE 2.0 (Klein Goldewijk *et al.*, 1994).

• *Feedbacks to Land Disappearance*. We noted above that some regions could not satisfy their requirement for range land during the simulation period under the Conventional Wisdom scenario. In reality, as critical land disappears a variety of feedbacks are set in motion which slow the dwindling of this land area such as switching to food imports or shifts in preferences for different foods. Some of these feedbacks should be included in the model.

• *Sensitivity Analysis*. As pointed out for the ADM, the LCM could also benefit from much more extensive testing and sensitivity analysis than was presented in this paper.

5.3 CONCLUSIONS

The two models presented in this paper plus other components of the IMAGE 2.0 model provide a method for linking agricultural demand with climate change and land suitability. This method provides insight into potential food shortages or self-reliance in various parts of the world, and can provide new information about the feasibility or desirability of measures to mitigate climate change. For the two model presented in this paper we conclude the following:

• For 1970 to 1990 , the LCM successfully simulates the total regional rates of deforestation and is within a factor of two of country-scale deforestation rates where the highest rates do occur. The model also captures some finer spatial detail of deforestation. These results show that there may be some validity to the main hypotheses of the land cover model, that regional demands for land can serve as a surrogate of regional and local demands for driving local land cover changes, and that land use rules can be used to represent driving forces of land conversions. Nevertheless, progress should be made to incorporate country-scale and local driving forces into the model.

• Results from a model experiment ("Conventional Wisdom Scenario") to year 2100 showed that future trends in land conversions could be strikingly different in different regions even though consistent (though different) assumptions were used for economic growth and population assumptions. In some regions, assumed technological improvements together with modest economic and population growth can lead to reduced demand for agricultural land and range land, whereas rapid increases in population and economic growth in other regions prevail over technology, leading to rapid diminishment of forest and savanna land.

• The model experiment also pointed out the great influence of assumed future technological improvements in crop yield on simulated land cover patterns. Special effort should be given to creating realistic scenarios of these technological improvements.

• The integration of agricultural and land cover calculations with climate change calculations is a first attempt to simulate coupled dynamics of land cover and climate for the whole world in geographic detail, and can provide new information about shifts in agricultural areas related to climate and the influence which changing land cover has on climate.

Acknowledgements

The authors are indebted to their colleagues at RIVM who contributed to the development of the models described herein: K. Klein Goldewijk, M.S. Krol, R. Leemans, J.G. van Minnen and A.M.C. Toet. We thank B. Lübkert-Alcamo for her detailed review of the manuscript and the reviewers for their comments. This research was supported by the Dutch Ministry of Housing, Physical Planning and the Environment, as well as the Dutch National Research Program on Global Air Pollution and Climate Change, Grant Numbers 851037, 851042, and 851045.

References

Addiscot, T.M., A.P. Whitemore and D.S. Powlson: 1991, *Farming, Fertilizers and the Nitrate Problem*, CAB International, Oxon, U.K., 170 pp.

Alcamo, J.M., G.J.J. Kreileman, M.S. Krol and G. Zuidema: 1994a, Modeling the global society-biosphere-climate system, Part 1: Model description and testing, *Wat. Air Soil Pollut.*, 76 (this volume).

Alcamo J.M. , G.J. van den Born, A.F. Bouwman, B.J. de Haan, K. Klein Goldewijk, O. Klepper, J. Krabec, R. Leemans, J.G.J. Olivier, H.J.M. de Vries and H. van der Woerd: 1994b, Modeling the global society-biosphere-climate system, Part 2: Computed scenarios, *Wat. Air Soil Pollut.*, 76 (this volume).

Blakeslee, L.L., E.O. Heady and C.F. Framingham: 1973, *World Food Production, Demand, and Trade*, Iowa State University Press, Iowa, pp. 31-74.

Bock, B.R. and G.W. Hugert: 1991, Fertilizer nitrogen management, in: R.F. Foilet, D.R. Kuney and R.M. Cruse (eds), *Managing Nitrogen for Groundwater Quality and Farm Profitability, Soil Sci. Soc. Am.*, Wisconsin, USA, pp. 139-164.

Bouwman, A.F., L. van Staalduinen and R.J. Swart: 1992, *The IMAGE Land Use Model to Analyze Trends in Land-Use Related Emissions.* Report no. 222901009, RIVM, Bilthoven, 149 pp.

Daugherty, A.B.: 1987, *Major Uses of Land in the United States*, U.S. Dept. of Agriculture, Agricultural Report Number 643, 35 pp.

Esser, G.: 1989, Global land-use changes from 1860 to 1980 and future projections to 2500, *Ecol. Mod.*, 44: 307-316.

Evenson R.E.: 1988, Technological opportunities and international technology transfer in agriculture, in: G. Antonelli and A. Quadrio-Curzio (eds), *The Agro-Technological System Towards 2000*, Elsevier Science Publishers B.V., Amsterdam, pp. 133-168.

FAO: 1971, *Agricultural Commodity Projection 1970-1980*, Vol. I and Vol. II, FAO, Rome.

FAO: 1992a, *AGROSTAT-PC, Computerized Information Series: User manual, Population, Land use, Production, Trade, Food balance sheets, Forest products*, Edition October 1992, FAO, Rome.

FAO: 1992b, *Forest Resources Assessment 1990, Tropical Countries*, FAO, Rome, 61 pp.

FAO: 1993, *Agriculture: Towards 2010*, FAO, Rome, 320 pp.

Fischer, G., K. Frohberg, M. Keyzer and K. Parikh: 1988, *Linked National Models: A Tool for International Food Policy Analysis*, Kluwer, Dordrecht, 214 pp.

Gohl, B.: 1981, *Tropical Feeds; Feed Information Summaries and Nutritive Values*, FAO, Rome, 529 pp.

de Haan, B.J., M. Jonas, O. Klepper, J. Krabec, M.S. Krol and K. Olendrzyński: 1994, A dynamic atmosphere-ocean model for integrated assessment of global change, *Wat. Air Soil Pollut.*, 76 (this volume).

Hall, D.O. and P.J. de Groot: 1987, Introduction: the biomass framework, in: D.O. Hall and R.P. Overend (eds), *Biomass Regenerable Energy*, John Wiley & Sons, Chichester, pp. 3-24.

Hesterman, O.B.: 1988, Exploiting forage legumes for nitrogen contribution in cropping systems, in: W.L. Hargrove (ed), *Cropping Strategies for Efficient Use of Water and Nitrogen*, ASA-CSSA-SSSA, Special Publication no. 51, Wisconsin, USA, pp. 155-166.

IPCC: 1990, *The IPCC Scientific Assessment*, WMO/UNEP, Cambridge University Press, 365 pp.

Klein Goldewijk, K., J.G. van Minnen, G.J.J. Kreileman, M. Vloedbeld and R. Leemans: 1994, Simulating the C carbon flux between the terrestrial environment and the atmosphere, *Wat. Air Soil Pollut.*, **76** (this volume).

Kreileman, G.J.J. and A.F. Bouwman: 1994, Computing land use emissions of greenhouse gases, *Wat. Air Soil Pollut.*, **76** (this volume).

Leemans, R.: 1992, Modelling ecological and agricultural impacts of global change on a global scale, *J. of Sci. & Ind. Res.*, **51**: 709-724.

Leemans, R. and A.M. Solomon: 1993, Modelling the potential change in yield and distribution of the earth's crops under a warmed climate, *Clim. Res.*, **3**: 79-96.

Leemans, R. and G.J. van den Born: 1994, Determining the potential global distribution of natural vegetation, crops, and agricultural productivity, *Wat. Air Soil Pollut.*, **76** (this volume).

Martinez, A.: 1990, *Fertilizer Use Statistics and Crop Yields*, Technical Bulletin; T-37, International Fertilizer Development Center, P.O. Box 2040, Muscle Shoals, Alabama 35662, USA, 35 pp.

Myers, N.: 1991, Tropical forests: present status and future outlook, *Clim. Cha.*, **19**: 3-32.

Olson, J., J.A. Watts and L.J. Allison: 1985, *Major World Ecosystem Complexes Ranked by Carbon in Live Vegetation: A Data base*, NDP-017, Oak Ridge National Laboratory, Oak Ridge, Tennessee, U.S.A., 164 pp.

Parry, M.L., T.R. Carter and N.T. Konijn (eds): 1988, *The Impact of Climatic Variations on Agriculture, Volume 1: Assessments in Cool Temperate and Cold Regions, Volume 2: Assessments in Semi-arid Regions*, Kluwer Academic Publishers, Dordrecht, 876 pp.

Pepper, W., J. Leggett, R. Swart, J. Wasson, J. Edmonds and E. Mintzer: 1992, *Emission Scenarios for the IPCC: An Update*, Unpublished report available from the Intergovernmental Panel on Climate Change (IPCC), 114 pp.

Prasad, K.K.: 1987, Woodfired heaters, in: D.O. Hall and R.P. Overend (eds), *Biomass Regenerable Energy*, John Wiley & Sons, Chichester, pp. 3-24.

Prentice, I.C., W. Cramer, S.P. Harrison, R. Leemans, R.A. Monserud and A.M. Solomon: 1992, A global biome model based on plant physiology and dominance, soil properties and climate, *Journal of Biogeography*, **19**: 117-134.

Prentice, I.C., M.T. Sykes and W. Cramer: 1993, A simulation model for the transient effects of climate change on forest landscapes, *Ecol. Mod.*, **65**: 51-70.

Roningen, V., J. Sullivan and P. Dixit: 1991, *Documentation of the Static World Policy Simulation (SWOPSIM) Modelling Framework*, Agriculture and Trade Analysis Division, Economic Research Service, U.S. Department of Agriculture, Staff Report No. AGES 9151, 198 pp.

Sanderson, F.H.: 1988, The Agro-Food Filiere: A macroeconomic study on the evolution of the demand structure and induced changes in the destination of agricultural outputs, in: G. Antonelli and A. Quadrio-Curzio (eds), *The Agro-Technological System Towards 2000*, Elsevier Science Publishers B.V., Amsterdam, pp. 186-211.

Scandizzo, P.L. and D. Diakosawas: 1987, *Instability in terms of trade of primary commodities 1900-1982*, FAO Economic and Social Development Paper nr. 64, FAO, Rome, 227 pp.

Skole, D. and C. Tucker: 1993, Tropical deforestation and habitat fragmentation in the Amazon: satellite data from 1978 to 1988, *Science*, **260**: 1905-1910.

Smil, V.: 1980, China's Energetics: A Systems Analysis, in: V. Smil and W. Knowland, *Energy in the Developing World*, Oxford Press.

Stern, P.C., O.R. Young and D. Druckman (eds): 1992, *Global Environmental Change: Understanding the Human Dimensions*, National Academy Press, New York, 308 pp.

Tisdall, S.L., W.L. Nelson, J.D. Beaton, and J.L. Harlin: 1993, *Soil Fertility and Fertilizers*, (**5th ed.**), MacMillan Publishing Company, New York, 549 pp.

Turner, B.L., R.H. Moss and D.L. Skole (eds): 1993, *Relating Land Use and Global Land-Cover Change: A Proposal for an IGBP-HDP Core Project*, IGBP Report No. 24, Stockholm, 56 pp.

Udo, R.K.: 1987, *The Human Geography of Tropical Africa*, Heinemann Educational Books, London. 244 pp.

UNCTAD: 1991, *Handbook of International Trade and Development Statistics 1991*, UNCTAD.

UNEP: 1992, *World Atlas of Desertification*, Edward Arnold, London.

de Vries, H.J.M., R.A. van den Wijngaart, G.J.J. Kreileman, J.G.J. Olivier and A.M.C. Toet: 1994, A model for calculating regional energy use and emissions to evaluate global climate change policies, *Wat. Air Soil Pollut.*, **76** (this volume).

Webster, C.C. and P.N. Wilson: 1980, *Agriculture in the Tropics*, (**2nd ed.**), Longman, London, 640 pp.

Williams, M.: 1992, Forests, in: B.L. Turner, W.C. Clark, R.W. Kates, J.F. Richards, J.T. Mathews, and W.B. Meyer (eds), *The Earth as Transformed by Human Action,* Cambridge University Press.

World Bank: 1989, *Social Indicators of Development 1989: data on diskette,* World Bank, Washington.

World Bank: 1992, *World Tables 1992, The International Bank for Reconstruction and Development,* World Bank, Washington, 669 pp.

WRI: 1986, *World Resources 1986,* Basic Books, New York, 353 pp.

WRI: 1988, *World Resources 1988-89,* Basic Books, New York, 372 pp.

WRI: 1990, *World Resources 1990-91,* Oxford University Press, New York, 383 pp.

WRI: 1992, *World Resources 1992-1993,* Oxford University Press, New York, 385 pp.

SIMULATING THE CARBON FLUX BETWEEN THE TERRESTRIAL ENVIRONMENT AND THE ATMOSPHERE

K. KLEIN GOLDEWIJK, J.G. VAN MINNEN, G.J.J. KREILEMAN,
M. VLOEDBELD AND R. LEEMANS.

National Institute of Public Health and Environmental Protection (RIVM),
P.O. Box 1, 3720 BA Bilthoven, The Netherlands.

Abstract. A Terrestrial C Cycle model that is incorporated in the Integrated Model to Assess the Greenhouse Effect (IMAGE 2.0) is described. The model is a geographically explicit implementation of a model that simulates the major C fluxes in different compartments of the terrestrial biosphere and between the biosphere and the atmosphere. Climatic parameters, land cover and atmospheric C concentrations determine the result of the dynamic C simulations. The impact of changing land cover patterns, caused by anthropogenic activities (shifting agriculture, de- and afforestation) and climatic change are modeled implicitly. Feedback processes such as CO_2 fertilization and temperature effects on photosynthesis, respiration and decomposition are modeled explicitly. The major innovation of this approach is that the consequences of climate change are taken into account instantly and that their results can be quantified on a global medium–resolution grid. The objectives of this paper are to describe the C cycle model in detail, present the linkages with other parts of the IMAGE 2.0 framework, and give an array of different simulations to validate and test the robustness of this modeling approach. The computed global net primary production (NPP) for the terrestrial biosphere in 1990 was 60.6 Gt C a^{-1}, with a global net ecosystem production (NEP) of 2.4 Gt C a^{-1}. The simulated C flux as result from land cover changes was 1.1 Gt C a^{-1}, so that the terrestrial biosphere in 1990 acted as a C sink of 1.3 Gt C a^{-1}. Global phytomass amounted 567.5 Gt C and the dead biomass pool was 1517.7 Gt C. IMAGE 2.0 simulated for the period 1970 - 2050 a global average temperature increase of 1.6 °C and a global average precipitation increase of 0.1 mm/day. The CO_2 concentration in 2050 was 522.2 ppm. The computed NPP for the year 2050 is 82.5 Gt C a^{-1}, with a NEP of 8.1 Gt C a^{-1}. Projected land cover changes result in a C flux of 0.9 Gt C a^{-1}, so that the terrestrial biosphere will be a strong sink of 7.2 Gt C a^{-1}. The amount of phytomass hardly changed (600.7 Gt C) but the distribution over the different regions had. Dead biomass increased significantly to 1667.2 Gt C.

Keywords: climate change, biogeophysical feedbacks, geographically explicit global C cycle model, CO_2 fertilization, soil respiration, land cover change.

1. Introduction

Many ecologists have begun to use their models to evaluate how elevated atmospheric CO_2 levels and climatic change might affect ecosystems (Melillo *et al.*, 1990; Leemans, 1992; Prentice and Solomon, 1991; Solomon and Shugart, 1993). However, most of the models were not developed specifically to address these global issues, but rather reflect the diverse interests and approaches of the scientists who developed them. The models differ in many respects, including the extent to which they are process based, the level of ecological organization (population, stand, community, ecosystem or biome), and the size of temporal and spatial dimensions.

We have been involved in a major model development to address the many and diverse aspects of global change. Global change is defined here as the whole range of anthropogenic influences on global biogeochemical, atmospheric and ecological processes. This therefore not only includes climatic change, but also the direct effects of

Water, Air, and Soil Pollution **76**: 199–230, 1994.
© 1994 *Kluwer Academic Publishers.*

changing atmospheric composition, changes and modifications of regional and global land cover, the impacts of these changes and their complex interactions. These topics have a high policy relevance and many international organisations have been called upon to evaluate the likelihood of global change (e.g. Houghton *et al.*, 1990; Houghton *et al.*, 1992). Much progress has been made with respect to understanding climate change (e.g. Mitchell *et al.*, 1990) and biogeochemical cycles (Houghton *et al.*, 1992), but, unfortunately, little progress has been made to date in understanding the land–use related changes (Turner *et al.*, 1993). These changes have pronounced consequences for future atmospheric C concentrations (Vloedbeld and Leemans, 1993).

The Integrated Model to Assess the Greenhouse Effect (IMAGE 2.0; Alcamo *et al.*, 1994, this volume) is especially developed to address all global change issues comprehensively. This is a logical consequence of the main aim of this model framework: To use state–of–the–art scientific models to assist policy makers in the development and evaluation of future scenarios to adapt to- or mitigate the negative effects of global change. The included models are therefore chosen on basis of their relevance in addressing global change issues and not only for being available. If necessary (and possible), models should be developed to fill gaps in research areas not covered. The whole modeling frame work consists currently of a series of integrated submodels that all deal with different parts of the earth system. The three major parts of IMAGE 2.0 are intended to simulate the relevant processes in the Energy/Industry, Terrestrial Environment, and the Atmosphere/ Ocean subsystems. In this paper we present the C Cycle model that is part of the Terrestrial Environment System (TES). The main objective of TES is to simulate greenhouse gas fluxes to and from the terrestrial biosphere. Secondary objectives are to assess the consequences of land cover change, to evaluate C sequestering potentials, and identify and quantify the important feedback processes within the terrestrial biosphere. It is linked with other components through climate, soil and topography (Atmosphere/Ocean System: Krol and van der Woerd, 1994; the Terrestrial Vegetation model: Leemans and van den Born, 1994), and land cover (Land Cover Change model: Zuidema and van den Born, 1994).

Although we have based our model on earlier, more traditional C models (especially those developed by Goudriaan and Ketner, 1984, and Goudriaan, 1992) and have adopted their basic approach by simulating the most important processes in major ecosystems or biomes, we differ largely in the dimensionality of simulated processes. The traditional models almost all characterize an ecosystem or biome by their global extent and model the relevant processes only globally, using presumed average parametrizations. Such an approach does not allow for a regional adjustment of important processes and a more process–based incorporation of land cover change. We have developed a different approach, where a C model is implemented on a global grid of 0.5° resolution. Each grid cell is characterized by a specific land cover type and a range of environmental parameters. This allows for simulations of the C cycle that are adapted to regional conditions. Our model should therefore be more suiTable to address global change issues at a wide range of scales (region, country, continent or globe).

In this paper we first give a description of the implementation of our C cycle model.

We will not only emphasize the basic approach, but will also elaborate on the implementations of a wide range of feedbacks that influence the C Cycle and thus atmospheric CO_2 concentrations. To evaluate the model, we will then present the results of a global simulation for the historic land cover patterns of 1970-1990. These results will be discussed against other field and model analyses obtained from the literature. We will further present a preliminary sensitivity analyses to determine the most sensitive model parameters, and finally we present some applications with transient simulation using different global change scenarios.

2. The Terrestrial C Cycle model

2.1   BACKGROUND

In IMAGE 2.0, the emissions of greenhouse gases from the terrestrial biosphere are calculated by two sub–models. Both models are part of the TES (see Figure 1 of Alcamo *et al.*, 1994). The Terrestrial C Cycle model is aimed at simulating the fluxes of C between the atmosphere and the terrestrial biosphere, and within the terrestrial biosphere, while the Land Use Emission model simulates all other greenhouse gases (cf. Kreileman and Bouwman, 1994). The Terrestrial C Cycle model computes C fluxes from and to the atmosphere stemming from natural processes and various anthropogenic influences such as deforestation or forestation.

These anthropogenic influences are very important for the total global C budget (Houghton *et al.*, 1990; Vloedbeld and Leemans, 1993). Understanding the significance of land cover change for the C Cycle is not possible without additional information on land use. Land cover changes are mostly driven by human activites, and land–use practices themselves have also major direct effects on environmental processes and systems (Turner *et al.*, 1993). Recent estimates of C fluxes related to land–use changes seem to converge towards values between 1 and 2 Gt C a^{-1}. (Detwiler and Hall, 1988; Houghton *et al.*, 1990). Compared with a global C flux estimate of 5.5 – 6.5 Gt C a^{-1} resulting from fossil fuel combustion, these fluxes from land conversions are surely significant.

2.2   MODEL DESCRIPTION AND IMPLEMENTATION

Some earlier traditional C cycle models are implementations using globally aggregated processes in few major ecosystems (e.g. Emanuel *et al.*, 1984). Other models do not simulate regional specific responses, but parameterize all processes by ecosystem type or biome, which are not spatially explicit (e.g. Houghton *et al.*, 1983; Goudriaan and Ketner, 1984; King *et al.*, 1992). Some biosphere models calculate monthly net exchange of CO_2 between the atmosphere and the terrestrial biosphere by latitudinal belts; the model of King (1986) is an extrapolation from various site-based biome models, while the model of Box (1988) is based on geographically varying processes.

Recently, several geographically more explicit models have been developed; the

TABLE 1

Parameters used in the Terrestrial Carbon Cycle model of IMAGE 2.0.

Landcover type	Initial NPP (g C/m2 yr)	Partitioning coefficients (unitless)				Lifetime (yr)							HF (unitless)	CF (unitless)
		Leaf	Branch	Stem	Root	Leaf	Branch	Stem	Root	Litter	Humus	Charcoal		
Agricultural land	400	0.8	0.0	0.0	0.2	1	1	1	1	1	30	500	0.3	0.05
Ice	0	0.5	0.1	0.1	0.3	1	10	50	10	5	100	500	0.6	0.05
Cool (semi)desert	50	0.5	0.1	0.1	0.3	1	10	50	10	5	100	500	0.6	0.05
Hot desert	50	0.5	0.1	0.1	0.3	1	10	50	10	3	50	500	0.6	0.05
Tundra	100	0.5	0.1	0.1	0.3	1	10	50	10	5	100	500	0.6	0.05
Cool Grass/Shrub	350	0.6	0.0	0.0	0.4	1	10	50	3	1	60	500	0.6	0.05
Warm Grass/Shrub	400	0.6	0.0	0.0	0.4	1	10	50	3	1	30	500	0.6	0.05
Xerophytic woods/scrub	350	0.3	0.2	0.3	0.2	1	10	50	10	1	50	500	0.4	0.05
Taiga	450	0.3	0.2	0.3	0.2	3	20	50	10	5	60	500	0.6	0.05
Cool Conifer Forest	550	0.3	0.2	0.3	0.2	3	20	50	10	3	50	500	0.6	0.05
Cool Mixed Forest	600	0.3	0.2	0.3	0.2	1	20	50	10	1	40	500	0.6	0.05
Temp. Deciduous Forest	600	0.3	0.2	0.3	0.2	1	20	50	10	1	40	500	0.6	0.05
Warm Mixed Forest	650	0.3	0.2	0.3	0.2	1	20	50	10	1	40	500	0.4	0.05
Trop. Dry Forest/Savanna	450	0.3	0.2	0.3	0.2	2	10	30	5	1	20	500	0.4	0.05
Trop. Seasonal Forest	800	0.3	0.2	0.3	0.2	2	10	30	8	1	20	500	0.4	0.05
Trop. Rain Forest	1000	0.3	0.2	0.3	0.2	2	10	30	8	1	20	500	0.4	0.05
Wetlands	700	0.3	0.2	0.3	0.2	2	10	30	8	1	100	500	0.6	0.05

HF = Humification factor
CF = Carbonization factor

regression-based Osnabrück Model (OBM, Esser, 1987; 1991), and the proces-based Terrestrial Ecosystem Model (TEM, Raich et al., 1991; McGuire et al., 1992; Melillo et al., 1993). Both use spatially referenced data on climate, soils and vegetation. The OBM model uses also prescribed land-use (change) information to simulate C pools and fluxes, but there are no feedbacks between decomposition and productivity. TEM does not take land-use into account but simulates C and N pools and fluxes on a monthly basis. It does take, however, some feedbacks into account (recycling of N to simulate transient response of NPP to elevated CO_2).

Our Terrestrial C Cycle model is implemented on a grid with a medium resolution of 0.5° longitude and latitude. This grid is chosen for all submodels within TES. Each grid cell is assumed to be homogeneous and characterized by a specific land cover type, climate, soil and topography. Land cover and climate change dynamically over time in IMAGE 2.0 simulations. Soil and topography are assumed to be constant. For other, advanced and off–line analyses the grid can also be linked to other relevant environmental data bases and coupled to an especially developed Geographic Information System for further analyses and mapping. The land cover types are based upon the potential vegetation and agricultural patterns (see Leemans and van den Born, 1994 and current (1970) patterns as defined by the global land cover data base of Olson et al. (1985). Land cover patterns are updated by the Land Cover Change model and fed into the C Cycle and Land Use Emission models. This allows for a dynamic simulation of the effects of changing land cover patterns on the C Cycle.

Each grid cell is divided into several major C pools (Figure 1). The definitions are similar to those by Goudriaan and Ketner (1984). The major pools are plants (Living biomass; partitioned into leaves, branches, stem and roots) and soil (Dead biomass; litter, humus and stable humus plus charcoal). The driving force of the C cycling among these pools is the Net Primary Production (NPP). NPP is gross primary production, (= the photosynthetically fixed C), minus the respiration loss of C within a plant. The partitioning of NPP among the different plant compartments is defined by fixed fractions (Table 1). All plant and soil compartments are characterized by a specific lifetime. The partition fractions, life times and initial NPP are defined uniquely for each compartment and land cover type (Table 1). The values are assumed to be representative for current (i.e. 1970) atmospheric CO_2 concentrations and are derived from an extended literature review (e.g. Atjay et al., 1979; Goudriaan and de Ruiter, 1983; Olson et al., 1985; Raich et al., 1991; and Waring and Schlesinger, 1985). We further assume that these initial values do not change within the decade to century time horizon of the IMAGE 2.0 simulations.

The predefined NPP values are characteristic for a steady state situation, where C uptake equals C release. We assume that this encompasses a valid initialization of the C cycle model for current conditions. Little data with a comprehensive global coverage exist for an improved initialization. We have adopted the Olson et al. (1985) land cover data base to define the distribution of initial land cover for current conditions. Using his C densities and extents for each land cover class generates global total C values comparable with our initial C values, indicating that our initialization is reasonable.

In steady state systems, the outflow from each compartment is a function of C content and

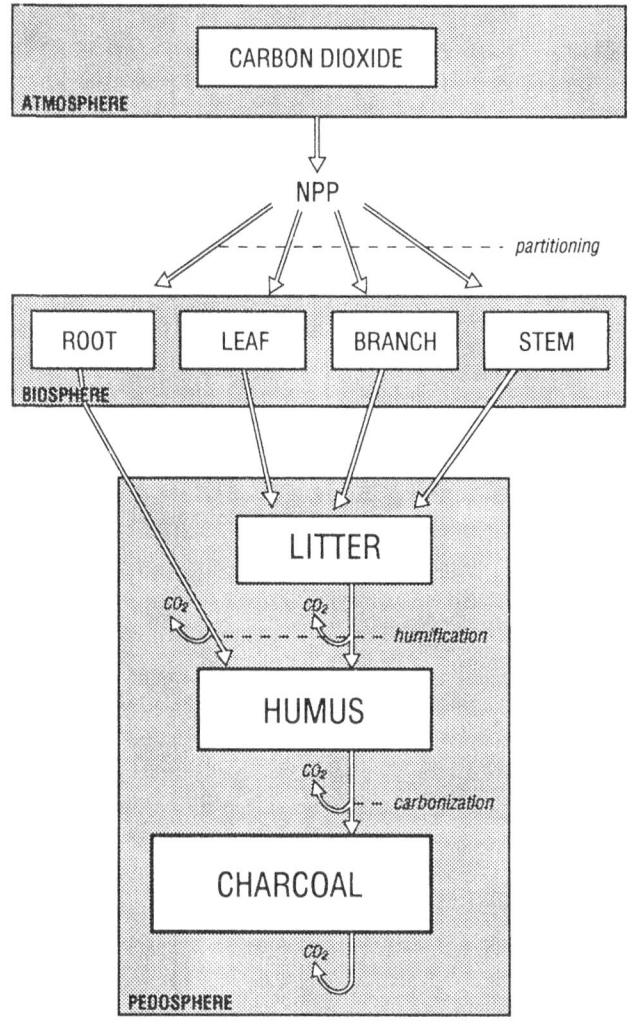

Figure 1. The components of the C Cycle model with the different compartments and their linkages. The C fluxes are characteristic for each Land Cover Type. (NPP stands for Net Primary Production.)

its average lifetime. The total global C cycle is in equilibrium under such conditions. However, we are more interested in the C content and the dynamics after disturbances and changes of one or more compartments of grid cells. In case of such changes, for example due to deforestation, NPP recovers towards its initial value following a logistic function (Cooper, 1983; Dewar, 1990; 1991; Dewar and Cannell, 1992). In formula:

$$y_{g,t} = \frac{1}{1 + b_l \cdot e^{-r_l(t-t_0)}}$$

1

where $y_{g,t}$ is the fractional increase in net primary production in grid cell g at time t (dimensionless, value between 0 and 1). The parameter b_l is a scaling factor for C storage for each land cover type l, depending on the initial C content of a land cover type (b_o) and the C content of the old-growth land cover type (b_m); it is calculated as $(b_m - b_o)/b_o$. The relative growth rate of C storage for each land cover type is represented by r_l. The initial NPP value of each land cover type is further corrected for environmental constraints or enhancements. This model structure provides an integrated structure, where several feedback processes acting on the C cycle are modeled intrinsically.

The actual NPP of a grid cell at a specific time ($NPP_{g,t}$, t C km^2 a^{-1}) becomes thus:

$$NPP_{g,t} = NPPI_l \cdot y_{g,t} \cdot \sigma_{g,t} \cdot \left(1 + \beta_{g,t} \cdot \ln(\frac{CO_{2,t}}{CO_{2,i}}) \right) \qquad 2$$

where $NPPI_l$ is the land cover type specific NPP (t C km^2 a^{-1}); $\sigma_{g,t}$ is the feedback parameter for direct effects of temperature on plant growth (dimensionless, see section 2.2.2.2); $\beta_{g,t}$ is the CO_2 fertilization factor (dimensionless, see section 2.2.2.1), that accounts for changing atmospheric CO_2 concentrations. The CO_2 concentration in the atmosphere at time t is represented by $CO_{2,t}$, while $CO_{2,i}$ is the initial CO_2 concentration (both in ppm). The feedback factors ($\sigma_{g,t}$ and $\beta_{g,t}$) are computed for every five year period in order to smooth their effect through time. However, $NPP_{g,t}$ is adjusted at a annual timestep, because the atmospheric CO_2 concentration changes gradually.

Net Ecosystem Production (NEP) defines the outcome of the total C budget for each cell. It is assembled from the partitioned NPP values minus the C loss by natural decay of all C pools. In natural undisturbed ecosystems the C-flux from each C pool equals its content divided by its average lifespan τ (Table 1). Leaves, branches and stemwood decay normally to the litter pool (L). The root biomass (B_4) is transferred gradually to the humus pool (H). NEP is positive when there is a net C flux from the atmosphere towards the biosphere. A positive NEP marks an actual sequestering of C within a grid cell. NEP is also influenced by environmental constraints. The most important feedback process here is the influence of moisture availability and temperature on the soil respiration rate. The actual NEP ($NEP_{g,t}$, t C km^2 a^{-1}) is defined as:

$$NEP_{g,t} = NPP_{g,t} - (1 - \lambda_l) \cdot \left(\frac{B_{4g,t}}{\tau(B_{4l})} + \phi_{g,t} \cdot \frac{L_{g,t}}{\tau(L_l)} \right)$$
$$- (1 - \delta) \cdot \phi_{g,t} \cdot \left(\frac{H_{g,t}}{\tau(H_l)} \right) - \phi_{g,t} \cdot \frac{K_{g,t}}{\tau(H_l)} \qquad 3$$

where the terms to the right of NPP represent the C flux due to soil respiration. Only a humification fraction λ_l (Table 1) is transferred to the humus pool. Humus, or soil organic matter, decays much slower than litter, and a carbonization fraction δ (Table 1) enters a charcoal pool of relatively stable carbon (K) that contains recalcitrant humus, charcoal and other forms of elementary carbon (Bouwman, 1989; Kortleven, 1963). $\Phi_{g,t}$

(dimensionless) is the factor that specifies the influence of temperature and moisture availability on decomposition processes (see section 2.2.2.3).

2.1.1   Land Cover Changes

An important feature of TES, and thus IMAGE 2.0, is the simulation of changes in land use through time. The simulation of these changes is regionally specific and it results in changes in land cover on the terrestrial grid. Socio–economic factors, such as demography and economic development (using the indicator gross national product, GNP), are used to estimate a demand for agricultural (food, fodder, meat and biofuels) and forestry (fuelwood and roundwood) products. These demands are satisfied in the Land Cover Change model (see Zuidema and van den Born, 1994 by combining current land cover patterns with agriculture and forestry potentials. The resulting land cover transformations, such as the establishment of new arable lands and pastures, define annual regional specific deforestation rates. Land cover change rates are thus not predefined and do not discriminate for any specific land cover type, as in the traditional C cycle models (e.g. Goudriaan and Ketner, 1984).

The simulated transient change in land cover has a great impact on all C pools of the terrestrial biosphere. The most common conversion is the conversion of forests to agricultural land or pastures. We assume that in the IMAGE 2.0 regions Latin America, Africa, India and East Asia most of the aboveground biomass (leave, branch, stem and litter) of warm mixed forest, tropical dry forest/savanna, tropical seasonal forest and tropical rain forest is burnt at the site. In all other regions and land cover types the phytomass is assumed to enter the humus pool. Only a small fraction (2-3%) does not burn and enters the charcoal pool. This fraction is consistent with recent observations (Robinson, personal communication, 1993; Fearnside, 1991), and is much lower than earlier estimates of Seiler and Crutzen (1980). The latter two assumed that c. 20% of the aboveground biomass could end up as charcoal after deforestation. Due to the long residence time of charcoal in the soil, we cannot justify such high estimate. It would lead to a strong and stable C sink, something which is not supported by experimental results (Robinson, personal communication, 1992; Fearnside, 1991). Typically, also a stemwood fraction remains unburnt (Fearnside, 1991). We assume that this fraction (10%) together with its related dead root biomass enters the humus pool directly. The land cover change thus leads to an instantaneous flux of C to the atmosphere and NEP will become strongly negative. Due to the high levels of the soil C pools, decomposition will also accelerate and remain at a relatively high level for several years (depending on the environmental and C characteristics of that grid cell). Later, NEP will level off but remains negative, so that converted land will act as a C source.

Another important conversion is the conversion of grasslands (pastures) into agricultural land (e.g. arable lands). For tropical grasslands, we assume that all aboveground biomass is burnt and only a very small fraction is converted to charcoal. For temperate grasslands we assume that all aboveground biomass enters the litter pool. NEP will also be negative after these conversion, but not so strong as after deforestation. Due to the relatively small changes in soil C pools, NEP will gradually decrease to zero with time.

The C cycle model is also suited to compute reversed land cover change processes, such as reforestation (forest to forest) or forestation (agricultural land or pastures to forests). For these conversions NEP will be driven by NPP (Equation 1). Gradually all C pools will become saturated with the C values characteristic for a steady state situation. These conversions lead to an uptake of C by the biosphere. NEP is positive and the converted land will act as a sink for CO_2. This is because soil decomposition rates tend to adjust slowly to the above ground changes, and it can take many years before an equilibrium state is reached, where C uptake equals C decomposition.

In the current version of the C Cycle model, the C dynamics are deterministic and characterized by the type of land cover conversions. Impacts such as shifting vegetation zones due to a changing climate (Leemans, 1992) or more effective use of moisture resources (Vloedbeld and Leemans, 1993) can be taken into account. However, the transient dynamics are driven by NPP characteristics only for a land cover type within a grid cell (Equation 1). In the future versions we would like to incorporate more realistic transient dynamics that also corrects NPP for ecological processes such as succession. The method that we are currently testing is related to the approach suggested by Smith and Shugart (1993).

2.2.2 Feedback processes

Feedback processes modify NPP and/or NEP. Some feedback processes increase NEP (negative feedback), while others decrease NEP (positive feedback). Feedback processes potentially can have a strong influence on the C cycle and thus on the final atmospheric C concentrations (Lashof, 1989; Vloedbeld and Leemans, 1993). The most important feedback processes with respect to the C cycle that are implemented in IMAGE 2.0, are: Impact of temperature change on photosynthesis and respiration (plant growth), response of soil respiration to climate change, CO_2 fertilization, and shifts in vegetation patterns due to changes in water use efficiency and climate.

2.2.2.1 *CO₂ fertilization*
Many studies suggest that high atmospheric CO_2 concentrations enhances plant growth (for a review see Bazzaz, 1990; Körner, 1993). This CO_2–fertilization effect is most pronounced if plants have plentiful supplies of nutrients and light. A CO_2–rich environment could allow plants to use less water per CO_2 unit during photosynthesis, and thus enhance growth. This is evident for plants with the C_3 photosynthetic pathway, such as forest trees and many herbs and weeds, and for crops like rice, wheat, potato and beans. The other photosynthetic pathway, C_4, is less affected (Mooney *et al.*, 1991). This group includes many grasses mainly in hot, dry tropical and subtropical regions, including crops like maize, sorghum and sugar cane.

CO_2 from the atmosphere diffuses into the plant through stomata, and under high atmospheric CO_2 concentrations the stomatal openings can be smaller to obtain effective internal CO_2 concentrations. This consequently also reduces evaporation rates and thus increases the Water Use Efficiency (WUE, Figure 2J). Increased WUEs are observed for species with C3 or C4 pathways (Carlson and Bazzaz, 1980; Bazzaz and Carlson, 1984; Miao *et al.*, 1992; Eamus, 1991; Strain and Cure, 1985). The increase in WUE has two

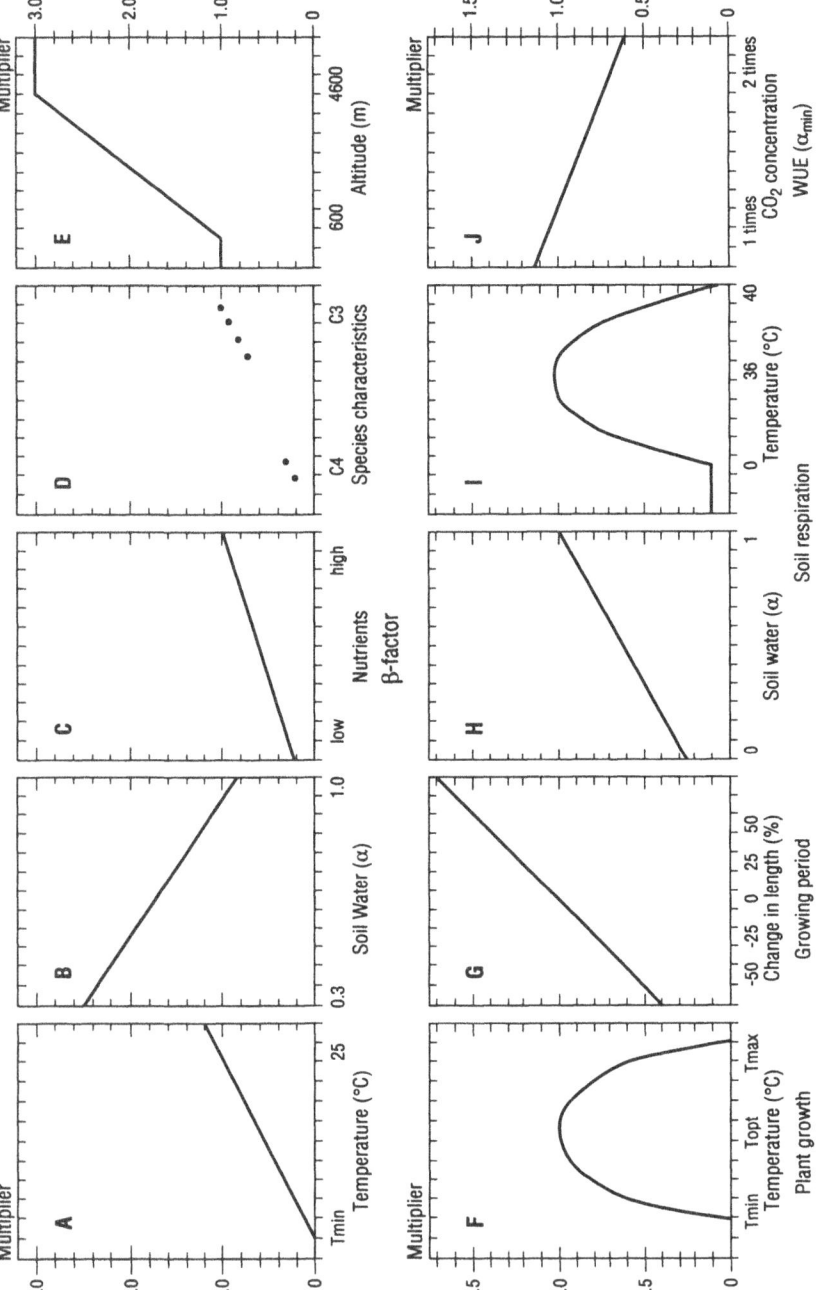

Figure 2: The influence of environmental factors on the feedback processes. The top diagrams represent the environmental modifications of the β factor, while the lower diagrams give the modifications for respectively plant growth, changes in growing period, soil respiration and water use efficiency.

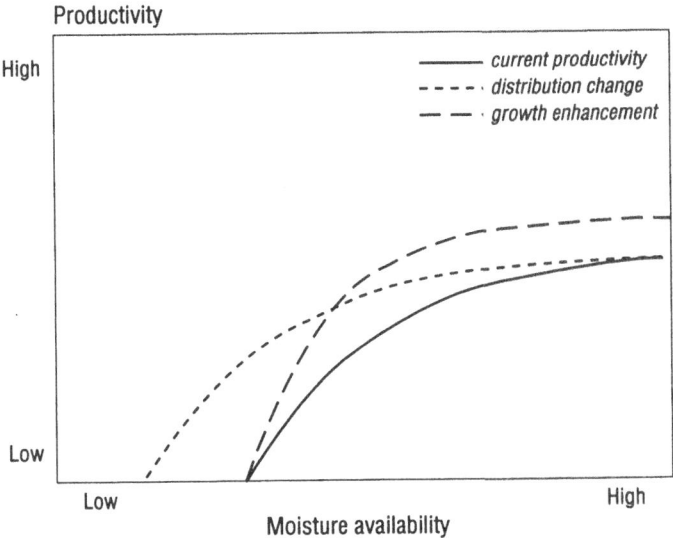

Figure 3: The two aspects of changing WUE under increasing atmospheric CO_2-concentrations (dotted lines). The first component increases productivity, while the second changes the distribution towards more xeric conditions.

effects (Figure 3). First, it somewhat enhances growth (Körner, 1993) and second it influences the distribution of a species. Under increased atmospheric CO_2 levels, a species can survive and grow under more xeric conditions (see section 2.2.2.4).

In this C Cycle model the relation between increased atmospheric levels of CO_2 and NPP is defined by the CO_2 fertilization factor, β (Equation 2). The main factors that control ß are temperature, soil moisture and nutrient availability, species characteristics, and altitude (Figure 2A-E). The determination of the grid– and time–specific ß factors is implemented using a multiplicative approach (Vloedbeld and Leemans, 1993):

$$\beta_{g,t} = (\gamma_{g,t}^T \gamma_{g,t}^W \gamma_g^N \gamma_{g,t}^C \gamma_g^A) \beta_i \qquad\qquad 4$$

with $\gamma_{g,t}^T$, $\gamma_{g,t}^W$, $\gamma_{g,t}^C$ the correction factors for respectively temperature, soil water and species characteristics for a grid cell g at time t, and γ_g^N, γ_g^A the correction factors for respectively nutrient availability and altitude of a grid cell g. The latter two correction factors are assumed to be constant during the length of the simulation. All γ's have no dimension. The initial ß factor (β_i) is set to 0.7 for a doubling of atmospheric CO_2 concentrations. This value is assumed to be appropriate for an annual C_3 crop species with little or no environmental constraints at 25°C (Vloedbeld and Leemans, 1993). The actual β value ($\beta_{g,t}$) is much lower for almost all cells during any simulation. For example, in 1970 the mean global β value is 0.28, which is in good agreement with values found by different other authors (e.g. Kohlmaier et al., 1989; Polglase and Wang, 1992).

The ambient temperature influences the magnitude of CO_2 fertilization. CO_2

fertilization does not occur below a certain minimum growing season temperature (T_{min}). We have implemented the temperature response by using the mean growing season temperature. We assumed $\gamma_{g,t}^T$ to be linear (Figure 2A) with an optimum and maximum temperature that is characteristic for each land cover type (cf. Vloedbeld and Leemans, 1993). Although some recent studies (e.g. Long, 1991) have shown that this temperature response shifts under changing atmospheric CO_2 concentrations, we have not included such a shift, because it is not yet clear what the actual response should be.

Due to increased WUE, CO_2 fertilization should be most pronounced under xeric conditions. In IMAGE 2.0 moisture availability is defined by using the Priestley–Taylor index α, which is the ratio between actual annual evapotranspiration and the annual equilibrium evapotranspiration. The α index is computed for each grid cell in the Terrestrial Vegetation model (Leemans and van den Born, 1994 and we have assumed a simple linear response curve (Figure 2B). The $\gamma_{g,t}^W$ is 1 in areas with little or no moisture stress, and it increases under more xeric conditions. This approach is consistent with measured values given by Morison (1985) and Körner (1993).

In nutrient–poor soils the C fertilization effect can be neglected. We assumed for this version of the C Cycle model that the availability of nutrients for plant growth only depends on soil characteristics. Nitrification, denitrification, decomposition, land cover type, atmospheric inputs and climatic constraints are not taken into account yet. Approaches such as those developed by McGuire *et al.* (1991) and Raich *et al.* (1991) are planned for future versions of our model. We have used the soil data from the FAO/UNESCO soil map of the world (FAO/UNESCO, 1974, as digitized by Zobler (1986)) to define different soil fertilities (see Leemans and van den Born, this volume). A 70% decrease of ß is assumed for soils with the lowest fertility, while no decrease will occur in fertile soils (Goudriaan and de Ruiter, 1983; Figure 2C).

This enhanced growth response is in C3 plants much larger than in C4 plants (Bazzaz, 1990; Mooney *et al.*, 1991). Furthermore, Körner (1983) ranked other plant physiognomic types, such as evergreen trees, woody plants, herbaceous plants, perennials and annuals, according to their specific CO_2 fertilization responses. We have defined specific responses for each land cover type on basis of the plant type composition of each land cover type (cf. Vloedbeld and Leemans, 1993 and Figure 2D).

Plants at high altitudes are strongly CO_2 limited, due to decreasing CO_2 pressures with increasing altitudes. The CO_2 fertilization effect at high altitudes is therefore stronger than at lower altitudes. We have used the global topography data base (Leemans and Cramer, 1991) to correct for differences in altitude. We assumed that no correction was needed for altitudes below 600 m and that the correction factor further increases with a value of 0.05 per 100 meter (Figure 2E).

2.2.2.2 *Impact of temperature change on plant growth* Changing temperatures influences the photosynthetic and respiration rates of plants directly. Photosynthesis starts at a minimum temperature (T_{min}), increases to an optimum value (T_{opt}) and decreases afterwards. Respiration increases exponentially with increasing temperatures. We have combined these two responses into a single response function (Figure 2F) where the correction factor for NPP$_i$ is set to 1 for T$_{opt}$. Gross photosynthetical rates equal the respiration rates at a certain temperature (T$_{max}$) This results in a balanced situation where net C fluxes within the plant are zero. Above this temperature no sustained plant growth is possible. The minimum, optimum and maximum temperature varies among habitat, species and altitudes (Larcher, 1980).

Another important temperature effect of climate change is the extension of the length of the growing season under a warming climate. A longer growth period results in a higher NPP. We assumed a linear response to the increase of the length of the growing period (Figure 2G). However, this could lead to very large NPP increases in severely dry of cold climates. Therefore we constrained the response to a maximum of 3 times the original value.

2.2.2.3 Response of soil respiration to climate change The most important factors that control decomposition rates of organic matter in soils are temperature and soil moisture content (Carlyle and Than, 1988; Parton *et al.*, 1987). Decomposition rates of organic material are always low below a threshold temperature and increase with rising soil temperature until an optimum value, after which decomposition declines again. However, even in extreme conditions some decomposition occurs always. Decomposition rates further increase nearly linearly with soil moisture content until saturation. We have implemented the functions using the mean temperature of the growing period and the α moisture index (Figure 2H+I). The combined effect is obtained by multiplying the individual values. Although theoretically this could easily result in small values, due to the lower limits of the individual response curve, we never obtain those.

2.2.2.4 Effect of climate change on Water Use Efficiency The potential vegetation of a grid cell is determined in the Terrestrial Vegetation model (Leemans and Van den Born, this volume). This potential vegetation is based on the BIOME model (Prentice *et al.*, 1992). BIOME defines climatic envelopes that mimic the ecophysiological response of coarse species aggregations (= Plant Functional Types, PFT). Assemblages of PFT create the major vegetation zones. These vegetation patterns change automatically when climate changes. This highly non–linear response differs per region and vegetation type and strongly influences the C dynamics of the terrestrial biosphere. Other important vegetation changes are driven by changes in WUE. With increasing atmospheric CO_2 concentrations, plants can grow under more xeric conditions (Figure 3). Using the climatic envelopes of the BIOME, the moisture constraints are adjusted accordingly (Vloedbeld and Leemans, 1993 and Figure 2J). The changing vegetation patterns are accounted for in the Land Cover Change model (Zuidema and van den Born, 1994 and enter through this model into the C cycle simulations.

2.2.3   Linkages
The Terrestrial C Cycle model exchanges information with the other components of IMAGE 2.0. It uses many of the climatic indices, which are computed by the Terrestrial Vegetation model (Leemans and van den Born, 1994 to define the magnitude of the

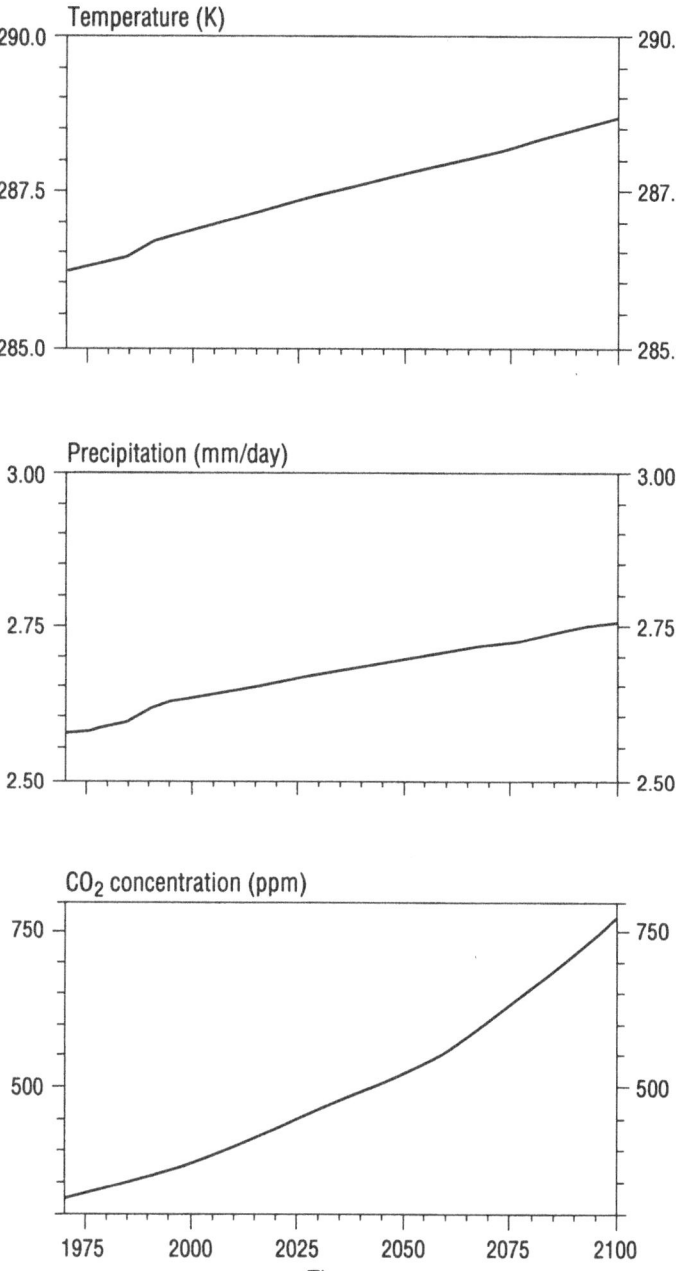

Figure 4: Global average temperature, precipitation and CO_2 concentration as simulated by IMAGE 2.0 for Conventional wisdom scenario 1970 -2100.

climate related feedback processes. It is further directly linked with the Land Cover Change model by using that model's global land cover updates as inputs for the C calculations. The outcome of the C cycle model are the fluxes of CO_2 from and to the atmosphere. These are used together with the CO_2 emissions from the Energy/Industry Emissions model (de Vries *et al.*, 1994 and the Atmospheric/Ocean model (Krol and van der Woerd, 1994 to determine the build–up of CO_2 concentrations in the atmosphere. These atmospheric CO_2 concentrations are subsequently used again to define the magnitude of the relevant feedbacks (Figure 4).

This comprehensive framework of linkages between the different components of IMAGE 2.0 allows for transient simulations of different global–change aspects, such as land cover, atmospheric composition and climate, that continuously influence the dynamics and characteristics of the terrestrial C cycle. The linkages, feedbacks and the geographic explicit simulation of environmental constraints are based on our current understanding of the functioning of the biosphere. Using process–related models will provide a more robust projection of future global change trends than the earlier analyses. Due to data, model, knowlegde and computational constraints, we probably cannot yet reduce the relatively high uncertainty of these projections, but will provide some preliminary scenario's.

2.3   SIMULATION FOR THE PERIOD 1970-1990.

An aggregated version of the Olson *et al.* data base (1985) defines the 1970 reference situation for land cover (cf. Leemans and van den Born, 1994). Initial C pools are assumed to be in an equilibrium state in 1900 and are initialized by running the C cycle model annually without any land cover change from 1900 (295 ppm atmospheric CO_2 concentration) until 1970 (325 ppm atmospheric CO_2 concentration), with a changing climate. Current climate is initialized with the global data base of Leemans and Cramer (1991). This initialization run allows all C pools to adapt to and stabilize under the different feedback mechanisms. The resulting sink in the terrestrial biosphere is 1.8 Gt C in 1970. We then ran the model further to 1990 using the regional land cover changes computed by the Land Cover Change model (Zuidema and van den Born, 1994).

The resulting values for NPP, biomasses and net C fluxes are compared with published data (Table 2). These comparisons are only indicative because most authors use different, and often incompatible classification schemes for land cover types. There are further large differences in the geographic regions covered and in the tabulated extent of land cover types. For example, many authors use the land cover class 'Tropical woodlands'. Tropical and woodlands are both ill–defined terms, which are used inconsistently in the literature. We have adapted the aggregated land cover classification based on the Olson *et al.* data base (1985) and all our Figures are based on the extent and patterns in this data base and we can provide digitized, geographic explicit results of the IMAGE 2.0 simulations to make intercomparisons between studies less complicated.

The initial NPP in 1970 of 58.5 Gt C a^{-1} is relatively high compared to most other estimates, which range from 46 – 61 Gt C a^{-1} (e.g. Whittaker and Marks 1975; Atjay *et al.*,

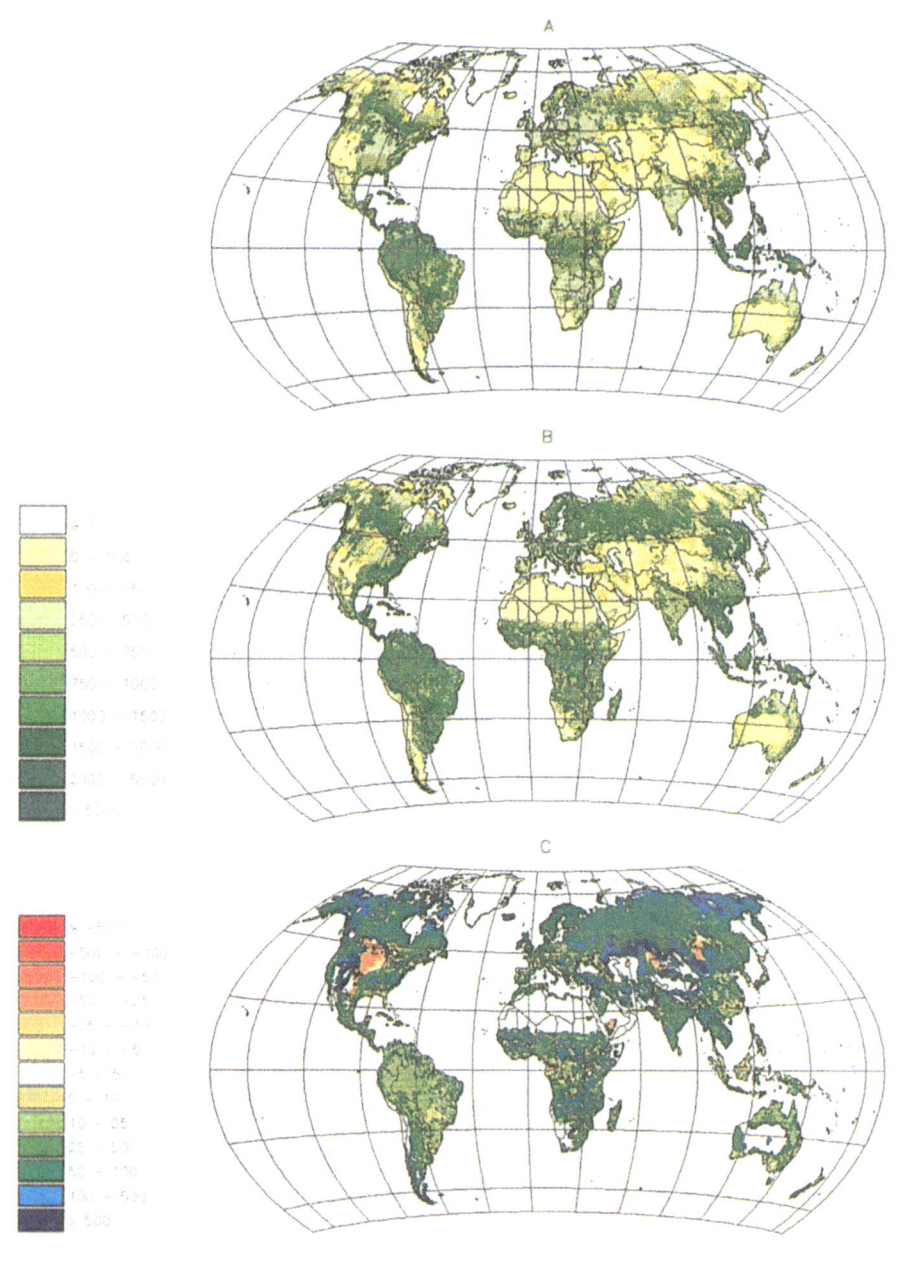

Figure 5: Net Primary Production (t C km^{-2} a^{-1}) in 1970 (A), 2050 (B) and the percentage difference between 1970 and 2050 (C).

TABLE 2
Summary of global and regional net primary production estimates.

Author	Region	Amount Pg C a⁻¹	Area 10¹² m²	Density g C m⁻² a⁻¹	Remarks
Whittaker & Likens (1973)	World	48.3	149.0	324	table 1.
Whittaker & Likens (1975)	World	52.8	149.0	354	
Atjay et al. (1979)	World	59.9	149.3	401	
Goudriaan & Ketner (1984)	World	61.9	121.1	511	fig. 9, p.184
Olson et al. (1985)	World	60.2	151.0	399	table 2., p.21
Esser (1987)	World	48.6	133.9	363	
Box (1988)	World	67.8	149.4	454	table 5., p.1118
Esser (1991)	World	45.0	133.9	336	table 31.10., p.702
King et al. (1992)	World	46.1	133.4	346	table 1., p.9
Seino & Uchijima (1992)	World	61.2	149.0	411	table 2., p.45
Polglase & Wang (1992)	World	59.8	122.3	489	table 2.
Melillo et al. (1993)	World	53.2	127.3	418	table 2, p.236
IMAGE 2.0 (1994)	World	58.5	134.8	434	
Raich et al. (1991)	South America	12.5	17.8	702	no agric. land and
IMAGE 2.0 (1994)	Latin America	15.7	20.7	756	wetland., table 6.
McGuire et al. (1992)	North America	7.0	20.8	339	no agric. land and
IMAGE 2.0 (1994)	Canada + USA	8.8	19.3	455	wetland., p.110
Trexler (1991)	USA	5.9	8.9	667	fig 3. + fig.6
IMAGE 2.0 (1994)	USA	4.3	9.4	456	
Kolchugina & Vinson (1993)	former USSR	4.4	14.0	311	only forest biomes
IMAGE 2.0 (1994)	CIS	7.8	21.9	354	

1979; Esser, 1991; King and Neilson, 1992; Olson *et al.*, 1985; Melillo *et al.*, 1993). The global estimate for phytomass in 1970 is 602.3 Gt C. This value is also in the upper range of values when compared with other estimates (500 to 650 Gt C: Goudriaan, 1992; King and Neilson, 1992; Schlesinger, 1977; Polglase and Wang, 1992). These high values can partly be explained by the somewhat high estimates of initial NPP values for some of the different land cover types combined with their extent, but are mainly caused by the absence of any land cover changes during the initialization period 1900 1970. After 1970, global phytomass values decline rapidly to an intermediate value of 567.5 Gt C (1990).

The calculated global amount of dead biomass in 1970 of 1498.8 Gt C is in good agreement with other estimates of Schlesinger (1977), Atjay *et al.* (1979), Goudriaan and Ketner (1984) and Polglase and Wang (1992). It increases slowly through time, due to the input of charcoal that results from burned biomass during land cover changes. The long lifetime of charcoal leads to a small C sink.

This C Cycle model also allows for a more regional comparison (Figure 5, Table 2, 3 and 4). Here we have also obtained good agreements for NPP, biomasses and C fluxes. For example, the 1970 NPP of 15.7 Gt C a⁻¹ for Latin America, is not significantly different

TABLE 3
Summary of global and regional living plus dead biomass estimates.

Author	Region	Amount Pg C	Area 10^{12} m^2	Density kg C m^{-2}	Remarks
Living biomass estimates.					
Schlesinger (1977)	World	826.9	147.0	5.6	table 2., p.56
Atjay et al. (1979)	World	560.0	149.3	3.8	
Goudriaan & Ketner (1984)	World	594.0	121.0	4.9	fig. 9, p.184
Olson et al. (1985)	World	561.3	151.1	3.7	table 2., p.21
Polglase & Wang (1992)	World	642.0	122.3	5.2	table 2.
Smith et al (1992)	World	737.2	151.1	4.9	table 2., p.310
Cramer & Solomon (1993)	World	604.0	135.5	4.5	table 3, p.106. Sparse
IMAGE 2.0 (1994)	World	602.3	134.8	4.5	agriculture
Kurz et al. (1992)	Canada	12.0	5.8	2.1	forest biomes, table 9.
IMAGE 2.0 (1994)	Canada	73.8	9.8	7.5	
Trexler (1991)	USA	18.2	8.9	2.0	only forests
IMAGE 2.0 (1994)	USA	34.7	9.4	3.7	
Kolchugina & Vinson (1993)	former USSR	88.0	14.0	6.3	forest biomes, fig. 5
IMAGE 2.0 (1994)	CIS	107.7	21.9	4.9	
Flint & Richards (1994)	India region	7.0	4.1	1.7	excl.Pakistan, Bhutan,
IMAGE 2.0 (1994)	India region	12.3	5.2	2.4	Nepal, Sikkim, Maldives
Dead biomass estimates.					
Schlesinger (1977)	World	1515.0	149.0	10.2	table 2., p.56
Atjay et al. (1979)	World	1636.0	149.3	11.0	
Post et al. (1982)	World	1395.0	129.6	10.8	table 2., p.158
Goudriaan & Ketner (1984)	World	1531.3	121.0	12.7	fig. 9, p. 184
Polglase & Wang (1992)	World	1342.0	122.3	11.0	table 2
Cramer & Solomon (1993)	World	1313.0	135.5	9.7	table 3, p.106. Sparse
IMAGE 2.0 (1994)	World	1498.8	134.8	11.1	agriculture
Kurz, et al. (1992)	Canada	76.4	5.8	13.2	table 9., page 40.
IMAGE 2.0 (1994)	Canada	175.9	9.8	17.9	
Trexler (1991)	USA	39.3	3.2	12.3	only forests
IMAGE 2.0 (1994)	USA	124.4	9.4	13.2	one third is forest area
Kolchugina & Vinson (1993)	former USSR	338.0	14.0	24.1	forest biomes, fig. 5.
IMAGE 2.0 (1994)	CIS	301.5	21.9	13.8	

TABLE 4
Living biomass simulations for all IMAGE 2.0 regions in 1970 and 2050 (Gt C).

	Canada	USA	Latin America	Africa	OECD Europe	Eastern Europe	CIS	Middle East	India+other S.Asian	China + C.P.countr.	East Asia	Oceania	Japan	Total
year 1970														
Agricultural land	0.2	1.7	1.7	1.5	0.6	0.3	0.8	0.1	1.1	0.9	0.7	0.5	0.1	10.2
Ice	0.0	0.0	0.0	0.0	0.0	0.0	0.0	0.0	0.0	0.0	0.0	0.0	0.0	0.0
Cool (semi)desert	0.0	0.2	0.5	0.0	0.0	0.0	0.6	0.0	0.0	0.2	0.0	0.0	0.0	1.5
Hot desert	0.0	0.1	0.5	4.1	0.0	0.0	0.2	1.4	0.1	0.4	0.0	2.5	0.0	9.3
Tundra	3.6	1.5	0.7	0.0	1.0	0.0	6.3	0.1	0.2	1.5	0.0	0.0	0.0	15.0
Cool Grass/Shrub	0.3	1.1	0.0	0.1	0.0	0.0	1.5	0.0	0.0	1.3	0.0	0.0	0.0	4.4
Warm Grass/Shrub	0.0	0.3	7.6	3.8	0.1	0.1	0.1	0.3	0.8	1.2	0.2	0.4	0.0	14.7
Xerophytic Woods/Scrub	0.0	1.5	15.6	17.7	1.3	0.0	0.1	0.8	1.6	0.0	0.0	8.0	0.0	46.6
Taiga	49.0	3.9	0.0	0.0	7.7	0.0	71.3	0.0	0.0	3.2	0.0	0.0	0.0	135.1
Cool Conifer Forest	16.4	8.5	0.0	0.0	1.9	0.9	12.7	0.0	0.5	3.9	0.0	0.0	0.6	45.5
Cool Mixed Forest	3.1	3.9	0.1	0.0	0.9	0.6	9.8	0.0	0.0	2.6	0.0	0.0	1.3	22.4
Temp. Deciduous Forest	0.0	2.8	0.1	0.0	0.3	0.3	2.0	0.2	0.0	6.2	0.0	0.0	0.1	11.8
Warm Mixed Forest	0.0	7.0	19.6	7.0	2.0	0.8	0.4	0.3	4.3	14.7	7.2	3.8	0.3	67.3
Trop. Dry Forest/Savanna	0.0	0.2	17.9	37.8	0.0	0.0	0.0	0.0	1.8	1.0	1.1	4.7	0.0	64.4
Trop. Seasonal Forest	0.0	0.0	33.6	22.0	0.0	0.0	0.0	0.0	3.5	3.0	5.0	0.3	0.0	67.5
Trop. Rain Forest	0.0	0.0	37.3	5.5	0.0	0.0	0.0	0.0	0.3	0.0	13.9	0.3	0.0	57.2
Wetlands	1.2	2.0	9.5	6.9	0.8	0.1	2.0	0.2	0.2	0.5	4.6	1.5	0.1	29.5
Total	73.8	34.7	144.6	106.3	16.6	3.2	107.6	3.4	14.3	40.4	32.8	21.8	2.6	602.2
year 2050														
Agricultural land	0.2	1.5	2.2	5.4	0.6	0.3	0.5	0.6	1.8	1.8	1.7	0.7	0.2	17.5
Ice	0.0	0.0	0.0	0.0	0.0	0.0	0.0	0.0	0.0	0.0	0.0	0.0	0.0	0.0
Cool (semi)desert	0.0	0.5	0.9	0.0	0.0	0.0	1.2	0.0	0.0	0.4	0.0	0.0	0.0	3.0
Hot desert	0.0	0.2	0.7	5.2	0.0	0.0	0.2	0.9	0.8	0.9	0.0	3.3	0.0	12.1
Tundra	6.7	2.7	0.9	0.1	1.7	0.0	10.8	0.1	0.1	2.3	0.0	0.1	0.0	25.3
Cool Grass/Shrub	0.2	0.8	0.9	0.5	0.7	0.3	4.2	0.0	0.6	6.0	0.2	0.0	0.0	14.4
Warm Grass/Shrub	0.0	0.2	10.5	13.6	0.2	0.0	0.3	0.1	2.4	2.0	1	0.5	0.0	30.7
Xerophytic Woods/Scrub	0.0	0.7	11.4	0.8	1.4	0.1	0.1	0.0	0.0	0.3	0.0	8.9	0.0	23.2
Taiga	68.3	6.4	0.0	0.0	10.9	0.2	103.7	0.0	0.0	0.3	0.0	0.1	0.0	190.0
Cool Conifer Forest	21.6	11.1	0.0	0.0	1.2	0.5	17.9	0.0	0.0	0.1	0.0	0.0	0.0	52.3
Cool Mixed Forest	4.8	5.7	0.3	0.0	1.0	0.6	21.6	0.0	0.0	0.0	0.0	0.0	0.0	33.9
Temp. Deciduous Forest	0.0	10.3	0.2	0.0	1.4	1.8	3.7	0.0	0.0	0.0	0.0	0.0	0.0	17.5
Warm Mixed Forest	0.0	6.1	18.0	0.0	1.6	0.7	0.4	0.0	0.0	0.0	1.4	1.5	0.0	29.6
Trop. Dry Forest/Savanna	0.0	0.0	18.2	2.0	0.0	0.0	0.0	0.0	0.0	0.0	0.9	4.8	0.0	25.9
Trop. Seasonal Forest	0.0	0.0	36.1	1.9	0.0	0.0	0.0	0.0	0.0	0.0	2.7	0.3	0.0	41.0
Trop. Rain Forest	0.0	0.0	42.5	0.2	0.0	0.0	0.0	0.0	0.0	0.0	4.9	0.2	0.0	47.7
Wetlands	1.9	2.4	11.3	8.2	1.3	0.1	3.3	0.2	0.3	0.7	5.1	1.8	0.2	36.7
Total	103.7	48.7	153.9	37.6	21.8	4.5	167.5	2.0	6	14.6	17.9	22.2	0.4	600.7

from the 12.5 Gt C a^{-1} estimate of Raich *et al.* (1991). The difference is completely explained by the differences in study area (Latin America versus South America). A good fit also applies to our estimate for North America of 8.8 Gt C a^{-1}, which is close to the value 7.0 Gt C a^{-1} given by McGuire *et al.* (1992). Our total NPP value of 7.8 Gt C a^{-1} for the former Soviet Union (CIS) is almost twice as high than the estimate of 4.4 Gt C a^{-1} by Kolchugina and Vinson (1993). This obvious difference is mainly caused by the exclusion of tundra and grasslands biomes by the latter estimate, while we accounted for all biomes. The tundra and grasslands biomes represent c. 30% of our NPP value. Our estimates for living and dead biomass pools for CIS are respectively 107.7 Gt C a^{-1} and 301.5 Gt C a^{-1}, while the estimates of Kolchugina and Vinson (1993) are 88.0 and 337.6 Gt C a^{-1} (Table 3). The differences for the separate Figures are probably caused by using different extent for the land cover (read vegetation) types, and maybe even amplified by actual differences in partitioning among the leaf–branch–stem components.

Deforestation rates are another indicator on the performance of the models. In IMAGE 2.0, deforestation rates are presented as a range between two different aggregations of forests and woodlands in the tropics. The lower value of the range includes only the classes 'Xerophytic Wood', 'Tropical Seasonal Forest' and 'Tropical Rain Forest'. The upper value includes also 'Tropical Dry Forest' and 'Savanna'. Literature values can be based on different classifications, but should be compatible with this wider range. The FAO (1993) estimated for the period 1970 - 1980 a global annual deforestation rate of 11.3 Mha a^{-1}. Meyers (1980) and Houghton *et al.* (1983) estimated respectively 13.8 Mha a^{-1} (only closed forests) and 11.4 Mha a^{-1}. The computed deforestation rate by IMAGE 2.0 ranged in this period also from 11.0 to 15.5 Mha a^{-1} and is consistent with the other literature estimates. However, the estimated deforestation rates for the period 1980-1990 are generally higher than those for the period 1970-1980. The rates range from 15.4 (FAO, 1993) to 17.2 Mha a^{-1} (WRI, 1992). We calculated a much lower range of 9.4 - 14.1 Mha a^{-1}, which is consistent with recent regional assessments (Skole and Tucker, 1993). The C Cycle model therefore uses an adequate land cover conversion approach and generates realistic C fluxes from land cover conversion. The deforestation rates are more thoroughly presented and discussed in the Land Cover Change Model paper (Zuidema and van den Born, 1994).

3. Sensitivity

A traditional sensitivity analyses provides the effect of possible combinations of input parameters and their interactions on selected output variables (Swartzman and Kaluzny, 1987; Leemans, 1991). Although some sensitivity can easily be assessed using the global atmospheric CO_2 concentration, a complete sensitivity analyses is still difficult to pursue in this phase of model development and testing. Input parameters involve the characteristics of vegetation classes (c. 19 x 17), global land cover (c. 61704 x 11) and feedback processes (c. 9 x 17 x 61704) and all parameters add up to a total number of c. 10^7. Even the most elaborate sampling schemes could not deal with that amount of parameters comprehensively. A sensitivity analyses of such a complex model would therefore always focus on selected aspects. Here we will limit our focus to the importance of initial conditions for the C fluxes of the different feedback processes.

A large problem was the initialization for the first simulated year. It is common to assume that initially the C cycle of the terrestrial biosphere is neither a net C source or sink. This assumption, however, is surely not valid for the starting year (1970) of the IMAGE 2.0 simulations. The influence of feedback processes and land cover changes were abundant in the decades preceding 1970. An equilibrium initialization for 1970 led to unrealistically rapid changes during the early phase of the simulation and illustrates the sensitivity to the initial conditions. We therefore start the C cycle model in equilibrium in 1900 so that at the start of the actual land cover simulation (1970) changes in fluxes for the different feedbacks processes are already accounted for and the terrestrial biosphere is not in equilibrium. This avoids making large initial errors in the net C flux due to the sensitivity to the initial state.

The land cover patterns are important for the determination of the C fluxes from the terrestrial biosphere throughout the whole simulation. These patterns are simulated dynamically by the Land Cover Change model (Zuidema and van den Born, 1994 , and this model has therefore a large influence on the C cycle. It determines for example the regional deforestation rates. Future projections of C fluxes from the terrestrial biosphere are therefore strongly dependent on the assumptions made in the Land Cover Change model.

We have chosen an aggregated version of the data base by Olson *et al.* (1985) as the template for land cover in 1970 (Leemans and van den Born, 1994). This data base defines the initial extent of each land cover class and therefore strongly determines the initial global C storage. Although the total global C values are generally in good agreement with other literature values (see discussion above), we have only used the extents and assigned well established initial NPP values and other characteristics to all land cover classes (Table 1). The actual C storage is computed dynamically during a simulation and depends on feedback processes and actual land cover status. However, some of the land cover classes span a wide range of climatic conditions. These classes (eg. agricultural areas and grasslands) are most sensitive to the environmental modification of the feedback processes. Cells with extreme climates for the land cover type are highly responsive to such modifications. The changes in NPP caused by changes in the growing period or temperature were especially significant in this respect, but, unexpectedly, the initial value of the β–factor or initial NPP values had little influence. The discrepancies between the Olson *et al.* (1985) data base and the climate data base could further result in changes of NEP up to 0.5 Gt a^{-1}.

One of the most important applications of IMAGE 2.0 is the quantification of different feedback processes. We have included changes in CO_2 fertilization (Figure 6), plant growth (Figure 7) and soil respiration (Figure 8), but also changes in growing period and land cover. CO_2 fertilization is the most significant negative feedback. Under increasing atmospheric CO_2 concentration, NPP increases and more C is stored in vegetation. The actual magnitude of this feedback is modeled as a function of environmental

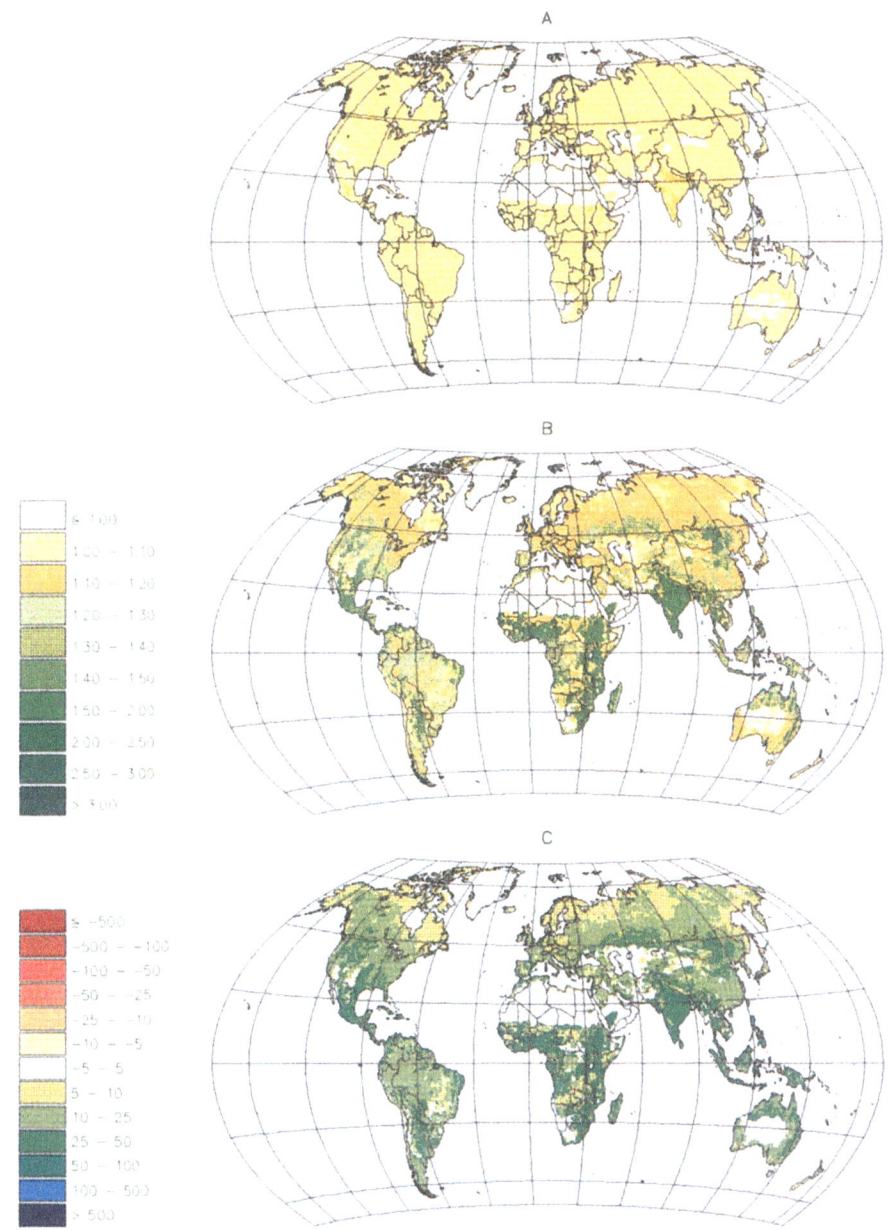

Figure 6: Total CO_2 fertilization effect $(1+\beta_{g,t} \cdot \ln(CO_{2,t}/CO_{2,i})$, Equation 2) in 1970 (A), 2050 (B) and the percentage difference between 1970 and 2050 (C).

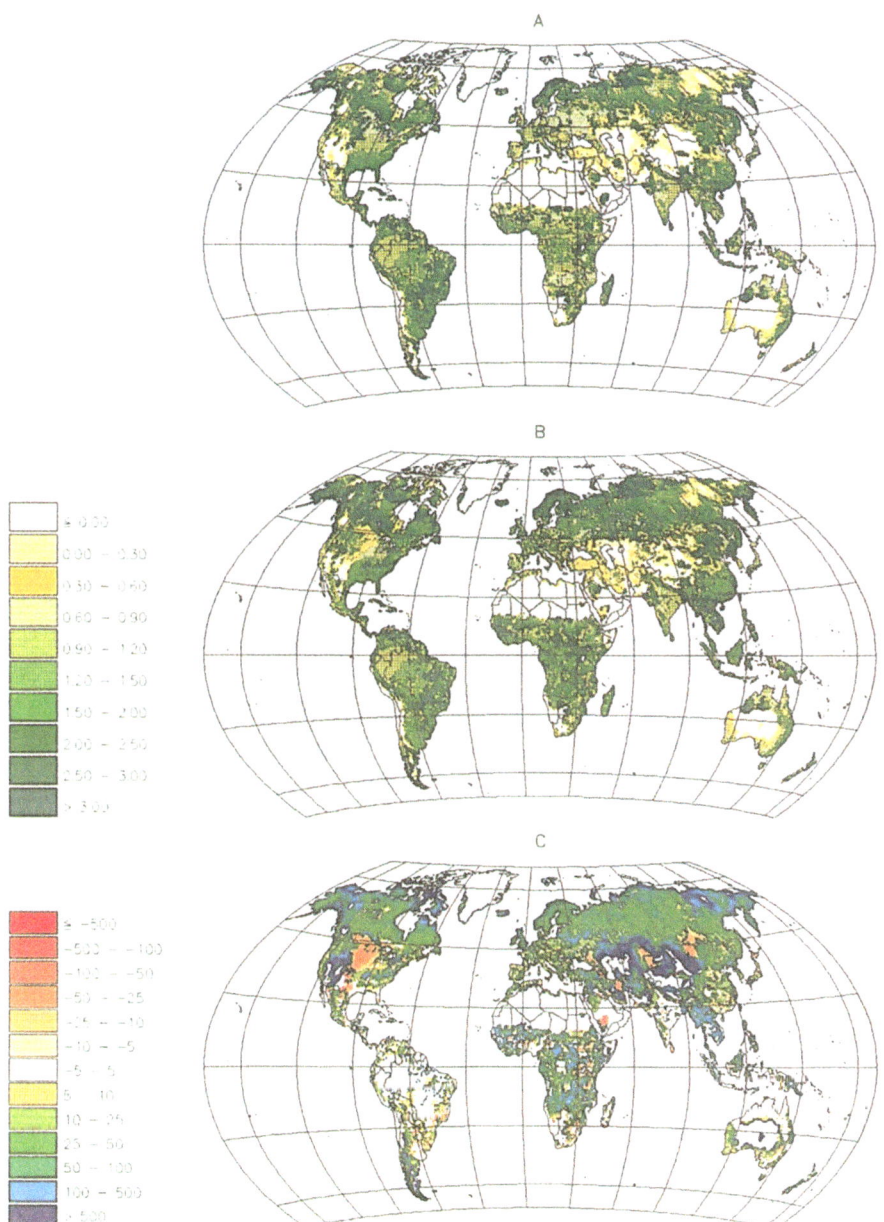

Figure 7: Temperature effect on photosynthetical rates ($\sigma_{g,t}$, Equation 2) in 1970 (A), 2050 (B) and the percentage difference between 1970 and 2050 (C). Cells where the land cover class changed during this period are not considered.

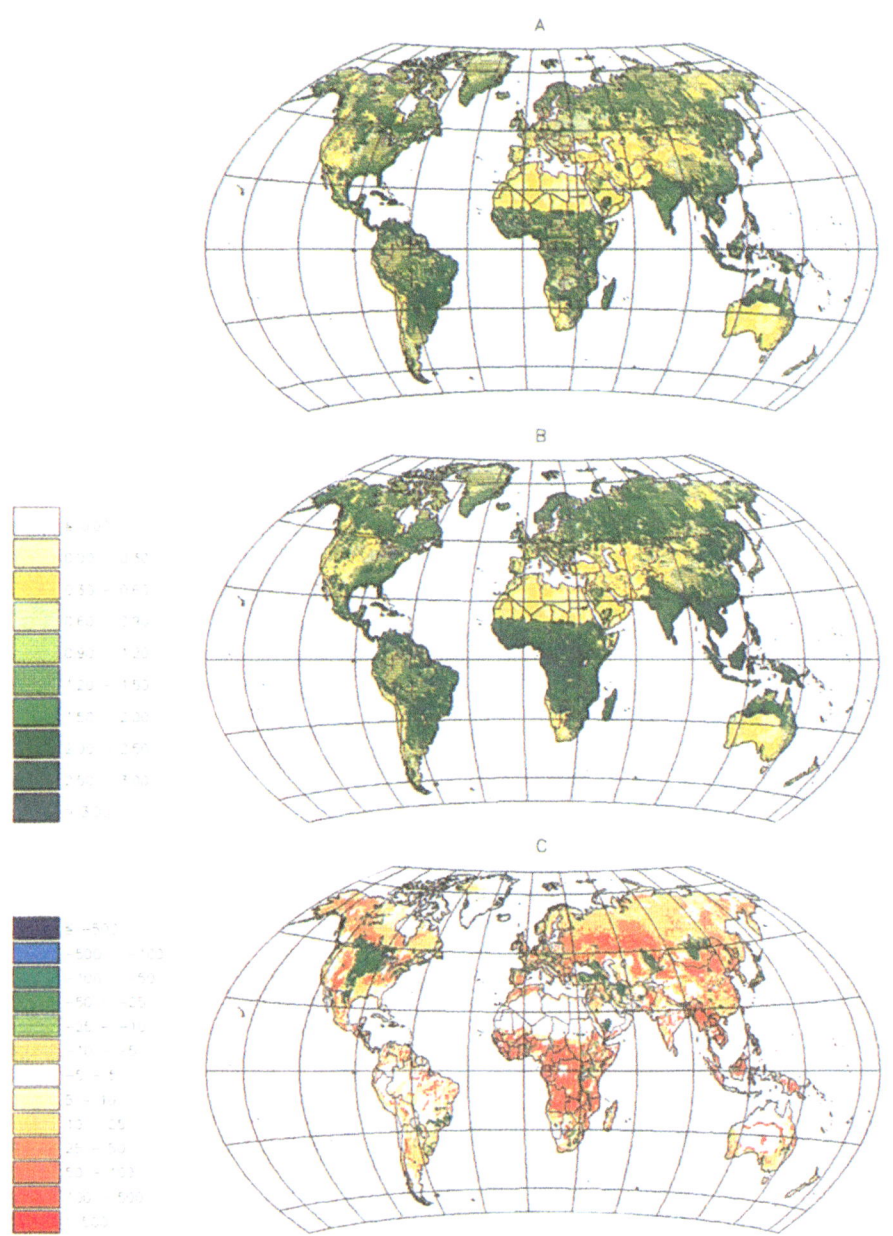

Figure 8: Effect of temperature and moisture on soil respiration ($\phi_{g,t}$, Equation 3) in 1970 (A), 2050 (B) and the percentage difference between 1970 and 2050 (C). Cells where the land cover class changed during this period are not considered.

TABLE 5

Accumulated net ecosystem production for each region between 1975 and 2100 (in Gt C).

year	1975	2000	2025	2050	2075	2100
Canada	2.3	12.23	27.8	57.2	88.4	119.0
USA	1.0	3.2	5.2	34.1	71.8	103.6
Latin America	1.4	9.4	23.9	43.1	66.7	92.1
Africa	1.1	8.0	19.4	37.1	54.2	61.7
OECD Europe	0.6	4.3	9.6	18.0	29.9	44.5
Eastern Europe	0.1	0.5	2.0	5.0	7.7	10.8
CIS	3.5	24.3	58.2	117.4	181.6	246.9
Middle East	0.2	1.0	2.6	3.9	4.6	5.8
India region	0.5	1.5	10.3	13.3	13.4	15.1
China region	1.0	4.9	7.0	23.7	32.2	46.8
East Asia	0.2	-0.8	-1.5	-0.3	1.3	3.4
Oceania	0.3	0.9	2.4	5.0	9.0	14.1
Japan	0.1	0.2	-0.5	-0.9	-1.0	-1.0
World	12.1	69.5	176.2	356.7	559.6	762.7

characteristics, such as temperature, soil moisture, nutrients, altitude and species characteristics and varies greatly for different regions (Figure 6). The importance of the different feedback processes and their interactions were preliminarily discussed by Vloedbeld and Leemans (1993) and Klein Goldewijk and Vloedbeld (1994) for Latin America. They concluded that land cover changes were the most important control of C fluxes. C fertilization, climatic effects and soil respiration only modified the outcome for Latin America. These conclusions are still valid, but the simulations presented here clearly illustrates the large regional differences of the feedback processes. The temperature effect on plant growth, for example, is most pronounced in the high latitude ecosystems (Figure 7).

4. Application to Conventional Wisdom Scenario

An illustration of the capabilities of the IMAGE 2.0 model incorporates the future projection of greenhouse gas emissions and associated climate change for a 'Conventional Wisdom Scenario'. This scenario specifies regional specific population, economic growth and energy use pathways (c.f. Alcamo et al., 1994). This scenario drives the land cover changes, and through the climatic and atmospheric composition feedbacks the C Cycle. The land cover changes for this scenario are large during the initial part of the simulation (1990-2025). The C emissions of land cover change are then just compensated by the C uptake of the terrestrial biosphere. After 2050 land cover changes decrease substantially in most regions. The main reason for this decrease is the lack of available non-agricultural land (especially in the regions India, China and Africa). Also, the sink function of the terrestrial biosphere is then strongly enhanced by the feedback mechanisms.

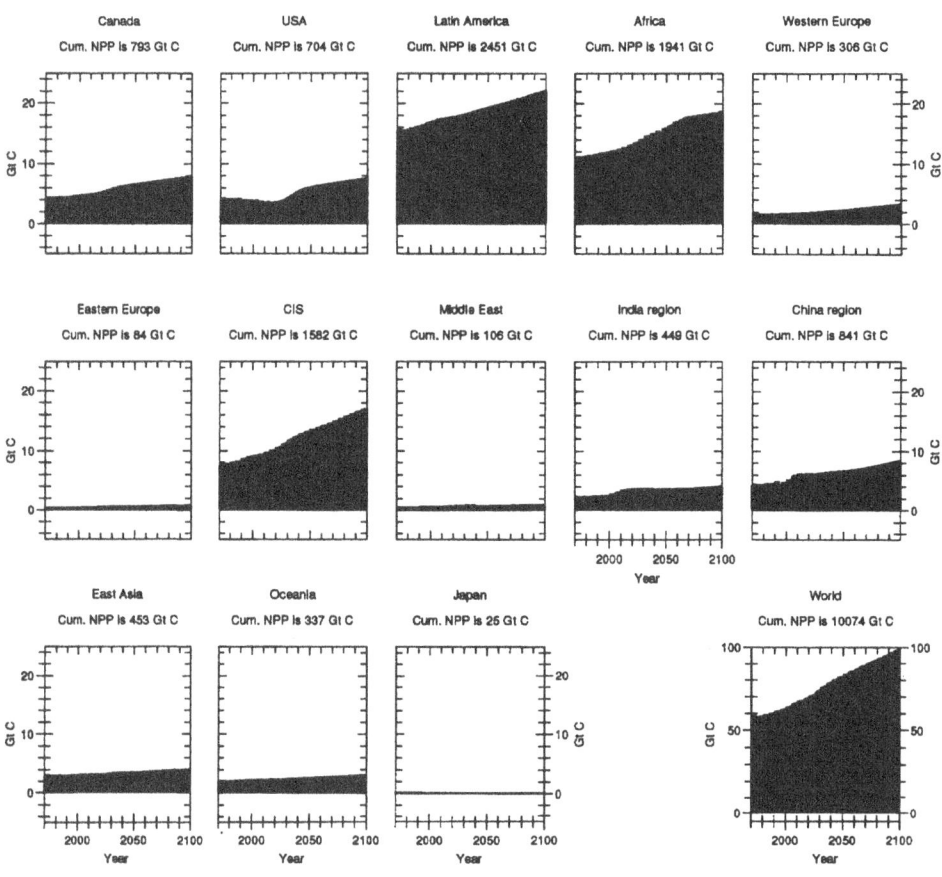

Figure 9: Development of the Net Primary Productivity between 1970 and 2050 for each region (Gt C a⁻¹).

The regional and global development of NPP and NEP is given in respectively Figures 9 and 10. The increase in net primary production is larger in high latitude, temperate and boreal regions, than in low latitude, tropical regions. There is even a pronounced tendency for the levelling of NPP in the latter regions. The temperature feedback on plant growth is most pronounced in Canada and CIS, which are now regions with temperature limited boreal forests. The more temperate regions, U.S.A. and Europe, also show the influence of these feedbacks, but only in a later stage. After 2050 the NEP increases remarkably in many regions. This is caused by the abandonment of agricultural land and a subsequent (natural) regrowth of forests. This process already starts early in the next century, but large C storages only develop slowly with a time lag of 25 - 100 years. This time lag is a direct consequence of our model approach (see above). Only Canada and CIS, and to a lesser extent Latin America and USA, hold vast forest areas throughout the simulation. These regions continuously sequester C (Table 5), a function that is enhanced by the feedback processes.

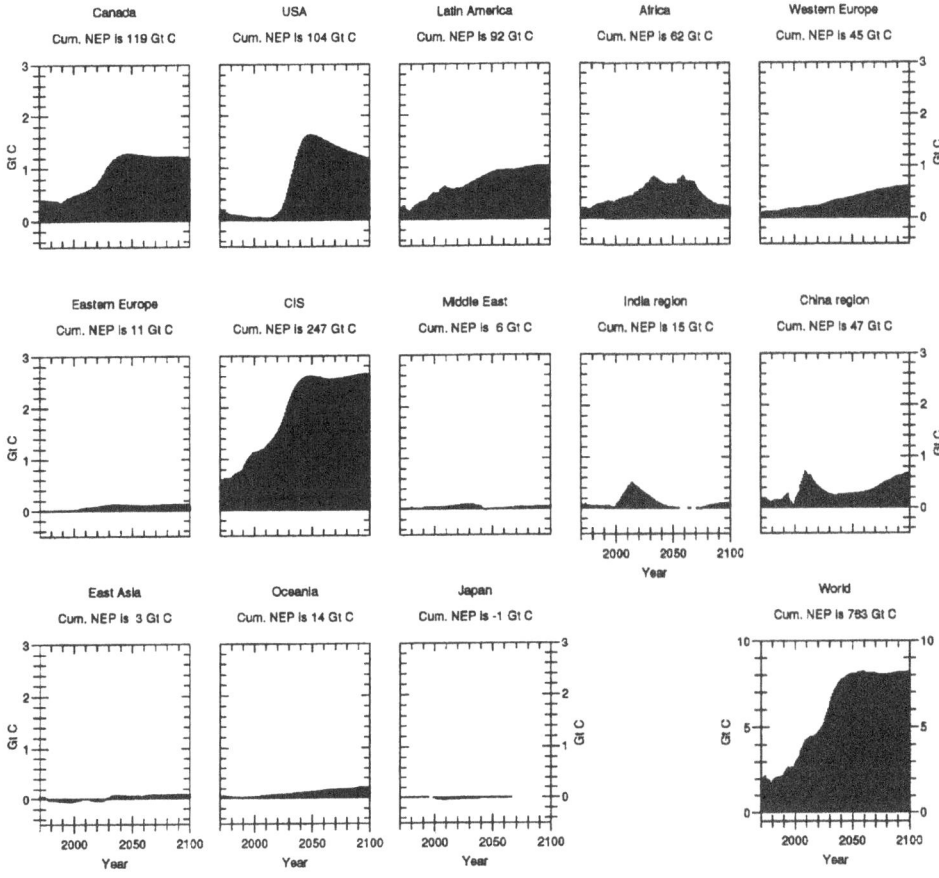

Figure 10: Development of the Net Ecosystem Productivity between 1970 and 2050 for each region (Gt C a^{-1}).

This simulated trend for the temperate and boreal regions is consistent with recent assessments. Kauppi *et al.* (1992) found increases in forest growth for most European forests from 1950 to 1980, while the ECE/FAO Temperate Forest Assessment (UN-ECE/FAO) presented increasing forest area's and growth rates in many temperate regions.

Surprisingly, the total global phytomass is 600.7 Gt C in 2050, which is somewhat lower than the 1970 value of 602.3 Gt C. However, the distribution over the different regions has dramatically changed. Tropical forests in Africa (loss of 68.7 Gt C), India (8.3 Gt C) and South East Asia (14.9 Gt C) almost completely disappeared. China would suffer large losses of temperate forest, despite the success of recent forestation projects (loss of 25.9 Gt C). The increase in phytomass occurs mainly in the temperate and boreal regions. CIS alone is already responsible of an increase in C storage since 1970 of 59.9 Gt C in 2050, and Canada accounts for 29.9 Gt C. Deforestation rates in Latin America decline considerably in this scenario, due to increased agricultural productivity. It remains the only region in this scenario where tropical forests remain an important sink of C (gain of

9.3 Gt C). This feature illustrates the capabilities of IMAGE 2.0 to generate regional specific assessments.

5. Conclusions

The IMAGE 2.0 Terrestrial C Cycle model is developed to calculate the C fluxes between the terrestrial biosphere and the atmosphere on a geographically explicit basis. Furthermore, a number of biogeophysical feedbacks have been implemented in order to examine the influence of climate change on the terrestrial C cycle.

- The IMAGE 2.0 C Cycle is very capable of simulating the most important aspect of the global C cycle. Its values are consistent with other analyses. The structure of this model allows for an improved scrutiny of the processes, linkages and feedbacks involved. The model is further capable of simulating geographically explicit changes in the C Cycle.
- Deforestation rates are a function of socio–economic drivers and are not prescribed. This allows for a more realistic simulation of land cover change and its impacts on the global C cycle.
- Although positive and negative feedback processes play a significant role, land cover changes are the most important factor in determining the C fluxes from the terrestrial biosphere to the atmosphere.
- The model allows for the development and evaluation of different scenarios. Complex scenarios with, for example, increased forestation rates or the use of biofuels to mitigate the buildup of atmospheric C can be addressed comprehensively. The consequence for agriculture and other anthropogenic land use can easily be addressed.

Acknowledgements

We thank S. Brown, A.M. Solomon and two anonymous reviewers for critical, but useful comments on earlier drafts of the manuscript. Further, we thank Joseph Alcamo, Jennifer Robinson and Gé Zuidema for fruitful discussions on the implementation and testing of this C Cycle model and the linkages with other parts of the IMAGE 2.0 framework. This reasearch is part of the Global Changes and Terrestrial Ecosystems (GCTE) project of the International Geosphere-Biopshere Programme (IGBP). It is supported by the Dutch National Research Programme on Climate Change and Global Air Pollution (NRP) under contract NOP 851042, and the Ministry of Housing, Physical Planning and the Environment under contract MAP 482507, MAP 481509 and MAP 482510 to RIVM.

References

Alcamo, J., G.J.J. Kreileman, M. Krol and G. Zuidema: 1994, Modeling the global society-biosphere-climate system, Part 1: model description and testing. *Wat. Air Soil Pollut.*, **76** (this volume).

Atjay, G.L., P. Ketner and P. Duvigneaud (eds): 1979, *Terrestrial Primary Production and Phytomass*, Wiley and Sons, pp. 129-187.

Bazzaz, F.A.: 1990, The response of natural ecosystems to the rising global CO_2 levels. *Annu. Rev. Ecol. Syst.*, **21**: 167-196.

Bazzaz, F.A. and R.W. Carlson: 1984, The response of plants to elevated CO2. I. Competition among an assemblage of annuals at two levels of soil moisture. *Oecologia*, **62**: 196-198.

Bouwman, A.F.: 1989, The role of soils and land use in the greenhouse effect. *Netherlands Journal of Agricultural Science*, **37**: 13-19.

Box, E.O.: 1988, Estimating the seasonal carbon source-sink geography of a natural, steady-state terrestrial biosphere. *Appl. Meteor.*, **27**(10): 1109-1124.

Carlson, R.W. and F.A. Bazzaz: 1980, The effects of elevated CO_2 concentrations on growth, photosynthesis, transpiration and water use efficiency of plants, in: J.J. Singh and A. Deepak (eds), *Environmental and Climatic Impact of Coal Utilization*, Academic Press, New York.

Carlyle, J.C. and U.B. Than: 1988, Abiotic controls of soil respiration beneath an eighteen-year-old Pinus radiata stand in South-eastern Australia. *Journal of Ecology*, **76**: 654-662.

Cooper, C.F.: 1983, Carbon storage in managed forests. *Can. J. For. Res.*, **13**: 155-166.

Cramer, W. and A.M. Solomon: 1993, Climatic classification and future global redistribution of agricultural land. *Clim. Res.*, **3**(1-2): 97-110.

Detwiler, R.P. and C.A.S. Hall: 1988, Tropical forests and the global carbon cycle. *Science*, **239**: 42-47.

Dewar, R.C.: 1990, A model of carbon storage in forests and forest products. *Tree Physiology*, **6**: 417-428.

Dewar, R.C.: 1991, Analytical model of carbon storage in the trees, soils and wood products of managed forests. *Tree Physiology*, **8**(3): 239-258.

Dewar, R.C. and M.G.R. Cannell: 1992, Carbon sequestration in the trees, products and soils of forest plantations: an analysis using UK examples. *Tree Physiology*, **11**: 49-71.

Eamus, D.: 1991, The interaction of rising CO_2 and temperatures with water use efficiency. *Plant, Cell Environm.*, **14**: 843-852.

Emanuel, W.R., G.G. Killough, W.M. Post and H.H. Shugart: 1984, Modeling terrestrial ecosystems and the global carbon cycle with shifts in carbon storage capacity by land use change. *Ecology*, **65**: 970-983.

Esser, G.: 1987, Sensitivity of global carbon pools and fluxes to human and potential climate impacts. *Tellus*, **39B**(3): 245-260.

Esser, G.: 1991, Osnabrück Biosphere model: structure, construction, results, in: G. Esser and D. Overdieck (eds), *Modern Ecology, Basic and Applied Aspects*, Elsevier, Amsterdam, pp.679-709.

FAO: 1993, *Forest Resources Assessment 1990 Tropical Countries*. 112, Food and Agricultural Organization of the United Nations, Rome, Italy.

FAO/UNESCO: 1974, *Soil Map of the World, 1:5,000,000*, Food and Agriculture Organisation of the United Nations.

Fearnside, P.M.: 1991, Deforestation in Brazilian Amazonia as a source of greenhouse gases. *Proceedings of Regional Conference on Global Warming and Sustainable Development: Perspectives from Developing countries*, Sao Paulo,.

Flint, E.P. and J.F. Richards: 1994, Trends in carbon content of vegetation in South and Southeast Asia with changes in land use, in: V.H. Dale (eds), *Effects of Land-Use Change on Atmospheric CO_2 Concentrations*, Springer-Verlag, New York, pp.201-299.

Goudriaan, J.: 1992, Biosphere structure, carbon sequestering potential and the atmospherci ^{14}C carbon record. *J. Exp. Bot.*, **32**(253): 1111-1119.

Goudriaan, J. and H.E. de Ruiter: 1983, Plant growth in response to CO_2 enrichment, at two levels of nitrogen and phosphorus supply. 1. Dry matter, leaf area and development. *Netherl. J. Agricult. Sci.*, **31**: 157-169.

Goudriaan, J. and P. Ketner: 1984, A simulation study for the global carbon cycle, including man's impact on the biosphere. *Clim. Change*, **6**: 167-192.

Houghton, J.T., B.A. Callander and S.K. Varney (eds): 1992, *Climate Change 1992. The Supplementary Report to the IPCC Scientific Assessment*, Cambridge University Press, 200 pp.

Houghton, J.T., G.J. Jenkins and J.J. Ephraums (eds): 1990, *Climate Change: The IPCC Scientific Assessment*,

Cambridge University Press, 365 pp.

Houghton, R.A., J.E. Hobbie, J.M. Melillo, B. Moore, B.J. Peterson, G.R. Shaver and G.M. Woodwell: 1983, Changes in the carbon content of terrestrial biota and soils between 1860 and 1980: a net release of CO_2 to the atmosphere. *Ecol. Monogr.*, **53**(3): 235-262.

Kauppi, P., K. Mielikäinen and K. Kuusela: 1992, Biomass and carbon budget of European forests, 1971 to 1990. *Science*, **256**: 70-74.

King, A.W.: 1986, *The seasonal exchange of carbon dioxide between the atmosphere and the terrestrial biopshere: extrapolation from site-specific models to regional models.*, Dissertation, 271 pp., Univeristy of Tennessee.

King, A.W., W.R. Emanuel and W.M. Post: 1992, A dynamic model of terrestrial carbon cycling response to land-use change. in: M. Kanninen (eds). *Carbon Balance of World's Forested Ecosystems: Towards a Global Assessment*, Joensuu, Finland.

King, G.A. and R.P. Neilson: 1992, The transient response of vegetation to climate change: a potential source of CO2 to the atmosphere. in: J. Wisniewski and A.E. Lugo (eds). *Natural sinks of CO_2*, Palmas Del Mar, Puerto Rico.

Klein Goldewijk, C.G.M. and M. Vloedbeld: 1994, The exchange of carbon dioxide between the atmosphere and the terrestrial biosphere in Latin America. *Advances in Soil Science*, in press.

Kohlmaier, G.H., E.-O. Sire, A. Janecek, C.D. Keeling, S.C. Piper and R. Revelle: 1989, Modelling the seasonal contribution of a CO_2 fertilization effect of the terrestrial vegetation to the amplitude increase on atmospheric CO_2 at Mauna Loa Observatory. *Tellus*, **41B**: 487-510.

Kolchugina, T.P. and T.S. Vinson: 1993, Carbon sources and sinks in forest biomes of the former soviet union. *Global Biogeochemical Cycles*, **7**(2): 291-304.

Körner, C.: 1993, CO_2 fertilization: The great uncertainty in future vegetation development, in: A.M. Solomon and H.H. Shugart (eds), *Vegetation Dynamics and Global Change*, Chapman and Hall, New York, pp.53-70.

Kortleven, J.: 1963, *Kwantitatieve aspecten van humusopbouw en humusafbraak*, Landbouwhogeschool, Wageningen. pp.

Kreileman, G.J.J. and A.F. Bouwman: 1994, Computing land use emissions of greenhouse gases. *Wat. Air Soil Pollut.*, **76** (this volume).

Krol, M.S. and H. van der Woerd: 1994, Simplified calculation of atmospheric concentration of greenhouse gases and other constituents for evaluation of climate scenarios. *Wat. Air Soil Pollut.*,**76**(this volume).

Kurz, W.A., M.J. Apps, T.M. Webb and P.J. McNamee: 1992, *The carbon budget of the Canadian forest sector: Phase I.* 0-662-19913-8, Northern Forestry Centre, Edmonton, Alberta, Canada.

Larcher, W.: 1980, *Physiological Plant Ecology*, Springer-Verlag , 303 pp.

Lashof, D.A.: 1989, The dynamics greenhouse: Feedback proceses that may influence future concentrations of atmospheric trace gases and climatic change. *Clim. Change*, **14**: 213-214.

Leemans, R.: 1991, Sensitivity analysis of a forest succession model. *Ecol. Mod.*, **53**: 247-262.

Leemans, R.: 1992, Modelling ecological and agricultural impacts of global change on a global scale. *J. Sci. Ind. Res.*, **51**: 709-724.

Leemans, R. and W. Cramer: 1991, *The IIASA database for mean monthly values of temperature, precipitation and cloudiness on a global terrestrial grid.* Research Report RR-91-18, International Institute of Applied Systems Analysis, Laxenburg, Austria.

Leemans, R. and G.J. van den Born: 1994, Determining the potential global distribution of natural vegetation, crops and agricultural productivity. *Wat. Air Soil Pollut.*, **76** (this volume).

Long, S.P.: 1991, Modification of the response of photosynthetic productivity to rising temperature by atmospheric CO_2 concentrations: Has its importance been underestimated? *Plant Cell Environm.*, **14**: 729-739.

McGuire, A.D., J.M. Melillo, L.A. Joyce, D.W. Kicklighter, A.L. Grace, B. Moore III and C.J. Vörösmarty: 1991, *Application of the terrestrial ecosystem model to estimate carbon and nitrogen dynamics for potential vegetation in North America*, The Ecosystems Center, Marine Biological Laboratory, Woods Hole, Massachuchetts, USA.

McGuire, A.D., M. Melillo, L.A. Joyce, D.W. Kicklighter, A.L. Grace, B. Moore III and C.J. Vorosmarty: 1992, Interactions between carbon and nitrogen dynamics in estimating net primary productivity for potential vegetation in North America. *Global Biogeochemical Cycles*, **6**: 101-124.

Melillo, J.M., T.V. Callaghan, F.I. Woodward and E. Salati: 1990, Effects on ecosystems, in: J.T. Houghton, G.J. Jenkins and J.J. Ephraums (eds), *Climate Change: The IPCC Scientific Assessment*, Cambridge University

Press, Cambridge, pp.283-310.

Melillo, J.M., A.D. McGuire, D.W. Kicklighter, B. Moore III, C.J. Vorosmarty and A.L. Schloss: 1993, Global climate change and terrestrial net primary production. *Nature*, **363**(6426): 234-239.

Meyers, N.: 1980, *Conversion of tropical moist forests*, National Academy of Sciences, USA.

Miao, S.L., P.M. Wayne and F.A. Bazzaz: 1992, Elevated CO_2 differentially alters the responses of co-occurring birch and maple seedlings to a moisture gradient. *Oecologia*, **90**: 300-304.

Mitchell, J.F.B., S. Manabe, V. Meleshko and T. Tokioka: 1990, Equilibrium climate change - and its implications for the future, in: J.T. Houghton, G.J. Jenkins and J.J. Ephraums (eds), *Climate Change: The IPCC Scientific Assessment*, Cambridge University Press, Cambridge, pp.131-172.

Mooney, H.A., B.G. Drake, R.J. Luxmoore, W.C. Oechel and L.F. Pitelka: 1991, Predicting ecosystems responses to elevated CO2 concentrations. *Bioscience*, **41**(2): 96-104.

Morison, J.I.L.: 1985, Sensitivity of stomata and water use efficiency to high CO_2. *Plant Cell Environm.*, **8**: 467-474.

Olson, J., J.A. Watts and L.J. Allison: 1985, *Major World Ecosystem Complexes Ranked by Carbon in Live Vegetation: A Database*. Carbon Dioxide Information Center, Oak Ridge, Tennessee, USA.

Parton, W.J., D.S. Schimel, C.V. Cole and D.S. Ojima: 1987, Analysis of factors controlling soil organic matter levels in Great Plains grasslands. *Soil Sci. Soc. Am. J.*, **51**: 1173-1179.

Polglase, P.J. and Y.P. Wang: 1992, Potential CO_2-enhanced carbon storage by the terrestrial biosphere. *Aust. J. Bot.*, **40**: 641-656.

Post, W.M., E.W. R., P.J. Zinke and A.G. Stangenberger: 1982, Soil carbon pools and world life zones. *Nature*, **298**: 156-159.

Prentice, I.C., W. Cramer, S.P. Harrison, R. Leemans, R.A. Monserud and A.M. Solomon: 1992, A global biome model based on plant physiology and dominance, soil properties and climate. *J. Biogeogr.*, **19**: 117-134.

Prentice, I.C. and A.M. Solomon: 1991, Vegetation models and global change, in: R.S. Bradley (eds), *Global Changes of the Past*, UCAR/Office for Interdisciplinary Earth Studies, Boulder, pp.365-383.

Raich, J.W., E.B. Rastetter, J.M. Melillo, D.W. Kicklighter, P.A. Steudler, B.J. Peterson, A.L. Grace, B. Moore III and C.J. Vörösmarty: 1991, Potential net primary productivity in South America: application of a global model. *Ecol. Appl.*, **1**(4): 399-429.

Schlesinger, W.H.: 1977, Carbon balance in terrestrial detritus. *Ann.Rev.Ecol.Syst.*, **8**: 51-81.

Seiler, W. and P. Crutzen: 1980, Estimates of gross and net fluxes of carbon between the biosphere and the atmosphere from biomass burning. *Clim. Change*, **2**: 207-247.

Seino, H. and Z. Uchhijima: 1992, Global distribution of net primary productivity of terrestrial vegetation. *J. Agr. Met.*, **48**(1): 39-48.

Skole, D. and C. Tucker: 1993, Tropical deforestation and habitat fragmentation in the Amazon: satellite data from 1978 to 1988. *Science*, **260**: 1905-1910.

Smith, T.M. and H.H. Shugart: 1993, The transient response of terrestrial carbon storage to a perturbed climate. *Nature*, **361**: 523-526.

Smith, T.M., J.F. Weishampel, H.H. Shugart and G.B. Bonan: 1992, The Response of Terrestrial C Storage to Climate Change: Modeling C Dynamics at Varying Temporal and Spatial Scales. in: J. Wisniewski and A.E. Lugo (eds). *Natural sinks of CO₂*, Palmas Del Mar, Puerto Rico.

Solomon, A.M. and H.H. Shugart (eds): 1993, *Vegetation Dynamics and Global Change*, Chapman and Hall, 338 pp.

Strain, B.R. and J.D. Cure: 1985, *Direct Effects of Increasing Carbon Dioxide on Vegetation*. United States Department of Energy, NTIS, Springfield, USA.

Swartzman, G.L. and S.P. Kaluzny: 1987, *Ecological Simulation Primer*, Macmillan Publishing Company, 370 pp.

Trexler, M.C.: 1991, *Minding the Carbon Store: Weighing U.S. Forestry Strategies to Slow Global Warming*, World Resources Institute, 81 pp.

Turner, D.P., J.J. Lee, G.J. Koerper and J.R. Barker: 1993, *The Forest Sector Carbon Budget of the United States: Carbon Pools and Flux under Alternative Policy Options*. EPA/600/3-93/093, US Environmental Protection Agency, USA.

UN-ECE/FAO: 1992, *The Forest Resources of the Temperate Zones. 1990 Forest Resource Assessment*. ECE/TIM/80, United Nations, New York.

Vloedbeld, M. and R. Leemans: 1993, Quantifying feedback processes in the response of the terrestrail carbon cycle to global change: The modelling approach of IMAGE-2.0. *Wat. Air Soil Pol.*, **70**: 615-628.

de Vries, B., R.A. van den Wijngaart, G.J.J. Kreileman, J.A. Olivier and S. Toet: 1994, A model for calculating regional energy use and emissions for evaluating global climate scenarios. *Wat. Air Soil Pollut.*, **76** (this volume).

Waring, R.H. and W.H. Schlesinger: 1985, *Forest Ecosystems. Concepts and Management*, Academic Press, Inc.,
    340 pp.
Whittaker, R.H. and G.E. Likens: 1973, *Carbon in the Biota*, United States Atomic Energy Commission.
Whittaker, R.H. and P.L. Marks: 1975, Methods of assessing terrestrial productivity, in: H. Lieth and R.H.
    Whittaker (eds), *Primary Productivity of the Biosphere*, Springer-Verlag, Berlin, pp.55-118.
WRI: 1992, *World Resources, 1992-1993*. World Resources Institute, Oxford University Press, New York.
Zobler, L.: 1986, *A World Soil File for Global Climate Modeling*. Scientific and technical Information Branch,
    New York.
Zuidema, G. and G.J. van den Born: 1994, Simulation of global land cover changes as affected by economic
    factors and climate. *Wat. Air Soil Pollut.*, **76** (this volume).

COMPUTING LAND USE EMISSIONS OF GREENHOUSE GASES

G.J.J. KREILEMAN AND A.F. BOUWMAN

National Institute of Public Health and Environmental Protection (RIVM)
P.O. Box 1, 3720 BA Bilthoven, The Netherlands

Abstract. A model has been developed to estimate the regional emission of greenhouse gases from land-use related sources. Driving forces for this model are the changing regional demand for food and wood products driven by demographic and economic developments (Zuidema *et al.*, 1994). To include the environmental conditions, which are essential factors determining the flux for certain sources, emissions are grid-based where possible. Grid-based explicit calculations are given for CH_4 emission from rice, wetlands, emissions from deforestation, savanna burning and agricultural waste burning and N_2O from natural soils, arable lands and deforestation. For a number of sources (landfills, domestic sewage treatment, termites, methane hydrates and aquatic sources) geographically explicit calculations are not yet possible because of data limitations. For most of the sources the global results of the calculations are in agreement with other scenario studies, although there are differences for a number of individual sources.

Keywords: animal excreta, animals, aquatic sources, arable lands, biomass burning, CH_4, CO, deforestation, emission, landfills, N_2O, rice cultivation, sewage, soils, termites, trace gases, wetlands

1. Introduction

Although the main source of global greenhouse gases is the world's energy system, a very significant part of these emissions originate from biotic, mostly land-based sources (Watson *et al.*, 1992). The percentage of non-fossil emissions of the most important gases may be about 25% for CO_2, 75% for CH_4 and > 70% for N_2O (Table 1). By taking into account the effect of different gases on radiative forcing, the non-energy emissions may contribute 40% to global warming (estimated from Shine *et al.*, 1990) for the period 1765 - 1990. Other reactive species, such as CO, NO_x and VOC also contribute indirectly to global warming and have major land-use related sources (Watson *et al.*, 1992).

In this paper, we describe a model for computing biotic emissions of important greenhouse gases. Because most of these biotic emissions are related to land use or land cover, we call it a "Land Use Emissions" model. This model is part of the Terrestrial Environment System of the IMAGE 2.0 model, whose goal is to provide a framework for modeling global changes in agricultural demand, vegetation potential, land cover transformations, and resulting biosphere-atmosphere exchanges of CO_2 and other greenhouse gases. The overall goal of the IMAGE 2.0 model is to relate - in a geographically-explicit way - human activity in the global energy and agricultural systems to modifications of climate and the biosphere. We believe that this quantitative overview of the global society-biosphere-climate system will have many different scientific and policy applications (Alcamo *et al.*, 1994a, b).

The specific goals of the Land Use Emissions model are:
·   To relate global land use to the flux of greenhouse gas emissions, such as the emissions

Water, Air, and Soil Pollution **76**: 231–258, 1994.

TABLE 1

Characteristics, sources and global emissions for the major greenhouse gases (Gt= gigaton, 1 Gt=10^{15} g, Tg= teragram, 1 Tg= 10^{12} g. Compiled from Watson et al. (1992) unless indicated otherwise.

	CO_2	CH_4	N_2O
Residence time (y)	120[1]	8-12	132
Concentration (ppm)	353	1.7	0.310
Annual increase (%)	0.5	1-2	0.2-0.3
Radiative absorption potential[2]	1	11	270
Global annual emission[3]	6.5-7.5 Gt C	460-600 Tg CH_4	11-17 T g N
% Biotic	20-30	75	>70
Sources	fossil fuels (6 Gt C)	rice paddies (20-150 Tg)	N-fertilizer use (<2 Tg)[4]
	deforestation and	wetlands (100-200 Tg)	natural soils (6.8 Tg)[5]
	shifting cultivation (0.6-2.6 Tg C)	ruminants (65-100 Tg)	land use change (0.4 Tg)[6]
		animal waste (20-30 Tg)	biomass burning (0.1-0.3 Tg)[7]
		termites (10-50 Tg)	animal waste (0.2-0.6 Tg)[6]
		landfills (20-70 Tg)	sewage (0.2-1.9 Tg)[6]
		oceans/lakes (6-45)	Oceans (2 Tg)
		domestic sewage treatment (25 Tg)	aquifers-irrigation (0.5-1.3 Tg)[6]
		biomass burning (20-80 Tg)	nitric acid (0.4-0.6 Tg)
		fossil (70-120 Tg) (coal mining,	adipic acid (0.4 Tg)
		gas exploitation; pet.industry)	automobiles (0.1-1.3 Tg)[6]
		CH_4 hydrates (0-5 Tg)	fossil fuel (<0.05)[6]
			atmospheric formation (?)[6]
			global warming (0.6)[6]
Sinks	atm. accumulation (3.5 Gt C)	atm. accumulation (28-37 Tg)	atm. accumulation (3-4.5 Tg)
	oceans (1.2-2.8 Gt C)	atm. chemistry (420-520 Tg)	atmosph. chemistry (7-13 Tg)
	biosphere (?)	soil oxidation (15-45 Tg)	soils (?)

[1] Isaksen et al. (1992); [2] for time horizon of 100 years; [3] The total budget or total sources is not necessarily the sum of the indicated individual sources; [4] Eichner (1990); [5] Bouwman et al. (1993); [6] Khalil and Rasmussen (1992); [7] Crutzen and Andreae (1990).

of CH_4 from rice fields and livestock. We note that the biotic emissions of CO_2 are not included in the Land Use Emissions model but are computed as part of Terrestrial Carbon model of IMAGE 2.0 (Klein Goldewijk *et al.*, 1993). This information will be used to estimate changes in emissions resulting from future changes in land use, and to evaluate strategies for reducing these emissions.

· To estimate the emissions resulting from biotic processes unrelated to human activity, such as N_2O emissions from soils in unmanaged forests.

· To estimate the biotic emissions (other than CO_2) from aquatic sources that are not accounted for elsewhere in IMAGE 2.0.

Our objective is to describe as many source categories as possible in a geographically-explicit way (Table 2) on a global grid of 0.5^0 x 0.5^0 latitude-longitude. This is not possible for some categories because of current data limitations. Future improvements in this direction are discussed later in the paper. Although grid-based calculations require extensive data collection and analysis efforts, they have the following benefits: (1) they can take into account the known relationships between local temperature and moisture conditions and the local rate of emissions, (2) they can take into account the relationship between land cover and emissions, (3) they allow for evaluation of local management strategies to reduce emissions, for example by controlling the rate of fertilizer input, or changing the practices of livestock management.

We first present background information of the greenhouse gases computed by the model, next model development, and then testing of calculations against 1990 reference data. Possible future improvements are discussed in the conclusions.

TABLE 2

Sources described in the IMAGE 2.0 Land Use Emission Model, type of gas emitted and the type of calculation and presentation. G = geographically explicit; R = regional total; W = global total

Source	Species	Type of calculation
Wetland rice fields	CH_4	G
Natural wetlands	CH_4	G
Landfills	CH_4	R
Domestic sewage treatment	CH_4	R
Animals	CH_4	R
Animal waste	CH_4, N_2O	R
Termites	CH_4	W
Methane hydrates	CH_4	W
Aquatic sources	CH_4, N_2O	W
Biomass burning	CH_4, CO, NO_x, N_2O, VOC	
-Deforestation		G
-Savanna burning		G
-Agricultural waste burning		G
Natural soils	N_2O	G
Agricultural fields	N_2O	G
Deforestation	N_2O	G

2. Background to land use emissions

In this paper we will discuss emission calculations for the gases CH_4 and N_2O. Emissions of CO, NO_x (NO and NO_2), and VOC (excluding CH_4), are discussed only briefly. Fluxes of CO_2 are the result of the Terrestrial Carbon Model as discussed in Klein Goldewijk *et al.*, (1993). Table 1 gives the major characteristics of CO_2, CH_4 and N_2O.

CH_4 has a great number of major and minor biotic sources. It is produced during microbial decomposition of organic material under anaerobic conditions. Natural wetlands, wet rice cultivation and landfill sites for solid waste dumping are places where anaerobic conditions prevail. CH_4 is also formed in the digestive tract of ruminating animals and in the guts of various insects, the major species being termites. The abiogenic process of biomass burning is a further source of CH_4. Between 20 and 30% of total sources of CH_4 is fossil or non-living CH_4; this fraction includes approximately 30 Tg a^{-1} of non-living CH_4 from wetlands and old peat layers (Cicerone and Oremland, 1988). The emissions from fossil sources therefore amount to 50-100 Tg a^{-1} (Table 1). Further inputs are from wetland rice cultivation (contribution of about 10% to the total source strength), natural wetlands (20%), biomass burning (10%), landfills for solid waste dumping (10%) and oceans and fresh water lakes another 5%; insects, primarily termites, may contribute 5% (Table 1). CH_4 accumulates in the atmosphere, and its major sinks are atmospheric oxidation by OH radicals and oxidation in soils. Estimates of these sinks are highly uncertain (Table 1).

For N_2O it is clear that most of this gas is formed in soils during nitrification and denitrification (Bouwman *et al.*, 1993). We do not know why N_2O is increasing. Probably the increase is caused by a great number of minor sources. These include (1) sources related to land use, such as N-fertilizers, animals manures, N-deposition and biomass burning; (2) fossil sources, such as fossil fuel combustion, industrial sources, automobiles; and (3) aquatic sources, including oceans and coastal waters, sewage treatment, freshwater systems, aquifers and irrigation. Recently even global warming and atmospheric formation were identified as potential sources (Table 1). The degree of uncertainty in global estimates of emissions is considerable. The uncertainty varies from source to source, and there are also regional differences. A number of attempts have been made to reduce the uncertainty by synthesizing statistical data on land use, biogeochemical processes and fluxes related to these processes (e.g. Cicerone and Oremland, 1988; Watson *et al.*, 1992; Khalil and Rasmussen, 1992). In other studies the atmospheric concentrations were used in modeling approaches to deduce seasonal and latitudinal variations in source strengths (Fung *et al.*, 1991; Taylor, 1992). Such modeling approaches can yield better insight in parameterizations to be used in models like IMAGE 2.0.

3. Model Description and Implementation

A list of the sources that are described in the Land Use Emissions Model with the type of calculation and representation in the model (geographically explicit or otherwise) is presented in Table 2. Below each of these sources will be discussed briefly.

3.1. CH$_4$ FROM RICE CULTIVATION

The harvested area of paddy rice, about 10% of the world's total cultivated area, has showed a steady increase of more than 1% a^{-1} during the last decades (FAO, 1991). About half of the cultivated rice area is covered by irrigated wetland rice fields, 30% by rain fed wetland rice fields, about 8% deep water rice and 15% is upland rice. Globally 60% of the harvested area is managed under a triple cropping system, 15% is double cropped and 25% is planted once a year (Matthews *et al.*, 1991). Upland rice fields do not produce significant amounts of CH$_4$ (Bouwman, 1991). The estimated global emission is 20-150 Tg CH$_4$ y^{-1}, but currently it is thought that the global contribution may be 60 Tg a^{-1} (Watson *et al.*, 1992). The uncertainty in this estimate is caused by scarcity of flux measurements, gaps in the knowledge of the global distribution of the various rice ecologies, soil types, and soil, water and crop management (levels, fertilizer type, application mode and timing of fertilization, both mineral and organic fertilizers, weeding and harvesting practices, water management and rice varieties) and associated CH$_4$ formation and oxidation (Bouwman, 1991).

The geographic basis for wetland rice fields is the Olson *et al.* (1983) land cover data base, corrected for the statistical information from FAO (1991). Although attempts have been made to relate CH$_4$ emission to temperature (see e.g. Aselmann and Crutzen, 1989), the knowledge of fluxes is still inadequate to apply such methods. The mean daily emission for temperate wetland rice fields used in one of the estimates of Aselmann and Crutzen (1989) is 310 mg m^{-2} day^{-1}, with a period of inundation of 130 days. As there is still no reliable method to estimate CH$_4$ emissions on the basis of rice ecology and climate, rice variety and soil and crop management, we use a mean global emission factor of 350 mg CH$_4$ m^{-2} day^{-1} and an inundation period of 130 days. This yields a global emission of about 60 Tg a^{-1}, which is in agreement with Watson *et al.* (1992). The percentage of each cell covered by rice is a result of calculations by the Land Cover Change Model (Zuidema *et al.*, 1994). The total rice area for each region is then calculated, and the total harvested area wetland rice is estimated as the total rice area minus the 1990 area of dryland rice (from Bouwman, 1991). Hence, we assume that the dryland rice area is constant in time and that changes in rice production occur in the wetland rice fields. This is consistent with, for example, EPA (1990).

3.2. CH$_4$ FROM NATURAL WETLANDS

The global wetland area is 530x10^6 ha (Matthews and Fung, 1987) to 570x10^6 ha (Aselmann and Crutzen, 1989), but the various estimates show regional discrepancies and definitional differences. The global emission from wetlands ranges between 40 and 160 Tg a^{-1}, most of this stemming from regions between 50°N and 70°N and between 10°N and 10°S (Matthews and Fung, 1987; Aselmann and Crutzen, 1989). Although there is evidence that CH$_4$ production and oxidation are stimulated by increasing temperatures, most studies (see e.g. Bartlett et al., 1992) are locale specific and cannot be used at the global scale. Also, the relationship between CH_4 fluxes and net primary production (NPP) has been tested in global approaches (Aselmann and Crutzen, 1989; Taylor et al., 1991). Unfortunately, to date field measurements are too scarce to verify the global validity of such relations. For estimating future emissions, therefore, the effect of temperature and NPP has been neglected.

Although there is considerable regional and global disagreement, the inventory of wetlands by Matthews and Fung (1987) clearly shows a better overall correspondence to the Olson et al. (1983) data base than the maps produced by Aselmann and Crutzen (1989), particularly in the cells with high fractional inundation. Therefore, the gridded wetland areas and mean annual emission for the different wetland types in Matthews and Fung (1987) have been adopted. From this data base the regional contributions are estimated. Because of the difficulties in describing changes in wetland areas (Zuidema et al., 1994) the global emission from wetlands is assumed constant in time at 111 Tg a^{-1}.

3.3. CH_4 FROM LANDFILLS AND DOMESTIC SEWAGE

Anaerobic decay of collected municipal and industrial organic matter that is dumped in landfills add 20-70 Tg CH_4 a^{-1} to the atmosphere globally (Watson et al., 1992). The largest current contribution comes from landfills in the industrialized world. Projections for the year 2000 suggest an increase in urban population and a doubling of waste generation and associated CH_4 emission in developing countries (Bingemer and Crutzen, 1987). Reduction of emissions from landfills may be achieved by gas harvesting or by recycling of solid wastes.

As no gridded global inventory of landfills and human waste production is available, the estimated country emissions of SEI (1992) are used to yield regional totals for the year 1990. The change in urban population is used to calculate past and future emissions from this source, assuming regionally-specific per capita waste productions that are constant in time.

Domestic sewage treatment has been identified as a significant global source of atmospheric CH_4. Watson et al. (1992) report a global estimate for 1990 of 25 Tg CH_4 a^{-1}. This estimate and its underlying factors are still very uncertain. Because of lack of data on the geographic distribution of this source, the global emission was simply related to the world population in agreement with Pepper et al. (1992).

3.4. CH_4 FROM ANIMALS

The global CH_4 emission from enteric fermentation in the digestive tract of animals may be 65 - 100 Tg y^{-1} (Crutzen et al., 1986). Approximately 80% is produced by cattle and buffaloes. About 65% of the CH_4 produced by domestic animals is emitted in the developing countries and Centrally Planned Asian countries. The geographic basis for the estimates is the regional average animal density for cattle and sheep and goats for the cells defined as grassland. Locally this may be different from the animal densities data bases produced by Lerner *et al.* (1988). Apart from the error made by assuming regional average densities, in many parts of the world animals do not graze in areas where grasslands are found. Hence, we recognize a number of potential errors made in presenting grid-based emissions from this source. The numbers of pigs, poultry, camels, buffaloes, horses, mules and asses are not represented on a grid-basis.

The estimates of future numbers of cattle, sheep, goats, pigs and poultry are based on estimates of future demand for meat and milk from Zuidema *et al.* (1994). Numbers of buffaloes, horses, mules and asses are added as regional constant emissions throughout the simulation period. We use estimates of CH_4 produced per head of cattle from Gibbs and Leng (1993), based on data for representative cattle types and leading to slightly different distributions compared to Crutzen *et al.* (1986). The contribution of the other animals (pigs, sheep, goats, buffaloes, horses, mules and asses is based on emission rates from Crutzen *et al.* (1986). The global source calculated thus is 79 Tg CH_4 a^{-1} for 1990 (see section 4.3). The consequence of using constant emissions per head, is that the emissions per unit of product will decrease as in the future animal productivity increases. This also applies to the CH_4 emissions from animal excreta.

3.5. CH_4 FROM ANIMAL EXCRETA

Recently animal excreta were identified as a potential source of CH_4 (Watson, 1992). The CH_4 emission from animal waste for dairy and non-dairy cattle, sheep and goats, pigs and poultry, is calculated from the 1990 regional estimated emission based on the 1990 manure usage presented by Gibbs and Woodbury (1993), who estimated CH_4 conversion factors for different livestock manure systems for a number of countries and regions of the world. Emissions for horses, mules and asses are calculated using the regional emission factor for the category "other animals" in Gibbs and Woodbury (1993). Since Gibbs and Woodbury (1993) give no estimates for buffaloes, the CH_4 emission factor for buffaloes is the global mean for the category "other animals". The global emission from animal waste for 1990 of about 14 Tg a^{-1} is somewhat lower than the estimate given in Watson *et al.* (1992). Manure usage systems are assumed constant in time with constant emission factors. The major contributors to waste CH_4 losses are Western Europe, USSR and the USA.

3.6. CH_4 FROM TERMITES

Estimates for this source range from 10 to 50 Tg a^{-1} (Watson *et al.*, 1992). The total population of termites shows an increasing trend, caused by extensions of the global cultivated area, burning and deforestation and possibly by increasing biomass production through CO_2 fertilization. Because of the lack of reliable data, we assume a constant global emission from termites of 20 Tg CH_4 a^{-1}, as proposed by Pepper *et al.* (1992).

3.7.   CH$_4$ FROM AQUATIC SOURCES AND DESTABILIZATION OF METHANE HYDRATES

Oceans and lakes may emit 6 - 45 Tg CH$_4$ a^{-1} (Table 1), mainly from sewage and other organic material. We assume a constant global source strength of 15 Tg CH$_4$ a^{-1}.

Methane hydrates may be released by warming (Cicerone and Oremland, 1988). Methane hydrates are most prevalent at depth in permafrost and in sea sediments. Our tentative global estimate of the CH$_4$ release from hydrates used of 5 Tg CH$_4$ a^{-1} (Pepper *et al.*, 1992) is assumed constant in time.

3.8.   CH$_4$, CO, NO$_X$, N$_2$O AND VOC FROM BIOMASS BURNING

The burning of biomass is a source of CH$_4$, CO, N$_2$O, VOC, many other trace gases and soot (Crutzen and Andreae, 1990). Biomass burning includes burning during deforestation, wildfires, burning of agricultural wastes and fuelwood, savanna fires, and shifting cultivation. There are several reports giving estimates for the global emissions caused by biomass burning. The emission factors for biomass burning were taken from the best documented global study of emissions from biomass burning (Crutzen and Andreae, 1990). They reported that $1 \pm 0.6\%$ of the released C evolves as CH$_4$, $10 \pm 5\%$ as CO, and $1.3 \pm 0.3\%$ as VOC. The N$_2$O-N losses amount to $0.7 \pm 0.3\%$ of the released N, emissions of NO$_x$ are 12.5 ± 5.3 % of the N released. The C/N ratios used were 0.01 for forests, 0.006 for savannas and 0.01 - 0.02 for agricultural wastes (Crutzen and Andreae, 1990).

For the emissions associated with deforestation the geographic basis is formed by the simulated deforestation (Zuidema *et al.*, 1994), resulting in emissions for 1990 of 14 Tg CH$_4$ a^{-1}, 251 Tg CO a^{-1}, 1 Tg NO$_x$ a^{-1}, 0.1 Tg N$_2$O-N a^{-1} and 14 Tg VOC a^{-1}. These estimates are higher than those made by Crutzen and Andreae (1990) (see also section 4.2).

We use the 5° x 5° grid distribution for savanna burning given by Hao *et al.* (1990) to calculate regional total emissions. The global emission for 1990 from savanna burning is about 16 Tg CH$_4$ a^{-1}, 284 Tg CO a^{-1}, 1 Tg NO$_x$-N a^{-1}, 0.1 Tg N$_2$O-N a^{-1} and 16 Tg a^{-1} of VOC. The scalar for savanna burning emissions in future predictions is the area of the land cover class of tropical dry/savanna.

Current agricultural waste burning is 650 (500 - 800) Tg C (Crutzen and Andreae, 1990), resulting in emissions of circa 8 Tg CH$_4$ a^{-1}, 141 Tg CO a^{-1}, 1.5 Tg NO$_x$-N a^{-1}, 0.1 Tg N$_2$O-N a^{-1} and 8 Tg VOC a^{-1}. Future estimates of the emissions from agricultural waste burning are related to the area of arable land in each region.

Shifting cultivation may be an important source of trace gases (Crutzen and Andreae, 1990), but it is not represented in the current model version, because of the scarcity of data of the regional distribution of this process and its effects on C cycling. The trace gas emissions resulting from fuelwood burning are discussed in de Vries *et al.* (1994).

3.9.   N$_2$O FROM NATURAL SOILS

A simple model for assessing the potential for N$_2$O production in natural soils was adapted from Bouwman *et al.* (1993). This model was based on concepts developed by Matson *et al.* (1989), who stressed the need for stratification of ecosystems to describe the variability of soil and environmental conditions responsible for N_2O fluxes. Such approaches lead to considerable improvements of regional estimates of N_2O emissions (Matson and Vitousek, 1990). The basis for this model is the observed strong relation between N_2O fluxes and the amount of nitrogen cycling through the soil-plant-microbial biomass system, as described in the conceptual 'hole in the pipe' model and relations between soil moisture and N_2O flux (Davidson, 1991). Five regulators of N_2O production were used: (1) input of organic matter (CARBON); (2) effect of soil temperature on decomposition, mineralization and nitrification (SOD) (3) soil fertility (FERT); (4) effects of soil moisture status on decomposition, mineralization and nitrification (H2O); and (5) effect of soil oxygen status on denitrification (O2). The monthly N2O potential was then calculated as:

$$N_2O\text{-}\ potential = [CARBON * SOD * FERT * H_2O * O_2]^{0.2} \qquad (1)$$

The simulated monthly N_2O potential was compared with reported measurements from different tropical and temperate ecosystems in Figure 1. About 65% of the variability of measured fluxes could be explained with this model. The regression equation was used to calculate emissions on a 0.5°x0.5° grid yielding a global annual N_2O emission from soils under natural vegetation of 6.7 Tg N_2O-N a^{-1} for 1990. The slight difference between this estimate and Bouwman *et al.* (1993) is caused by the different scalar for input of organic matter (NPP estimates *vs* the normalized difference vegetation index in Bouwman *et al.* 1993), different climate data bases used and minor differences in the soil water budget model. However, the resulting emissions for broad ecosystems are consistent with other estimates (Bouwman, A.F.; Van Der Hoek, K.W. and Olivier, J.G.J.: 'Uncertainty in the global source distribution of N_2O. Hereafter referred to as Bouwman *et al.*, in prep.).

3.10. N₂O FROM ARABLE LANDS

The fertilizer induced emission is calculated as 1.25% of the mineral fertilizer N (see 'best' estimate in Bouwman *et al.*, in prep.). This estimate yields a global fertilizer induced emission for 1990 of 0.9 Tg N_2O-N a^{-1} for 1990. The global fertilizer induced emission is lower than the most recent IPCC scenario estimate of 2 Tg a^{-1} (Pepper *et al.*, 1992), that was based on N_2O losses of 2.6% of fertilizer applied. Its geographic distribution of emissions of this estimate seems more realistic than other estimates, e.g. Eichner (1990) (Bouwman *et al.*, in prep.). The regional nitrogen fertilizer use (Table 3) for the period 1970-1990 is from FAO (1991), while future projections are from Pepper *et al.* (1992).

3.11. ENHANCED SOIL N₂O EMISSIONS FOLLOWING DEFORESTATION

Human activity in tropical forests may change the emission rates of biogenic trace gases (Keller *et al.*, 1986). Burning may make nitrogen and other nutrients available in the residual material, mainly in the ash and undecomposed material. During the rains in

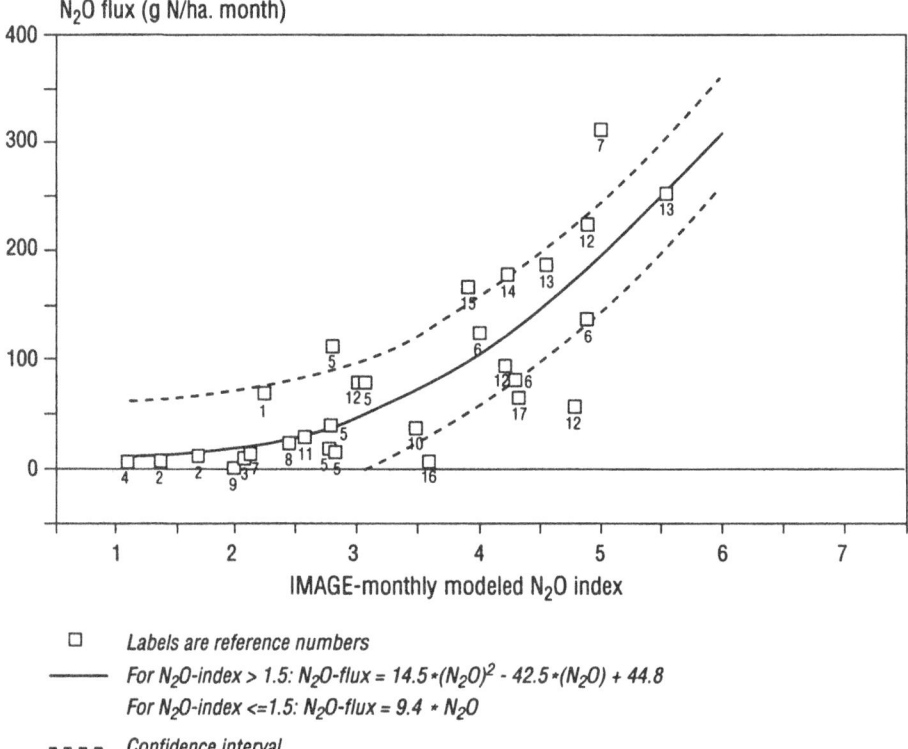

Figure 1: Relation between modeled N2O index and monthly flux measured for different sites, regression line and upper and lower confidence levels. The markers indicate the simulation results obtained with data derived directly from the databases for locations in the following ecosystems:
1 – Semi-arid grassland, Colorado, USA (Mosier *et al.*, 1981); 2 – Semi-arid grassland, Colorado, USA (Parton *et al.*, 1988); 3 – Temperate forest, Germany (Seiler and Conrad, 1981); 4 – Temperate chaparral, Wisconsin, USA (Cates and Keeney, 1987); 5 – Tall grass prairie; Grass; Temperate forest, Wisconsin, USA (Goodroad and Keeney, 1984); 6 – Temperate forest, New York, USA (Duxbury *et al.*, 1982); 7 – Tropical rain forest, Manaus, Brazil (Keller *et al.*, 1983); 8 – Temperate forest, various sites near Mainz, Germany (Schmidt *et al.*, 1988); 9 – Temperate coniferous forest, Massachusetts, USA (Bowden *et al.*, 1990); 10 – Marsh, Louisiana, USA (Smith *et al.*, 1983); 11 – Tropical savanna, Venezuela); Hao *et al.*, 1988); 12 – Tropical rainforest, Manaus, Brazil; Tena, Ecuador; Puerto Rico, USA (Keller *et al.*, 1986); 13 – Tropical rain forest, various sites, Costa Rica (Matson and Vitousek, 1987); 14 – Tropical rain forest, Manaus, Brzail (Luizao *et al.*, 1989); 15 – Tropical rain forest, Brazil (Livingston *et al.*, 1988); 16 – Tropical rainforest (2 sites) and varzea, Manaus, Brazil (Matson *et al.*, 1990); 17 – Tropical dry forest, Chamela, Mexico (Vitousek *et al.*, 1990); 18 – Tropical rainforest, Manaus, Brazil (Keller *et al.*, 1988).

savanna areas following the fires this may lead to prolonged enhancement of N_2O emissions (Anderson and Poth, 1989). Because of the scarcity of field observations, the post-burn effects of savanna burning on N_2O emission have not been included in this version of IMAGE 2.0.

Deforestation may lead to accelerated decomposition of litter, root material and loss of

TABLE 3

Future use and growth of N-fertilizers for the periods 1990-2025 and 2025-2100 (from Pepper et al., 1992). Gg = 10^9 g

Region	Use 1990	Growth 1990-2025	Use 2025	Growth 2025-2100	Use 2100
	Gg a^{-1}	(% a^{-1})	Gg a^{-1}	(% a^{-1})	Gg a^{-1}
Canada	1158	1.1	1692	-0.2	1425
USA	10141	1.1	14821	-0.2	12481
Latin America	3543	3.2	10629	0.6	16223
Africa	2063	3.2	6189	0.6	9446
OECD Europe	8585	1.1	12547	-0.2	10566
Eastern Europe	3205	1.3	5008	0.1	5408
CIS	8738	1.3	13653	0.1	14745
Middle East	2509	3.2	7527	0.6	11489
India + S. Asia	10299	3.2	30897	0.6	47159
China + CP countries	20536	1.0	29337	0.1	30804
East Asia	3375	3.2	10125	0.6	15454
Oceania	499	1.1	729	-0.2	614
Japan	612	1.1	894	-0.2	753
Total	75263	1.9	144048	0.3	176567

part of the soil organic matter in the first years after disturbance causing a pulse of N_2O emissions (Keller et al., 1993). For effects of deforestation on soil N_2O fluxes we included the calculations proposed by Bouwman et al. (in prep.) based on N_2O pulses during 10 years after clearing (based on Keller et al., 1993). In the first year after disturbance, the fluxes amount to 5 times the flux of the original ecosystem. They decrease linearly to the level of the new ecosystem in the 10th year, and this is usually lower than the flux from the original forest. In our calculations we assumed that the pulses of N_2O only occur in tropical rain and seasonal forests, with a rate of clearing of about 50,000 km^2 a^{-1}. We do not know if the enhanced N_2O emission will also occur after clearing of other vegetation types, such as xerophytic woods and savannas. The global emission for 1990 is about 0.2 Tg N_2O-N a^{-1}. This is about half of the estimated 0.4 Tg N_2O-N a^{-1} estimated by Bouwman et al. (in prep.) who assumed a deforestation rate of about 150,000 km^2 a^{-1}.

3.12. N_2O FROM ANIMAL WASTE

Khalil and Rasmussen (1992) recently referred to animal wastes as a source of N_2O. Their estimate of the global emission from this source is 0.2 - 0.6 Tg N_2O-N a^{-1}, based on N_2O from confined animals. We use the mean global nitrogen excretion per animal as proposed by Bouwman (1993). The N_2O losses for animal excreta are taken tentatively 0.5% (as in Bouwman, 1993, equal to the mean loss for all mineral fertilizers reported by Eichner, 1990). The estimated global amount of N excreted by animals is 120 Tg N a^{-1} for 1990, yielding an emission of 0.6 Tg N_2O-N a^{-1}. As the N in excreta is higher than the global use of N-fertilizers (76 Tg N a^{-1}; FAO, 1991), this may be an underestimate. The geographic basis and calculation of future numbers of animals is identical to the distribution of animals (see above).

3.13. N_2O FROM AQUATIC SOURCES

The estimated global oceanic N_2O emission contribution is 2.0 (1.4 - 2.6) Tg N a^{-1} (Watson *et al.*, 1992). Coastal waters and freshwater systems may also be N_2O sources, although Seitzinger (1990) suggested that they are not significant at the global scale.

N_2O may be formed below the rooting zone and N_2O in soil solution may leach to the groundwater and it may evolve to the atmosphere by degassing from groundwater entering surface streams. Ronen *et al.* (1988) estimated that N-contaminated aquifers may contribute 0.5 - 1.1 Tg N_2O-N a^{-1}, assuming that 1% of the world's aquifers is contaminated.

A global estimate for the aquatic sources of 2.2 Tg a^{-1} is used (Pepper *et al.*, 1992), which includes an oceanic contribution and emissions from aquifers. Data are still lacking to produce regional estimates for these sources.

4. Discussion of results

In the following sections we discuss the results of the application of the Land use Emission Model to a "Conventional Wisdom Scenario". The basic driving forces of this scenario were taken from on the IPCC IS92a scenario (Pepper *et al.*, 1992). It includes World Bank projections of population, increasing to 11.3 billion by 2100, a 2.9% a^{-1} economic growth between 1990 and 2025, and 2.3% a^{-1} for the period 2025-2100. Further scenario assumptions relate to energy consumption, industrial emissions and gradual phase out of emissions of CFCs.

We will discuss regional changes in land use first. These results are also discussed in Zuidema *et al.* (1994), but for a good understanding of the emission calculations a short summary of their results is presented in section 4.1. In section 4.2 we will discuss global model results of emissions and make a comparison with reference estimates for IPCC scenario IS92a described in Pepper *et al.* (1992). We will present regional developments in wetland rice cultivation and the population of dairy cattle and associated CH_4 emissions in section 4.3, and for the African continent we will discuss the simulation results for land cover, biomass burning emissions of CH_4 and soil emissions of N_2O in section 4.4.

4.1   REGIONAL LAND USE CHANGES

The changes in global land cover are the result of the Agricultural Demand Model (ADM) and the Land Cover Model (LCM) described in Zuidema *et al.* (1994). In ADM the demand for crop, animal and forest products are estimated. The LCM simulates grid-scale changes in global land cover by reconciling regional demands for land (based on results of the ADM) with the local potential of the land. For this a number of simple land use rules were formulated. The final output is the change in croplands, rangelands and forests. We will briefly discuss regional land use for 1990, 2025 and 2050 (Table 4).

The change in agricultural land is the result of changing demand and increasing productivity. Obviously, the balance of demand and productivity leads to both increases (Africa and all Asian countries except CIS) and decreases (North America, Latin America, Europe, CIS and Oceania) of agricultural land (including arable land and grasslands). In all regions, except the Middle East and Japan, there is an expansion of the areas of grazing land (types 6 and 7). In general, the redundant agricultural land in temperate regions is converted to conifer or mixed forests or temperate deciduous forest. In tropical regions with expanding agricultural areas and grazing land there is considerable deforestation. In a number of regions deforestation is very rapid. This leads to a near complete loss of tropical forests (types 14-16, Table 4) in Africa (in the year 2050), the Middle East (in about 2030), India and other South Asian countries (in about 2005) and China and other centrally planned countries (in the year 2005). Table 4 shows an increase in global agricultural area and grazing lands, while about 2/3 of the dry savannas are lost, and about 1/3 of tropical seasonal and rain forests are cleared. There are only minor changes in the areas of temperate and cool conifer and mixed forests.

Global losses of tropical dry savannas have important repercussions for the simulated savanna burning activities. In Africa the near complete loss of tropical forest has major consequences for emissions from biomass burning (deforestation mainly).

4.2.   GLOBAL EMISSION RESULTS

The emission calculations resulting for the year 1990 for CH_4 and N_2O are compared with a number of reference estimates of corresponding sources for a number of selected sources (Table 5). From the reference reports either the methodology has been adopted, giving by definition the same result, or emission factors have been adapted to arrive at equal global totals, as in the case of rice cultivation. In a number of cases our results are slightly different from those of reference studies.

Major differences are found for all emissions that are related to deforestation. Zuidema et al. (1994) estimate a deforestation rate of 154,800 km^2 for the period 1970 - 1980, while for 1980 - 1990 the estimated annual rate is 141,000 km^2. Official inventories of deforestation by FAO show that rates may have been 113,000 km^2 and 154,000 km^2 a^{-1} for 1970-1980 and 1980-1990, respectively (Alcamo et al., 1994a). Reference estimates for burning during deforestation from Crutzen and Andreae (1990) are based on the FAO deforestation estimates for the early 1980's. Therefore, our estimates of sources associated with deforestation are somewhat higher than the reference estimates in Table 5.

The changes in land cover for Africa (Figure 3a and 3b for 1990 and 2050, respectively) show that arable land and pastures increase at the cost of the forest area. In 2050 expansion of the agricultural area sharply decreases and in 2060 it stops, since then most of the forest has been converted to other uses. This has important implications for those African and also global sources that are directly related to deforestation.

Global emissions from the sources of greenhouse gases described in the Land Use

TABLE 4

Simulated regional land use for 1990 and 2050 from the Land Cover Model developed by Zuidema et al. (1993). Areas in 1000 km².

Land cover type	Canada	USA	Latin America	Africa	OECD Europe	Eastern Europe	CIS	Middle East	India+ S.Asia	China+ CP countries	East Asia	Oceania	Japan	WORLD
Land cover distribution for 1990														
1 Agricultural land	1306	4144	3471	3247	1703	779	2939	707	2739	2468	1591	1398	168	26661
2-5,17 Various types [1]	4047	1774	2496	9954	2884	21	6614	2831	544	3745	384	1742	33	37073
6-8 Grass/shrub, woods/shrub [2]	39	1846	6271	7401	432	83	3777	2879	1199	3601	78	3884	17	31505
9 Taiga	3305	481	3	0	567	14	6845	10	8	231	0	7	0	11470
10 Cool conifer forest	1022	713	0	0	77	26	763	5	5	244	0	0	26	2881
11 Cool mixed forest	122	166	47	0	69	40	714	0	0	141	0	0	113	1411
12 Temp. deciduous forest	0	77	50	0	29	84	156	31	5	298	0	0	2	732
13 Br.leav./warm mixed forest	0	225	821	299	275	130	91	88	180	665	181	151	25	3131
14 Tropical dry/savanna	0	8	2259	6836	0	0	0	125	217	120	139	881	0	10585
15 Tropical seasonal forest	0	0	2667	1811	0	0	0	0	270	309	362	36	0	5456
16 Tropical rain forest	0	0	2651	408	0	0	0	0	0	0	832	15	0	3906
Land cover distribution for 2050														
1 Agricultural land	662	2688	2790	8060	1290	675	1418	2319	2829	2879	2066	1130	352	29156
2-5,17 Various [1]	4047	1774	2499	9951	2884	21	6614	2394	534	3737	384	1742	31	36615
6-8 Grass/shrub, woods/shrub [2]	129	2073	6717	11414	769	162	4285	1962	1801	5032	413	4126	2	38884
9 Taiga	3401	481	3	0	516	11	6848	0	0	55	0	7	0	11321
10 Cool conifer forest	1186	720	0	0	40	14	877	0	0	32	0	0	0	2868
11 Cool mixed forest	415	405	47	0	55	59	1492	0	0	0	0	0	0	2473
12 Temp. deciduous forest	0	599	50	0	264	145	280	0	5	54	0	5	0	1402
13 Br.leav./warm mixed forest	0	688	987	0	218	91	86	0	0	29	55	161	0	2316
14 Tropical dry/savanna	0	8	2289	383	0	0	0	0	0	0	119	892	0	3691
15 Tropical seasonal forest	0	0	2717	136	0	0	0	0	0	3	215	36	0	3107
16 Tropical rain forest	0	0	2635	12	0	0	0	0	0	0	315	15	0	2978

[1] Including ice, cool (semi-) desert, hot desert, tundra and wetlands; [2] Including cool grass/shrub, warm grass/shrub and xerophytic woods/shrub

TABLE 5

Estimated global emissions for 1970, 1990 and 2050, and reference estimates for 1990 and 2050. Reference estimates for 1990 are from Watson et al. (1992) unless indicated otherwise. Reference estimates for 2050 are from Pepper et al. (1992), scenario IS92a. Emissions are expressed as Tg CH_4 a^{-1} and Tg N_2O-N a^{-1}. For each source the section is indicated where calculations are described. The range of uncertainty of the various estimates is not indicated here, but can be found in Table 1.

Source (section)	Emission (Tg a^{-1})				
	1970 IMAGE	1990 IMAGE	1990 REF.	2050 IMAGE	2050 REF.
a. Sources of CH_4					
Wetland rice fields (3.1)	53	59	60	52	87
Natural wetlands (3.2)	111	111	111 [a]	111	115
Landfills (3.3)	24	36	30	81	93
Domestic sewage treatment (3.3)	17	25	25	47	47
Animals (3.4)	66	79	80	161	173
Animal waste (3.5)	12	14	14 [b]	28	54
Termites (3.6)	20	20	20	20	20
Aquatic sources (3.7)	15	15	15	15	15
Methane hydrates (3.7)	5	5	5	5	5
Biomass burning (3.8)					34
- Deforestation [c,d]	16	14	6 [e]	12	14
- Savanna burning	17	16	13 [e]	6	-
- Agricultural waste burning	7	8	9 [e]	9	-
b. Sources of N_2O					
Soils, incl. background emission from arable lands (3.9)	6.6	6.7	6.8 [f]	8.0	6.1 [g]
Fertilizer induced (3.10)	0.4	0.9	2.2 [g,h]	2.0	4.2 [g,h]
Deforestation (3.11)[d]	0.2	0.2	0.4 [i]	0.1	1.1
Animal waste (3.12)	0.5	0.6	0.4 [j]	1.5	-
Aquatic sources (3.13)	2.2	2.2	2.0-2.2 [g]	2.2	2.2

[a] Matthews and Fung (1987); [b] Gibbs and Leng (1993); [c] estimate for deforestation exludes shifting cultivation, which contributes about 10 Tg CH_4 a^{-1} (Crutzen and Andreae, 1990); [d] For 1970 no values are available, since this calculation involves the history of deforestation over a period of some years. For deforestation effects on soil N_2O emission as well as for direct biomass burning the year 1975 is presented; [e] Crutzen and Andreae (1990); [f] Bouwman et al. (1993); [g] Pepper et al. (1992); [h] Estimate of total N_2O loss, including fertilizer induced and background losses from arable land. From this number the amount of circa 1 Tg N_2O-N a^{-1} needs to be subtracted to arrive at the 1990 fertilizer induced loss; [i] Bouwman et al. (in prep.); [j] Khalil and Rasmussen (1992).

Emissions Model show different patterns for CO_2, NO_x, CO and VOC on the one hand and for CH_4 and N_2O on the other hand (Figure 2). The first group shows a near constant global emission up till about the year 2020, and a decreasing trend towards a lower constant level after 2060. Emissions of N_2O and CH_4 increase from 1970 onwards up till 2050 (CH_4) and 2080 (N_2O), when a near constant level is reached. The first group of gases mainly stems from biomass burning. Both deforestation and savanna burning, as the dominant contributors in the first decades of the simulation, show a decreasing trend at the

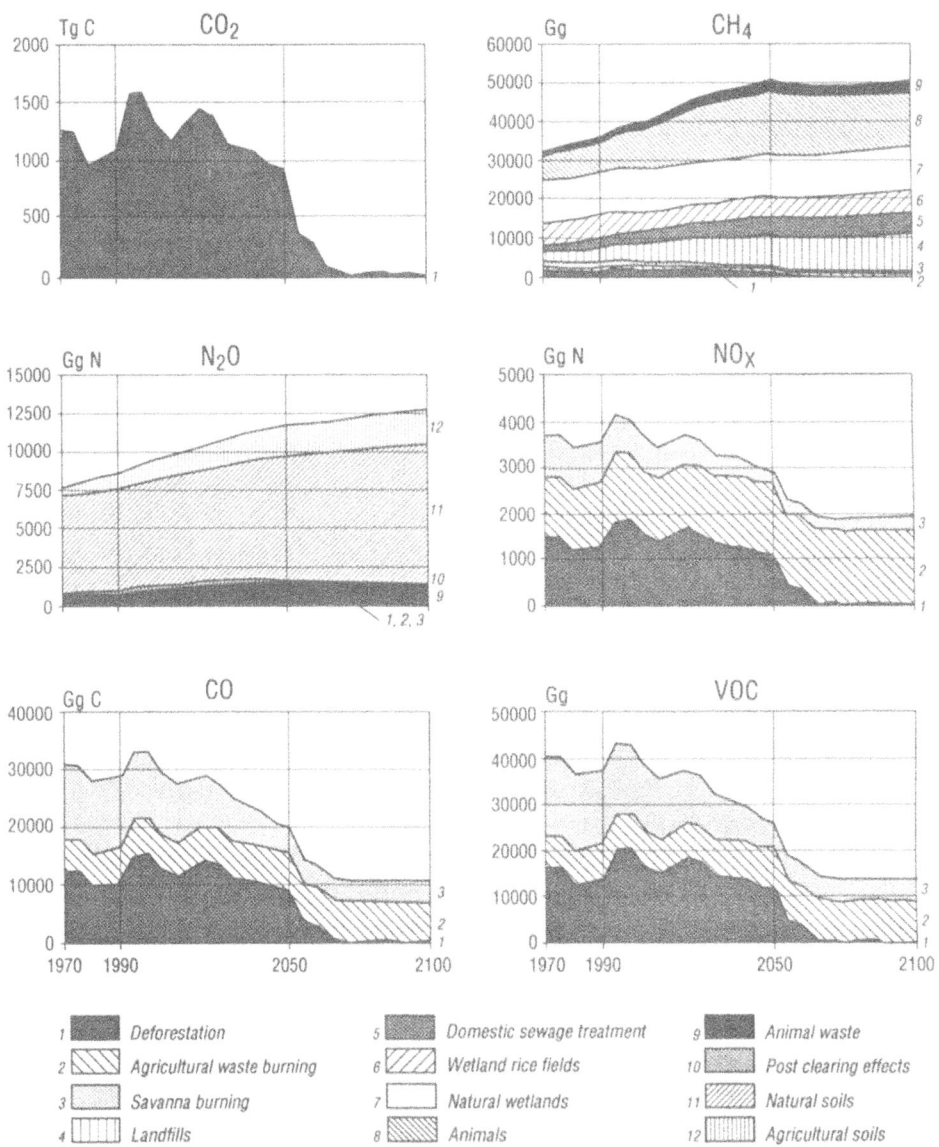

Figure 2: Global land-use related emissions of CO_2, CH_4, N_2O, NO_x, CO and VOC for the period 1970 - 2100, indicated for the sources described in the IMAGE Land Use Emissions Model.

global scale (Figure 2), determining the overall trend for this group of gases. The reason for the simulated decrease in deforestation, that is particularly evident from about the year 2000 onwards, is discussed in Zuidema *et al.* (1994).

CH_4 has a great variety of sources (Table 1), and its trend is determined by emissions

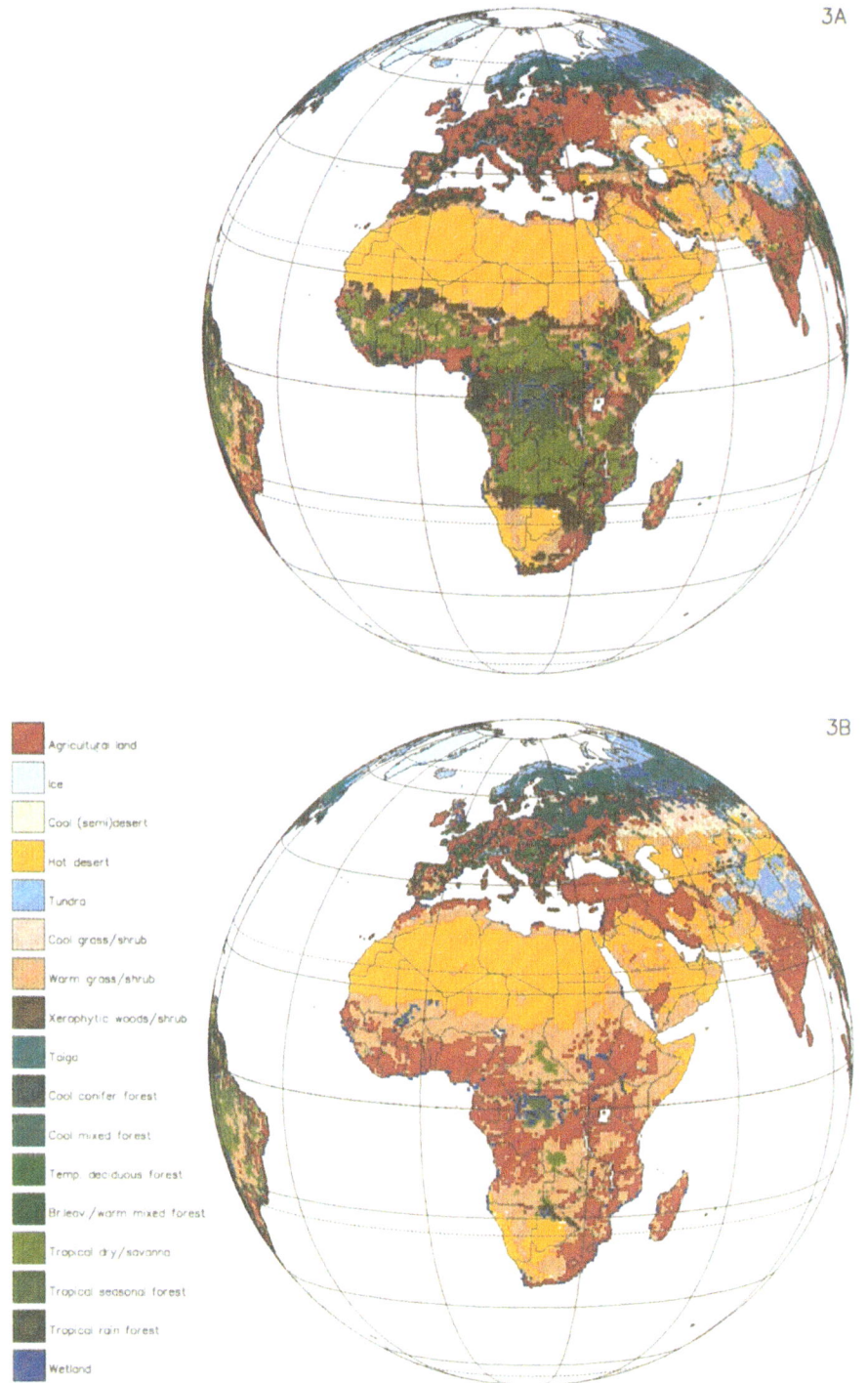

3A

3B

Agricultural land

Ice

Cool (semi)desert

Hot desert

Tundra

Cool grass/shrub

Warm grass/shrub

Xerophytic woods/shrub

Taiga

Cool conifer forest

Cool mixed forest

Temp. deciduous forest

Brleav./warm mixed forest

Tropical dry/savanna

Tropical seasonal forest

Tropical rain forest

Wetland

Figure 3a-3b: Selected simulation results for the African continent.- Land cover for Africa for (a) 1990 and (b) 2050 .

from animal population and by landfilling (Figure 2).

For N_2O all sources, including the natural soils, show an increasing trend. Since the area of tropical forests where highest emissions occur, decreases world-wide, the major cause of increase is the predicted climate change. As shown in section 3.9, the dynamic calculation of current emissions was done on the basis of a limited number of flux measurements. Many studies have concentrated on the effect of a number of environmental factors such as temperature and soil moisture on N_2O fluxes, but it is very difficult to assess their combined effect in nature. Therefore, validation of the results of the dynamic N_2O model is a difficult task. The projected increase between 1970 and 2100 is about 2.6 Tg N_2O-N a^{-1} or 1.1 Tg N a^{-1} per degree of global mean temperature rise. For comparison, Khalil and Rasmussen (1992) presented an estimate of 0.5 Tg N_2O-N a^{-1} per degree of global mean temperature increase from ice-core data.

The regional results for N_2O emissions from soils and fertilizers (Table 8) show similar patterns of near constant soil emissions in tropical regions (Latin America, Africa, India and other South Asian countries and East Asia). Post clearing effects are most important in regions with high deforestation rates, i.e. in Latin America, Africa, India and other South Asian countries and East Asia. The post-clearing N_2O increase caused by deforestation in Watson et al. (1992) is based on enhanced emissions from the total of the 2×10^6 km² of forests that were converted to pasture in the period 1940-1980. However, Bouwman et al. (in prep.) showed that this source is probably much less important, because the N_2O pulse probably lasts only about 10 years (Keller et al., 1993). Bouwman et al. (in prep.) assumed a total deforestation rate of about 150,000 km² a^{-1}. In our calculations we assumed that the pulses of N_2O only occur in tropical rain and seasonal forests, with a rate of clearing of about 50,000 km², because we do not know if the enhanced N_2O emission will also occur after clearing of other vegetation types, such as xerophytic woods and savannas. The global result of our calculations is a flux of about half of the estimate by Bouwman et al. (in prep.).

Fertilizer induced emissions form a significant contribution in 1970-1990 in Canada, USA, OECD Europe, CIS, China and other Centrally Planned Asian countries (Table 8). Fertilizer induced emissions show a major increase in the developing countries (Latin America, Africa, India and other South Asian countries and East Asia) between 1990 and 2100, although in some regions the natural soil source remains dominant. On the global scale the fertilizer contribution increases from 0.9 to 2.2 Tg N a^{-1} between 1990 and 2100 (Table 5 and 8).

4.3. REGIONAL RESULTS

Regional results for wetland rice (Table 6) clearly show that the regions "India and other South Asian countries" and "China and other Centrally Planned Asian countries" are the major rice producing regions. Obviously, the increase in production per unit harvested area - amounting to about 100% between 1990 and 2050 in India and 50% in China - more than compensates for the increased demand triggered by population growth (see Zuidema et al., 1994). Between 1990 and 2025 this leads to considerable decreases in the

TABLE 6

Calculated areas of wetland rice (in 1000 km^2) and associated CH$_4$ emission (in Gg CH$_4$ a^{-1}) for 1970, 1990 and 2050

Region	1970		1990		2050	
	Area	CH$_4$ emission	Area	CH$_4$ emission	Area	CH$_4$ emission
Canada	0	0	0	0	603	27
USA	7288	332	11645	530	16047	731
Latin America	56972	2592	53600	2439	54483	2479
Africa	9027	411	30507	1388	181113	8241
OECD Europe	3052	139	4056	185	4543	207
Eastern Europe	1669	76	1323	60	1781	81
CIS	3455	157	5050	230	4213	192
Middle East	7235	329	8942	407	78673	3580
India + S. Asia	477446	21724	543641	24736	382920	17423
China + CP countries	393192	17890	404450	18402	268200	12203
East Asia	167759	7633	218626	9947	131507	5984
Oceania	514	23	1527	69	2298	105
Japan	29816	1357	20129	916	7632	347
Total	1157426	52663	1303494	59309	1134039	51599

TABLE 7

Calculated numbers of dairy cattle (in 1000 heads) and associated emission of CH$_4$ (in Gg CH$_4$ a^{-1}) for 1970, 1990 and 2050.

Region	1970		1990		2050	
	Dairy cattle	CH$_4$ emission	Dairy cattle	CH$_4$ emission	Dairy cattle	CH$_4$ emission
Canada	2554	138	1379	75	1065	58
USA	12001	650	10127	549	9810	532
Latin America	23896	1197	37689	1888	42950	2152
Africa	21005	687	32200	1053	161026	5266
OECD Europe	36283	2329	31484	2021	36486	2471
Eastern Europe	13144	833	12495	792	13005	825
CIS	39400	2498	41700	2644	20798	1319
Middle East	7759	254	8756	286	21240	695
India + S. Asia	25074	710	42084	1191	269478	7626
China + CP countries	1249	56	3648	162	24314	1082
East Asia	122	3	704	20	11237	318
Oceania	5129	278	4009	217	3106	168
Japan	1198	65	1336	72	2797	152
Total	188814	9699	227611	10971	619312	22661

TABLE 8

Calculated N_2O emission from 'natural' soils (SOIL), fertilizer induced emission (FERT) and deforestation on N_2O (DEFOR) for 1970, 1990 and 2050. Fertilizer emissions based on the scenario of fertilizer use presented in Table III. N_2O emission in Gg N a^{-1}.

Region	1970			1990			2050		
	SOIL	FERT	DEFOR	SOIL	FERT	DEFOR	SOIL	FERT	DEFOR
Canada	108	4	0	111	14	0	156	19	0
USA	316	92	0	292	127	0	405	171	0
Latin America	2334	17	40	2389	44	49	2744	170	0
Africa	1670	10	18	1715	26	24	1972	99	113
OECD Europe	100	87	0	107	107	0	144	144	0
Eastern Europe	32	34	0	35	40	0	54	70	0
CIS	227	58	0	259	109	0	397	191	0
Middle East	69	5	0	73	31	0	82	120	0
India+ S. Asian co.	431	24	28	433	129	20	552	495	0
China + CP Asia co.	438	47	6	463	257	6	574	385	0
East Asia	624	10	77	607	42	74	692	162	2
Pacifica	221	2	2	225	6	0	255	8	0
Japan	22	9	0	24	8	0	26	10	0
Total	6593	397	171	6735	941	173	8053	2046	115

[*] For 1970 no values available, since this calculation involves the history of deforestation over a period of some years. Instead of 1970, for this source the year 1975 is presented

wetland rice area. On the basis of the global mean emission rate used here the projected emissions will decrease from 60 Tg CH_4 a^{-1} in 1990 to 47 Tg CH_4 a^{-1} in 2025. From 2025 onwards the increase in productivity is less than the growth in demand for rice, leading to an increase of emissions from wetland rice fields to about 52 Tg CH_4 a^{-1} in 2050 and 59 Tg CH_4 a^{-1} in 2100. This is different from Pepper *et al.* (1992), who used a global productivity growth in rice cultivation of some 50% between 1990 and 2050; for the period 2050 - 2100 they assumed that rice yields will improve by only 10%. The resulting emissions estimates by Pepper *et al.* (1992) are presented in Table 5.

The projections of agricultural demand (Zuidema *et al.*, 1994) lead to an increase in the number of dairy cattle by a factor 2.7 between now and 2050 (Table 7). After 2050 the global number of dairy cattle starts decreasing due to levelling off of the demand and due to the projected rise in productivity per head of cattle. This leads to an increase of emissions from dairy cattle from 11 Tg CH_4 a^{-1} in 1990 to 23 Tg CH_4 a^{-1} in 2050 and 19 Tg CH_4 a^{-1} in 2100. The increase of the emission between 1990 and 2050 is less than the increase in the animal population, because the emission per head is constant while productivity increases (Table 7). The estimated total CH_4 emission from all animals increases from 79 Tg CH_4 a^{-1} in 1990 to 161 Tg CH_4 a^{-1} in 2050 and then decreases to 141 Tg CH_4 a^{-1} in 2100. Global estimates for CH_4 from animal wastes for the same period (Table 5 and Figure 2) show a similar pattern. It should be noted that the emissions from animals and their waste are based on constant emission factors. Effects of changing diets and changing waste management have not been included.

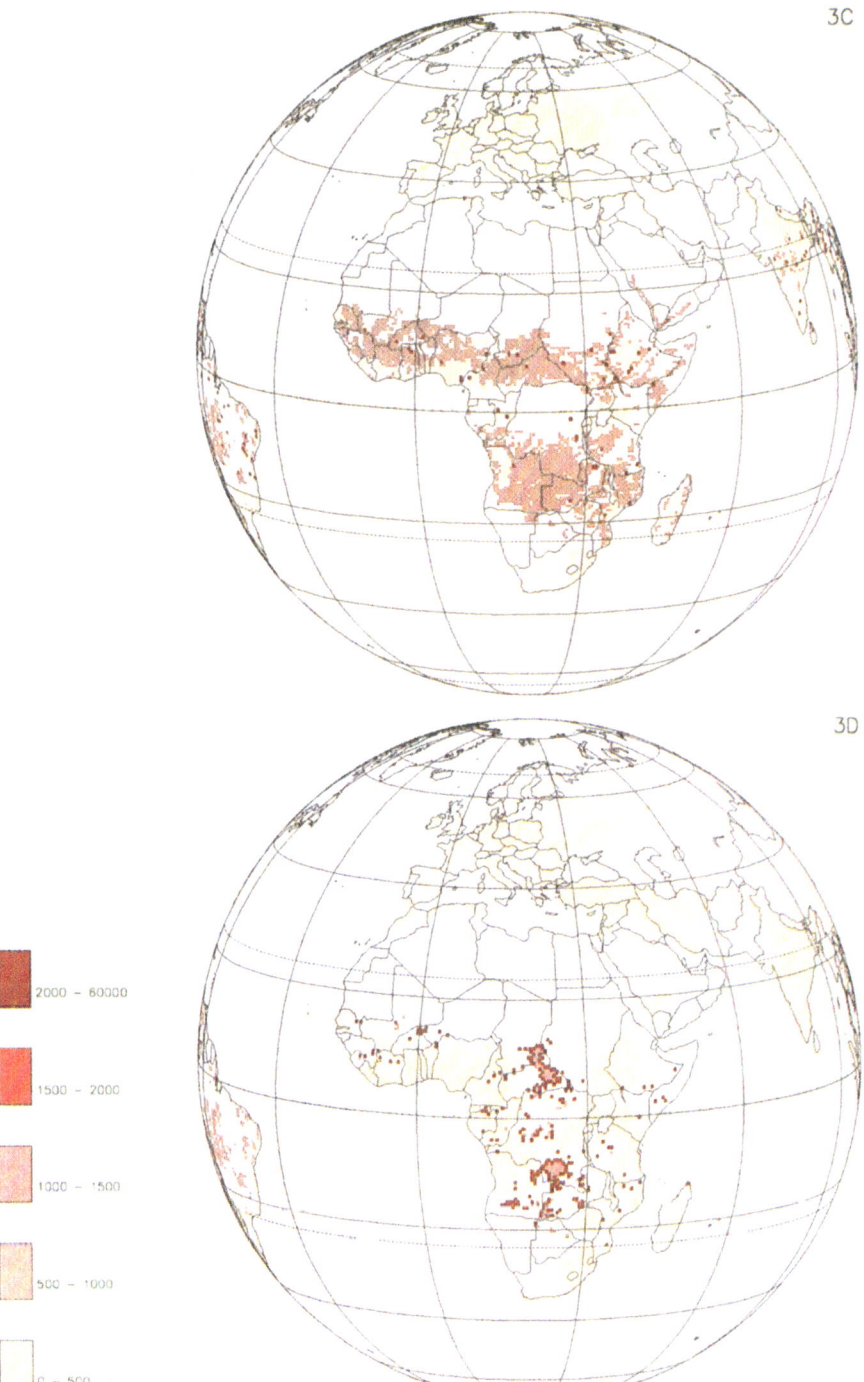

3C

3D

2000 – 60000

1500 – 2000

1000 – 1500

500 – 1000

0 – 500

Figure 3c-3d: Selected simulation results for the African continent. - Emission of CH_4 (kg CH_4 km^{-2} a^{-1}) from biomass burning (including deforestation, savanna burning and agricultural waste burning) for Africa for (c) 1990 and (d) 2050 . The estimated CH_4 emission from this source is 14 Tg CH_4 a^{-1} for 1990 and 15 Tg CH_4 a^{-1} in 2050. In 1990 the major contributor to this emission is savanna burning (70%), while in 2050 about 80% is from deforestation.

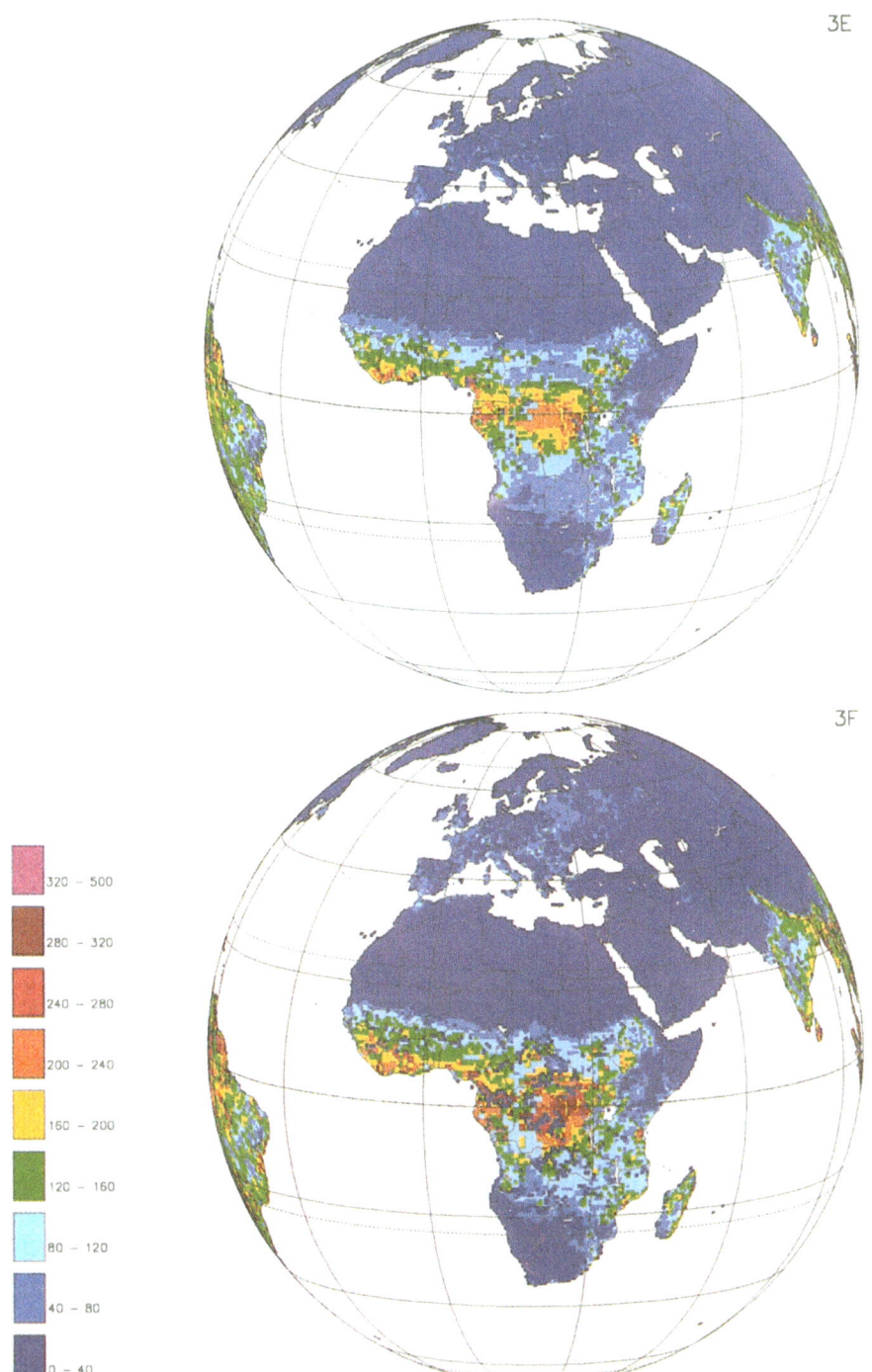

320 — 500

280 — 320

240 — 280

200 — 240

160 — 200

120 — 160

80 — 120

40 — 80

0 — 40

Figure 3e-3f: Selected simulation results for the African continent. - Emission of N_2O (kg N_2O-N km^{-2} a^{-1}) from natural soils for (e) 1990 and (f) 2050. The estimated emission from the African continent amounts to 1.7 Tg N_2O-N a^{-1} in 1990 and 2.0 Tg N_2O-N a^{-1} in 2050.

4.4. Results for Africa

African emissions of CH_4 from biomass burning (deforestation, savanna burning and agricultural waste burning) (Figure 3c, 1990; Figure 3d, 2050) illustrate a decrease in source strength from 15 Tg CH_4 a^{-1} in 1975 to 14 Tg CH_4 a^{-1} in 1990, 15 Tg CH_4 a^{-1} in 2050 and 3 Tg CH_4 a^{-1} in 2100. This is caused by rapid forest conversion, whereby after the year 2060 the emissions associated with deforestation decrease to zero because then all African forest areas have been converted to other uses (Figure 3b). The major contributor to biomass burning in the period 1970 - 2010 is savanna burning, while in the period 2010 - 2060 deforestation is most important (Figure 3c and d). At the global scale savanna burning and deforestation are of equal importance up till 2060 (Figure 2), when deforestation shows a fast decrease caused mainly by the diminishing forest area in Africa.

The model results for Africa for the natural or 'background' N_2O emission are presented in Figure 3e and 3f for 1990 and 2050, respectively. This example combines effects of the significant changes in land use caused by increasing population pressure with climatic impacts on N_2O emission. The estimated natural emission for Africa increases from 1670 Gg N_2O-N a^{-1} in 1970 to 1972 Gg a^{-1} in 2050. This suggests that - although the area of tropical forests decreases - changing climate patterns lead to a very slow increase in emission. The scenario for fertilizer use implies an increase from 10 Gg N_2O-N a^{-1} in 1970 to 26 Gg N a^{-1} in 1990, 99 Gg N a^{-1} in 2050 and 118 Gg N a^{-1} in 2100 (Table 8). The calculations suggest that the African deforestation enhancement of N_2O emissions amounts to 18 Gg N_2O-N a^{-1} in 1975, 24 Gg N_2O-N a^{-1} in 1990 and 113 Gg N a^{-1} in 2050 (Table 8). After the year 2060 deforestation and expansion of the agricultural area stops, as noted above.

5. Conclusions

The Land Use Emissions Model estimates the emissions on the basis of trends in land cover. It is based on current estimates for the various sources and species listed in Table 1 and 5. We have shown that - given the limitations of the available data bases and other statistical data - it is possible to present grid-based estimates for a number of different sources, including CH_4 emission from rice, wetlands, emissions of CH_4, CO, NO_x, N_2O and VOC from deforestation, savanna burning and agricultural waste burning and N_2O from natural soils, N-fertilizer and deforestation. For some of the sources (landfills, domestic sewage treatment, termites, methane hydrates and aquatic sources) geographically explicit calculations are not yet possible because of data limitations.

The model results show that calculations of future emissions are extremely sensitive to assumptions of demographic and economic development and the land's productivity (crop yields, animal productivity). Slight differences in the balance between demand and production of a product cause important differences in land use (Zuidema et al., 1994) and associated emissions. The major part of the disagreement between our results and those of e.g. Pepper et al. (1992) can be explained from differences in assumptions related to the productivity.

Improvements in emission calculations can be achieved in a number of different ways:
More reliable land cover distributions. The currently available global data bases of
land cover, e.g. Olson *et al.* (1983), have many disadvantages in terms of the definition
of land cover types. In particular Olson's description of agricultural systems and
grazing lands in these data bases is weak.

At present it is not possible to combine the above data bases with e.g. the grid-
based classification of rice-ecologies made by Matthews (1991) due to regional
discrepancies with Olson *et al.* (1983). Another weak point in the Olson *et al.* (1983)
data base are the natural wetlands. The areas with high fractional inundation in
Matthews and Fung (1987) corroborated with the Olson *et al.* (1983) wetlands, but
certainly a major part of the wetlands is not well represented.

In IMAGE 2 wetlands are constant, i.e. they can not be converted to other uses.
However, in practice wetland areas may strongly influence land conversions. In some
places wetlands may be protected areas, while in other areas they may be preferentially
converted to wetland rice. Under certain conditions it may not be economically
justifiable to reclaim wetlands for agriculture because of the high cost involved.

Implementation of grid-based calculations. Emission calculations for a number
sources are not geographically explicit. These sources include landfills, termites,
methane hydrates, and the aquatic sources of N_2O. Fung *et al.* (1991) attempted to
produce grid-based estimates for landfill emissions. It was decided to adopt country
estimates without being geographically explicit, because there is still considerable
debate on the order of magnitude of the source strength.

The uncertainty associated with CH_4 emissions from termites and methane
hydrates may also be reduced in the near future. Also, a better description of the
processes in oceans, continental shelves, coastal waters and fresh water systems is
required for a better understanding of the N_2O budget. Although the data base of
measurements is expanding, we are still far from making reliable and geographically
explicit estimates for the aquatic sources.

Inclusion of all biomass burning processes. The disadvantage of the inventory of
biomass burning produced by Hao *et al.* (1990) is that not all burning processes are
included and that it includes tropical biomass burning only. A more complete
inventory including estimates of deforestation, shifting cultivation, savanna burning,
agricultural waste and firewood burning, and wildfires, is still in preparation (NASA,
see Pacyna, 1992) and will form an important improvement. Using the NPP to
distribute emissions associated with savanna burning, deforestation and agricultural
waste burning is a further potential improvement.

Implementation of dynamic models for trace gas fluxes. Dynamic models are required
to predict future emissions under changing climate and land use. One example of such
a dynamic model, the N_2O from natural soils, is described above. Other sources for
which similar approaches may be implemented include the natural wetlands. A further
potential improvement is the WISE data base (Batjes and Bridges, 1992) that will
combine soil and agronomic factors in a model to estimate fluxes needed to produce a

more reliable inventory of CH_4 fluxes from wetland rice fields.

· *Use of demographic and socio-economic driving forces.* In many cases the driving forces of land use processes are poorly represented. Obvious examples of improvements include: (i) the use of population densities to allocate sources such as landfills; (ii) the use of relationships between GNP and waste produced per capita to estimate future landfill emissions; (iii) in future versions other ways to estimate CH_4 emissions from animal manure should be considered, e.g. coupling economic development to the manure usage systems in livestock.

6. Acknowledgements

This investigation was supported by the Dutch National Research Programme on Global Air Pollution and Climate Change (NOP contracts Nº 851037, 851042 and 851044), and the Netherlands Ministry of Housing, Physical Planning and Environment (MAP projects Nº 482505, 482507 and 482508). Two anonymous reviewers provided extensive comments that led to improvement of the paper. We are grateful to Joe Alcamo for fruitful discussions on the implementation of the Land Use Emissions Model and for his comments on earlier versions of this text.

References

Alcamo, J., Kreileman, G.J.J., Krol, M., Zuidema, G.: 1994a, Modeling the global society-biosphere-climate system, Part I: Model description and testing, *Water, Air and Soil Pollution*, **76** (this volume).

Alcamo, J., Van Den Born, G.J., Bouwman, A.F., De Haan, B., Klein Goldewijk, K., Klepper, O., Leemans, R., Olivier ,J.G.J, Toet, A.M.C., de Vries, H.J.M., Van de Woerd, H.:1994b, Modeling the global society-biosphere-climate system, Part II: Computed scenarios, *Water, Air and Soil Pollution*, **76** (this volume).

Anderson, I.C. and Poth, M.A.: 1989, Semiannual losses of nitrogen as NO and N2O from unburned and burned chaparral, *Global Biogeochemical Cycles*, **3**: 121-135.

Aselmann, I. and Crutzen, P.J.: 1989, Global distribution of natural freshwater wetlands and rice paddies: their net primary productivity, seasonality and possible methane emissions, *Journal of Atmospheric Chemistry*, **8**: 307-358.

Bartlett, K.B., Crill, P.M., Sass, R.L., Harriss, R.C. and Dise, N.B.: 1992, Methane emissions from tundra environments in the Yukon-Kuskokwim Delta, Alaska, *Journal of Geophysical Research*, **97**: 16645-16660.

Batjes, N.H. and E.M. Bridges (Eds.): 1992, *World inventory of soil emission potentials*. Proceedings of the International Workshop organized in the framework of the Netherlands National Research Programme on Global Air Pollution and Climate Change (NOP). WISE report No. 2, International Soil Reference and Information Centre, Wageningen.

Bingemer, H.G. and Crutzen, P.J.: 1987, The production of methane from solid wastes, *Journal of Geophysical Research*, **92**: 2181-2187.

Bouwman, A.F.: 1991, Agronomic aspects of wetland rice cultivation and associated methane emissions, *Biogeochemistry*, **15**: 65-88.

Bouwman, A.F.: 1993, Global distribution of nitrous oxide: analysis of the biogenic sources, in: Van Amstel, A. (Ed.), *Proceedings of the International Workshop on Methane and Nitrous oxide. Methods for National Inventories and Options for Control*. RIVM report 481507003. p. 261-272, National Institute of Public Health and Environmental Protection, Bilthoven, The Netherlands.

Bouwman, A.F., Fung, I., Matthews, E. and John, J.: 1993, Global analysis of the potential for N_2O production in natural soils, *Global Biogeochemical Cycles*, **7**: 557-597.

Bowden, R.D., Steudler, P.A., Melillo, J.M. and Aber, J.D.: 1990, Annual nitrous oxide fluxes from temperate forest soils in the northeastern United States, *Journal of Geophysical Research*, **95**: 13997-14005.

Cates, R.L. and Keeney, D.R.: 1987, Nitrous oxide emission from native and reestablished prairies in southern Wisconsin, *The American Midland Naturalist*, **117**, 35-42.

Cicerone, R.J. and Oremland, R.S.: 1988, Biogeochemical aspects of atmospheric methane, *Global Biogeochemical Cycles*, **2**: 299-327.

Crutzen, P.J. and Andreae, M.O.: 1990, Biomass burning in the tropics: impact on atmospheric chemistry and biogeochemical cycles, *Science*, **250**: 1669-1678.

Crutzen, P.J., Aselmann, I. and Seiler, W.: 1986, Methane production by domestic animals, wild ruminants, other herbivorous fauna and humans, *Tellus*, **38B**: 271-284.

Davidson, E.A: 1991, Fluxes of nitrous oxide and nitric oxide from terrestrial ecosystems, in: Rogers, J.E. and Whitman, W.B. (Eds), *Microbial Production and Consumption of Greenhouse Gases: Methane, Nitrogen Oxides and Halomethanes*. p.219-235, American Society of Microbiology, Washington, D.C.

Duxbury, J.M., Bouldin, D.R., Terry, R.E. and Tate III, R.L.: 1982, Emissions of nitrous oxide from soils, *Nature*, **298**: 462-464.

Eichner, M.J: 1990, Nitrous oxide emissions from fertilized soils: summary of available data, *Journal of Environmental Quality*, **19**: 272-280.

EPA: 1990, Greenhouse gas emissions from agricultural systems, Volume I and II, *Proceedings of the Workshop on Greenhouse Gas Emissions from Agricultural Systems*. Prepared for IPCC-RSWG-AFOS, EPA Office of Policy Analysis, Washington, September 1990

FAO: 1991, *Agrostat PC*. Computerized Information Series 1/3. Land use. FAO Publications Division, FAO, Rome.

Fung, I., John, J., Lerner, J., Matthews, E., Prather, M., Steele, L.P. and Fraser, P.J.: 1991, Three-dimensional model synthesis of the global methane cycle, *Journal of Geophysical Research*, **96**: 13033-13065.

Gibbs, M.J. and R.A. Leng: 1993, Methane emissions from livestock, in: Van Amstel, A. (Ed.), *Proceedings of the International Workshop on Methane and Nitrous Oxide. Methods for National Inventories and Options for Control*. RIVM report 481507003. p. 73-79, National Institute of Public Health and Environmental Protection, Bilthoven, The Netherlands.

Gibbs, M.J. and J.W. Woodbury: 1993, Methane emissions from livestock manure, in: Van Amstel, A. (Ed.), *Proceedings of the International Workshop on Methane and Nitrous Oxide. Methods for National Inventories and Options for Control*. RIVM report 481507003. p. 81-91, National Institute of Public Health and Environmental Protection, Bilthoven, The Netherlands.

Goodroad, L.L. and Keeney, D.R.: 1984, Nitrous oxide emission from forest, marsh and prairie ecosystems, *Journal of Environmental Quality*, **13**: 448-452.

Hao, W.M., Liu, M.H. and Crutzen, P.J.: 1990, Estimates of annual and regional releases of CO_2 and other trace gases to the atmosphere from fires in the tropics, based on the FAO statistics for the period 1975-1980, in: Goldhammer, J.G. (Ed.), *Fire in the Tropical Biota. Ecological Studies*, **84**: p 440-462. Springer Verlag, Berlin.

Hao, W.M., Scharffe, D.S, Crutzen, P.J. and Sanhueza, E.: 1988, Production of N_2O, CH_4 and CO_2 from soils in the tropical savanna during the dry season, *Journal of Atmospheric Chemistry*, **7**: 93-105.

Isaksen, I.S.A., Ramaswamy,V., Rodhe, H. and Wigley, T.M.L.: 1992, Radiative forcing to climate, in: Houghton, J.T., Callander, B.A. and Varney, S.K. (Eds.), *Climate Change 1992. The Supplementary Report to the IPCC Scientific Assessment*, p. 49-67, University Press, Cambridge.

Keller, M., Goreau, T.J., Wofsy, S.C.,Kaplan, W.A., McElroy, M.B.: 1983, Production of nitrous oxide and consumption of methane by forest soils, *Geophysical Research Letters*, **10**: 1156-1159.

Keller, M., Kaplan, W.A. and Wofsy, S.C.: 1986, Emissions of N2O, CH4 and CO2 from tropical forest soils, *Journal of Geophysical Research*, **91**: 11791-11802.

Keller, M., Kaplan, W.A., Wofsy, S.C. and Da Costa, J.M.: 1988, Emission of N_2O from tropical soils: response to fertilization with NH_4^+, NO_3^-, and PO_4^{3-}, *Journal of Geophysical Research*, **93**: 1600-1604.

Keller, M., Veldkamp, E., Weitz, A.M. and Reiners, W.A.: 1993, Pasture age effects on soil-atmosphere trace gas exchange in a deforested area of Costa Rica, *Nature*, **365**: 244-246.

Khalil, M.A.K. and Rasmussen, R.A.: 1992, The global sources of nitrous oxide, *Journal of Geophysical Research*, **97**: 14651-14660.

Klein Goldewijk, K., Van Minnen, J.G., Kreileman, G.J.J., Vloedbeld, M., Leemans, R.: 1994, Simulating the carbon flux between the terrestrial environment and the atmosphere, *Water, Air and Soil Pollution*, **76** (this volume).

Lerner, J., Matthews, E., and Fung, I.: 1988, Methane emission from animals: a global high resolution database, *Global Biogeochemical Cycles*, **2**: 139-156.

Livingston, G.P., Vitousek, P.M. and Matson, P.A.: 1988, Nitrous oxide flux and nitrogen transformations across a landscape gradient in Amazonia, *Journal of Geophysical Research*, **93**: 1593-1599.

Luizao, F., Matson, P., Livingston, G.,Luizao, R. and Vitousek, P.: 1989, Nitrous oxide flux following tropical land clearing, *Global Biogeochemical Cycles*, **3**: 281-285.

Matson, P.A. and Vitousek, P.M.: 1987, Cross-ecosystem comparisons of soil nitrogen and nitrous oxide flux in tropical ecosystems, *Global Biogeochemical Cycles*, **1**: 163-170.

Matson, P.A. and Vitousek, P.M.: 1990, Ecosystem approach for the development of a global nitrous oxide budget.Processes that regulate gas emissions vary in predictable ways, *Bioscience*, **40**: 667-672.

Matson, P.A., Vitousek, P.M. and Schimel, D.S.: 1989, Regional extrapolation of trace gas flux based on soils and ecosystems, in: Andreae, M.O. and Schimel, D.S. (Eds.), *Exchange of trace gases between terrestrial ecosystems and the atmosphere*. p. 97-108, Dahlem workshop report, Wiley and Sons, Chichester, New York.

Matson, P.A., Vitousek, P.M., Livingston, G.P. and Swanberg, N.A.: 1990, Sources of variation in nitrous oxide flux from Amazonian ecosystems, *Journal of Geophysical Research*, **95**: 16789-16798.

Matthews, E. and Fung, I.: 1987, Methane emission from natural wetlands: Global distribution, area, and environmental characteristics of sources, *Global Biogeochemical Cycles*, **1**: 61-86.

Matthews, E., Fung, I. and Lerner, J.: 1991, Methane emission from rice cultivation: geographic and seasonal distribution of cultivated areas and emissions, *Global Biogeochemical Cycles*, **5**: 3-24.

Mosier, A.R., Stillwell, M., Parton, W.J. and Woodmansee, R.G.: 1981, Nitrous oxide emissions from a native shortgrass prairie, *Soil Science Society of America Journal*, **45**: 617-619.

Olson, J.S., Watts, J.A. and Allison, L.J.: 1983, *Carbon in live vegetation of major world ecosystems*. ORNL 5862. Environmental Sciences Division Publication No.1997. Oak Ridge National Laboratory, Oak Ridge, Tennessee. National Technical Information Service. U.S. Dept. Commerce.

Pacyna, J.: 1992, Report of the Second Workshop of the Global Emissions Inventory Activity (GEIA), Lillestrom, June 1992.

Pepper, W., Leggett, J., Swart, R., Watson, J., Edmonds, J. and Mintzer, I.: 1992, *Emission scenarios for the IPCC: an update. Assumptions, methodology and results*, 115 p, prepared for IPCC Working Group I, May 1992.

Parton, W.J, Mosier, A.R. and Schimel, D.S.: 1988, Rates and pathways of nitrous oxide production in a shortgrass steppe, *Biogeochemistry*, **6**: 45-48.

Ronen, D., Magaritz, M. and Almon, E.: 1988, Contaminated aquifers are a forgotten component in the global N2O budget, *Nature*, **335**: 57-59.

Ryden, J.C: 1981, N_2O exchange between a grassland soil and the atmosphere, *Nature*, **292**: 235-237.

Schmidt, J., Seiler, W. and Conrad, R.: 1988, Emission of nitrous oxide from temperate forest soils into the atmosphere, *Journal of Atmospheric Chemistry*, **6**: 95-115.

SEI: 1992, *National Greenhouse Gas Accounts*. Current anthropogenic sources and sinks. 37p., Stockholm Environment Institute, Stockholm, Sweden.

Seiler, W. and Conrad, R.: 1981, Field measurements of natural and fertilizer induced N_2O release rates from soils, *Journal Air Pollution Control Association*, **31**: 767-772.

Seitzinger, S.P.: 1990, Denitrification in aquatic sediments, in Revsbech, N.P. and Sorensen (Eds.), *Denitrification in soil and sediment*, p. 301-322, Plenum Press, New York.

Shine, K.P., Derwent, R.G., Wuebbles, D.J., Morcrette, J.-J.: 1990, Radiative forcing to climate, in Houghton, J.T., Jenkins, G.J. and Ephreanus, J.J. (Eds.), *Climate Change. The IPCC Assessment*, p. 45-68, Cambridge University Press, Cambridge.

Smith, C.J., Delaune, R.D. and Patrick Jr, W.H.: 1983, Nitrous oxide emission from Gulf Coast wetlands, *Geochimica et Cosmochimica Acta*, **47**:, 1805-1814.

Taylor, J.A.: 1992, A global three-dimensioal lagrangian tracer transport modelling study of the sources and sinks of nitrous oxide, *Mathematics and Computers in Simulation*, **33**: 597-602.

Taylor, J.A., G.P. Brasseur, P.R. Zimmerman and Cicerone, R.J.: 1991, A study of the sources and sinks of methane and methylchloroform using a global three-dimensional lagrangian tropospheric tracer transport model, *Journal of Geophysical Research*, **96**: 3013-3044.

Vitousek, P.M., Matson, P., Volkman, C., Mass, J.M. and Garcia, G.: 1990, Nitrous oxide flux from dry tropical forests, *Global Biogeochemical Cycles*, **3**: 375-382.

de Vries, H.J.M., Olivier, J.G.J., Van den Wijngaart, R., Kreileman, G.J.J., and Toet, S.: 1994, A model for calculating regional energy use and emissions to evaluate global climate change policies, *Water, Air and Soil Pollution*, **76** (this volume).

Watson, R.T., Meira Filho, L.G., Sanhueza, E. and Janetos, A.: 1992, Sources and sinks, in: Houghton, J.T., Callander, B.A. and Varney, S.K. (Eds.), *Climate change 1992. The Supplementary Report to the IPCC Scientific Assessment*, pp 25-46, University Press, Cambridge.

Zuidema, G., Van Den Born, G.J., Alcamo, J. and Kreileman, G.J.J.: 1994, Simulating changes in land cover as affected by economic and climate factors, *Water, Air and Soil Pollution*, **76** (this volume).

ATMOSPHERIC COMPOSITION CALCULATIONS FOR EVALUATION OF CLIMATE SCENARIOS

M.S. KROL AND H.J. VAN DER WOERD

National Institute of Public Health and Environmental Protection (RIVM)
P.O. Box 1, 3720 BA Bilthoven, the Netherlands

Abstract. The future radiative forcing by non-CO_2 greenhouse gases depends strongly on the behavior of the OH radical, which represents the primary sink for CH_4, CO and H(C)FCs in the atmosphere. We present a simple model to describe the changes in the concentration of the main greenhouse gases. The focus is on the description of the atmospheric chemistry of OH and the important tropospheric oxidant and greenhouse gas O_3. Changes in the equilibrium concentrations of these oxidants will change the trends in the concentrations of greenhouse gases, especially CH_4. The model is applied to the 1992 IPCC emissions scenarios, as well as to an IMAGE 2.0 scenario, based on "Conventional Wisdom" assumptions. We find the following major results: for the central estimate of emissions assuming no additional policies (IS92a), the concentration of CH_4 keeps rising at rates similar to those observed over the last decades; results for the other IS92 scenarios range from stabilization early in the next century (IS92d) to an ever increasing rate of accumulation of CH_4 in the atmosphere (IS92f), even though these scenarios assume no policy interventions. The IMAGE 2.0 Conventional Wisdom scenario is similar to IS92a before the year 2025; afterwards the expansion of agricultural area significantly decreases the emissions of hydrocarbons and NO_x from savanna burning, not represented in the IS92 scenarios. This leads to stable levels of atmospheric CH_4 after 2025.

Keywords: atmospheric chemistry, CH_4, OH, O_3, emissions scenarios, integrated modeling.

1. Introduction

Atmospheric processes play a key role in the greenhouse effect. The radiative properties of trace gases and their concentrations are among the main factors determining the temperature of the atmosphere and at the earth's surface. The concentrations of the trace gases change due to a number of processes: emissions from the biosphere, deposition at the surface, transport processes within the troposphere and between the troposphere and stratosphere. This paper concentrates on chemical processes in the atmosphere that change the composition of the atmosphere, concentrating on radiatively active gases, as calculated by the Atmospheric Composition model. The aim of this model is to simulate the long-term evolution of the concentrations of greenhouse gases in the global atmosphere, for usage in an integrated model, IMAGE 2.0.

IMAGE 2.0 consists of a set of global models covering issues related to climate change such as energy use, land use and the greenhouse gas emissions related to these activities, the C cycle both on land and in the ocean and heat balances in both atmosphere and ocean (Alcamo *et al.*, 1994a). The input of the Atmospheric Composition model are the emissions of greenhouse gases and ozone precursors and are provided by the Energy and Industry Emission models, the Land Use Emission model and the Ocean Biosphere/Chemistry model (for the CO_2 flux into the ocean). Another source of input is the Zonal Atmospheric Climate model which, amongst others, simulates the atmospheric

Water, Air, and Soil Pollution **76**: 259–281, 1994.

climatic data relevant for the chemical calculations. The main output of the Atmospheric Composition model are the concentrations of the greenhouse gases used in the Zonal Atmospheric Climate model for the radiative forcing calculations. Furthermore the atmospheric concentration of CO_2 is used in the C cycle both on land (in parameterizations of fertilization effects) and in the ocean (in the gas-exchange parameterization).

The main greenhouse gases are H_2O, CO_2, O_3, CH_4, N_2O and CFCs. From these gases O_3 and CH_4 are most important in the chemistry of the troposphere. This is in contrast to H_2O, the most powerful absorber of radiation, whose concentration is determined by climatic processes rather than by chemical processes. Some key chemical processes, like the formation of the OH radical, do depend on the concentration of H_2O, however. By comparison CO_2 is chemically inert and is distributed almost homogeneously over the atmosphere. N_2O and CFCs are also inert to chemical reactions in the troposphere, but do have an atmospheric sink due to photodissociation in the stratosphere.

In this paper we describe the 0^+-D (box) model of atmospheric composition used in IMAGE 2.0. We call it a 0^+-D model rather than a 0-D model to emphasize that tropospheric processes have been parameterized using results from spatially explicit model studies (Hough, 1991; Prather and Spivakovsky, 1990; Thompson et al., 1989). This leads us to expect the model to simulate the tropospheric concentrations of greenhouse gases more accurately than an ordinary box model.

The model has been used to simulate the impacts of the 1992 IPCC scenarios on atmospheric composition. These scenarios are updates of the 1990 IPCC scenarios, defined in terms of economic and population growth and the availability of resources. All six scenarios assume either existing or proposed restricted climate policies only. This in contrast to the 1990 IPCC scenarios, which were partly defined by the year at which the atmospheric CO_2 concentration would double. In addition a simulation was made using the IMAGE 2.0 Conventional Wisdom scenario, built on assumptions similar to the IPCC scenarios, but including the main feedbacks of global change (Alcamo et al., 1994b). Another IMAGE 2.0 scenario, 'Ocean Realignment', investigates the sensitivity to changes in the ocean circulation.

First we will discuss the general modeling principles and give a description of the compounds and the processes affecting the concentrations of greenhouse gases (par. 2). In par 3. the validity of this model approach is tested, followed by the main results of model calculations for the six 1992 IPCC scenarios (par. 4). Paragraph 5 reports on results of using the model within the entire IMAGE 2.0 framework. Here the feedbacks of climate change on land use/cover and greenhouse gas emissions, and their effect on the atmospheric composition is discussed.

2. The Atmospheric Composition Model of IMAGE 2.0

The model simulating the composition of the atmosphere in IMAGE 2.0 is a one box model of the troposphere, with some stratospheric data added. The model aims at calculating the concentrations of greenhouse gases with an accuracy satisfactory for climate calculations in situations not differing too much from the present day situation.

The concentrations of greenhouse gases are represented by their tropospheric yearly average. This representation is satisfactory for long-lived greenhouse gases that are distributed homogeneously, like CO_2, CH_4, N_2O and CFCs. Tropospheric O_3 and compounds involved in the CH_4 chemistry, like OH, NO_x, CO and NMHCs, are distributed much more heterogeneously. This implies that the troposphere is a composition of regions with distinctly different atmospheric chemistry and corresponding response to emission scenarios. In the Atmospheric Composition model these chemically different regions do not appear as separate boxes. They are integrated to derive the globally averaged description of the atmospheric processes only. This integration is based on the 'chemically coherent region' approach.

2.1. The Local Atmospheric Chemistry Coupling

We mainly based our parameterizations on the results of the multi-1-D model of the Goddard Space Flight Center (GSFC) by Thompson *et al.* (1989) which explicitly describes the global atmosphere as a set of chemically coherent regions. A chemically coherent region is defined as a region in which the contaminant levels can be distinguished from levels in adjacent regions. As a consequence the concentrations of the trace gases and their chemical coupling are (almost) only a function of altitude. In particular the concentrations of the oxidants OH and O_3 are assumed to be in equilibrium with the concentrations of longer-living species such as CH_4 and CO and the emissions of source gases such as NO_x. The approach of neglecting transports is reasonable for some trace gases and an oversimplification for others.

For trace gases like CH_4 with a long lifetime, concentrations will be close to the tropospheric average concentration everywhere (except close to the sources), which implies that local concentrations depend more on global emissions than local emissions. For gases with short residence times like OH and NO_x the net transport between the regions will be small since these gases are involved in fast chemical reactions before they have time to be transported to the boundary of the region. The trace gases that are simulated less well by the coherent region approach are those that have a lifetime long enough to allow transport from one coherent region to another, but short enough to allow significant differences between the concentrations in the different regions. These include CO and O_3. For these gases the transport between regions is an important effect which is not taken into account.

Thompson *et al.* (1989) distinguish between 5 'typical' regions: an urban region, a clean continental region and a marine region, all in the latitude belt from 30°-60°N, a low latitude region (30°N-30°S) and a southern hemisphere mid latitude region (30°-60°S) (see Figure 1). The polar regions are not explicitly described in the model because they marginally contribute to the global emissions and oxidation by O_3 and OH.

Neglecting (horizontal) transport between the chemically coherent regions, each region can be described by one vertical column atmosphere (1-D). Thompson *et al.* (1989) made detailed 1-D model calculations to describe the change in total tropospheric amount of *x*

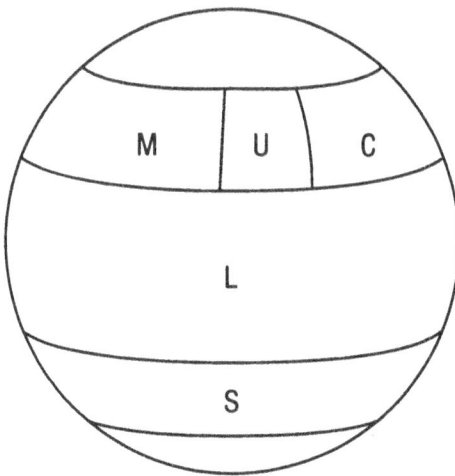

Figure 1: Schematic representation of the chemically coherent regions used in Thompson et al. (1989). The regions are: urban (U), clean continental (C) and marine (M) all in the mid latitudinal belt of the Northern hemisphere, low latitude (L), the tropical belt, and the Southern hemispheric mid latitudinal belt (S)

(OH and O_3) as function of numerous atmospheric conditions y. The changes are expressed as sensitivity coefficients $dlnx/dlny$, where $dlnx$ denotes the relative change in the quantity x. Here y stands for the emission rate of CH_4, CO or NO_x, for the column O_3 or for the H_2O mixing ratio. From the five quantities influencing OH considered in Thompson *et al.* (1989), the emissions of CH_4 and CO directly influence its sink while the emissions of NO_x, H_2O and column O_3 directly influence its source. For our box model we integrated the calculated responses of each of the 5 chemically coherent regions into one global sensitivity coefficient.

2.2.   A GLOBALLY AVERAGED BOX MODEL

Box models of atmospheric chemistry use mass-balances in simulating atmospheric concentrations of greenhouse gases. In a mass balance the difference between the sources and sinks of a greenhouse gas is computed, and converted to an increase or decrease of the atmospheric concentration of the gas. The sources taken into account are not only direct emissions from energy use or land-use but also atmospheric production by oxidations (for CO and CO_2). The sinks include atmospheric oxidation, photodissociation and transport losses (deposition and transport to the stratosphere). The concentrations of atmospheric oxidants are important in determining the oxidation sinks for CH_4, HCFCs and HFCs. The concentrations of these oxidants are usually parameterized using sensitivity coefficients.

In our parameterization we explicitly use the present day distributions of emissions, concentrations and reaction rates and the locally different trends in emissions and concentrations. In this way the 0^+-D model differs from other box models that often use averaged values for all quantities and process-describing parameters considered, as would be found for a homogeneous gas mixture. The difference is, above all, important in describing the evolution of the concentrations of OH and O_3 in the troposphere.

In Thompson *et al.* (1989) sensitivity coefficients are given for a range of values for the atmospheric conditions y, thus representing the nonlinear behavior of the atmosphere as a function of its inputs. The dependence of OH on hydrocarbon emissions is found to be nonlinear: higher emissions also lead to longer lifetimes of the hydrocarbons, which adds to the linear reaction of OH concentrations. The sensitivities of tropospheric O_3 are much more linear than the sensitivities of OH, but they also vary from region to region.

Nonlinearity is most apparent for the dependence of OH on NO_x emissions. Here the sensitivity coefficient increases with increasing NO_x emissions as the emissions are low. But for higher emission levels (corresponding to concentrations above 5-10 ppbv) the sensitivity even gets negative. This is what happens in the polluted urban region, where the emission rates for NO_x are two orders of magnitude larger than those for a southern hemispheric situation. In Appendix 1 a full description is presented of the influence of NO_x emission on local and global oxidant levels.

Reaction rates also require a careful translation to the global scale. The atmospheric oxidation rate of CH_4 (and HCFCs, HFCs) depends on the concentration of OH and on temperature. The oxidation proceeds faster at higher temperatures. In the troposphere the tropical region is not only warmer than the mid- and high-latitude regions, but it also shows higher concentrations of OH. Therefore simply taking mass-averaged values for both OH concentration and temperature would lead to an underestimation of the amount CH_4 oxidized by as much as 30 %. In our parameterization we follow the global averaging proposed in Prather and Spivakovsky (1990). For each reaction an effective globally-averaged OH concentration and an effective globally averaged tropospheric temperature are defined, taking into account the reaction kinetics and three-dimensional fields of OH and temperature. We also took from this work an effective globally averaged OH concentration for the CO oxidation rate, which depends on pressure. In using the average reaction rates we implicitly assume that possible changes in the distribution patterns of OH and temperature are less important than the changes in the average values of these data.

2.3. PARAMETERIZATIONS FOR THE 0+-D MODEL

The evolution of the greenhouse gas concentration is calculated from the global balance between production and loss terms. The production generally follows directly from the emissions scenarios, or alternatively from the emission modules of IMAGE 2.0. The loss term for CH_4 and CO depends mainly on the OH equilibrium concentration. This concentration depends also on production, described as a separate quantity coupled to the O_3 concentration, and loss through reaction with CH_4 and CO. This set of strongly coupled differential equations forms the heart of the 0+-D model of IMAGE 2.0

CO_2: Carbon dioxide is inert in the atmosphere. Sources of atmospheric CO_2 are emissions from energy use and industry and land use (land clearance and biomass burning) (Klein Goldewijk *et al.*, 1994; de Vries *et al.*, 1994). A small additional atmospheric source is

TABLE 1

Values for the parameters in the concentration equations of the compounds assumed not to be in equilibrium. f_X denotes the conversion factor from emissions to concentrations (in their respective units), k_{X+OH} the reaction rate for the oxidation of X by OH in $cm^{-3} \cdot a^{-1}$ (WMO, 1992), l_X stands for the transport losses to both stratosphere and biosphere (deposition) in a^{-1} (WMO, 1992) and lft_X for the average atmospheric residence time (lifetime) in years (WMO, 1992), p_X and Em_X denote the atmospheric concentration and emission of the compound, and are both given in the base year 1990 (IPCC, 1992).

compound X	p X(1990)	Em_X(1990)	f_X	lft_X	k_{X+OH}	l_X
CO_2	354 ppmv	7.4 Pg	0.469	-	-	-
CH_4	1.72 ppmv	506 Tg	$3.90 \cdot 10^{-4}$	-	$1.23 \cdot 10^{-7}$	0.0091
CO	0.095 ppmv	1164 Tg	$2.22 \cdot 10^{-4}$	-	$5.93 \cdot 10^{-6}$	0.473
N_2O	308 ppbv	12.9 TgN	0.212	150	-	-
CFC-11	272 pptv	298 Gg	0.0413	55	-	-
CFC-12	471 pptv	362 Gg	0.0490	116	-	-
CFC-113	70 pptv	147 Gg	0.0316	110	-	-
CFC-114	18.5 pptv	13.1 Gg	0.0347	220	-	-
CFC-115	5 pptv	6.5 Gg	0.0384	550	-	-
CCl_4	109 pptv	119 Gg	0.0385	47	-	-
HCFC-22	113 pptv	138 Gg	0.0723	-	$9.05 \cdot 10^{-8}$	0.0042
CH_3Cl	600 pptv	4000 Gg	0.124	-	$8.79 \cdot 10^{-7}$	0.02
CH_3CCl_3	140 pptv	378 Gg	0.0469	-	$2.20 \cdot 10^{-7}$	0.0213

provided by the atmospheric oxidation of hydrocarbons. The sinks for CO_2 are uptake by vegetation and by the ocean. The Atmospheric Composition model adds up the fluxes and converts the net flux into a concentration change.

CH_4: The only source of atmospheric methane is direct emission into the troposphere. The emission is converted linearly into a global increase term for the concentration in the troposphere (here taken to be the atmosphere up to the 100 hPa level), implying instantaneous and uniform mixing in one year. The sinks of CH_4 include transport to the stratosphere and deposition/soil uptake, accounting for a sink of 10 and 30 Tg a^{-1} respectively in 1990 (IPCC, 1992). These loss terms are taken to be proportional to the CH_4 concentration. The main sink is the oxidation of CH_4 by OH, proportional to both the CH_4 and the OH concentrations and accounting for a loss of approximately 420 Tg a^{-1} (IPCC, 1990). The reaction rate is calculated as the global average (following Prather and Spivakovsky (1990)) of the local reaction rate as given in Vaghjiani and Ravishankara (1991). The sum of the sources and sinks describes the change in concentration (p) of X per time step, and is given by:

$$dpX = f_X \cdot Em_X - (l_X + k_{X+OH} \cdot pOH) \cdot pX \qquad (1)$$

where $X = CH_4$; parameters are given in Table 1.

CO: Carbon monoxide competes with CH_4 for the consumption of OH. The resulting oxidation is the main sink of CO, accounting for approximately 2000 Tg a^{-1}. The loss is proportional to the concentrations of CO and OH; the local reaction rate is taken from Atkinson (1985). Another sink is soil uptake (approximately 300 Tg a^{-1}) and possibly transport to the stratosphere (100 Tg a^{-1}) (IPCC, 1992); these sinks are assumed to be proportional to the atmospheric concentration of CO. The sources for CO included are threefold: direct emissions, atmospheric oxidation of NMHCs and atmospheric oxidation of CH_4; 1990 values for these three sources are, in approximation, 1440 (IPCC, 1992), 390 (Roemer, 1991) and 600 (IPCC, 1992) Tg a^{-1}. The source from NMHC is included in the direct emissions of CO in this model, taking into account an average yield of 0.651 (mass/mass), as calculated by the 2-D TNO-Isaksen model (Roemer, 1991). Mass is converted into concentrations in a similar way as for CH_4. For the source from CH_4 we assume that 80% of the oxidized CH_4 ends up as CO, following Logan et al. (1981). This leads to:

$$dpCO = f_{CO} \cdot \left(Em_{CO} + 0.651 \cdot Em_{NMHC} \right) + 0.8 \cdot k_{CH_4+OH} \cdot pOH \cdot pCH_4$$
$$- \left(l_{CO} + k_{CO+OH} \cdot pOH \right) \cdot pCO$$

(2)

OH: Hydroxyl has a very short atmospheric lifetime, which means that its globally averaged concentration can be approximated by an equilibrium concentration. The production of OH is used as a separate quantity in the model, denoted by prod OH. The loss of OH is taken to be caused by CH_4 and CO alone. Small net losses due to NMHC oxidation and reaction with other atmospheric radicals or H_2 are neglected. The steady state assumption now leads to:

$$pOH = \frac{prodOH}{k_{CH_4+OH} \cdot pCH_4 + k_{CO+OH} \cdot pCO}$$

(3)

The production of OH depends on the concentrations of its precursors, namely the excited oxygen atom and tropospheric H_2O. The sensitivity of OH to tropospheric O_3 (and to H_2O) is based on the production of excited (O^1D) by photolysis of O_3 followed by OH production:

$$O(^1D) + H_2O \rightarrow 2 OH$$

(4)

However, an increase by 1 % in O_3 results only in approximately 0.5 % increase in OH production because O_3 also reacts with other species like NO_2 and because of repartitioning within the HO_x chemical family (Liu et al., 1987; Prather and Spivakovsky, 1990). The concentration of excited oxygen atoms is represented by the concentration of tropospheric O_3 (trop O_3) and by the column O_3 (col O_3), determining the UV radiation levels required for the photodissociation of O_3 in the troposphere.

Furthermore, NO_x significantly affect OH production by recycling OH in oxidation processes of hydrocarbons. These dependencies are linearized in the model as described in the Appendix 1. The NO_x influence on OH is split into two parts: one via the influence of NO_x on tropospheric O_3 and one directly, modeling the impact of for instance the reaction

$$HO_2 + NO \rightarrow OH + NO_2 \tag{5}$$

The direct influence of NO_x on OH is modeled as the difference of the total influence of NO_x on OH (sensitivity coefficient 0.4, see Table A.3) and the indirect influence via tropospheric O_3 (sensitivity coefficient dln trop O_3/dln Em NO_x = 0.064 and dln OH/dln trop O_3 = 0.5), yielding 0.368. In total the OH production is given by:

$$dln\, prodOH = 0.5 \cdot dln trop O_3 - 0.7 \cdot dln\, col O_3$$
$$+ 0.5 \cdot dln pH_2O + 0.368 \cdot dln\, Em_{NO_x} \tag{6}$$

O_3: Ozone is, second to H_2O, the most important greenhouse gas which is not homogeneously distributed over the troposphere. The two sources of trop O_3 are flux from the stratosphere and reaction of single oxygen $O(^3P)$ which is produced by photolysis of NO_2, the product of Reaction (5). In the model, the concentration of O_3 in the troposphere is treated similar to the production of OH: changes in the concentration of tropospheric O_3 are being related to changes in the concentrations of its precursors and to changes in the levels of the radiation responsible for its dissociation. The precursors included are CH_4 and CO and NO_x (only as emissions); the UV radiation level is parameterized by the change in column O_3. The sensitivity coefficients relating the relative changes are obtained from Prather (1989) or from Thompson et al. (1989). After taking the appropriate global averages (see Appendix 1), this leads to:

$$dln\, trop O_3 = 0.13 \cdot dln pCH_4 + 0.15 \cdot dln pCO$$
$$+ 0.064 \cdot dln Em_{NO_x} + 0.2 \cdot dln\, col O_3 \tag{7}$$

The column O_3 is related to the stratospheric concentration of active Cl by empirical fitting to the TOMS data. A threshold level of 1.33 ppbv of Cl and an ozone depletion rate of 3.4 % ppbv^{-1} for Cl levels exceeding the threshold level is used:

$$col O_3 = \min\left(1, 1 - 0.034 \cdot \left(p Cl_{strat} - 1.33\right)\right) \tag{8}$$

NO_x: The model only accounts for emissions (not concentrations) of nitrogen oxides (NO and NO_2). These emissions are input variables for the model to parameterize the impact on the production of OH (Equation 6) and on the concentration of tropospheric O_3 (Equation 7).

NMHC: Non-methane hydrocarbons are oxidized in the troposphere in a few days or weeks. NMHCs consume some 10% of the total OH radicals (Prather, 1989). The relatively fast oxidation is not described in the 0^+-D model; instead, NMHC is only treated as an additional source of CO. The yield of CO from the NMHC emissions is added to the direct emissions of CO, and is assumed to be 0.65 kg/kg, the yield found for present day conditions by the TNO-Isaksen model (Roemer, 1991). The linkage between NMHC oxidation to CO and tropospheric O_3 formation is neglected.

H_2O: In the integrated framework of IMAGE 2.0 the change in tropospheric water vapour is taken directly from the atmospheric climate calculations in the Zonal Climate Model (de Haan *et al.*, 1994). In this model radiative effects and heat transport are simulated on a two dimensional grid (latitude *vs.* height).

N_2O: Nitrous oxide is emitted from the surface and depleted by photodissociation in the stratosphere. The emissions, in units of mass, are converted into concentration increases assuming that N_2O is homogeneously distributed over the troposphere. The lifetime lft_{N2O} is estimated to be 150 years (IPCC, 1990). The current best estimate of the atmospheric lifetime of N_2O (WMO, 1992; IPCC, 1992) is not consistent with the update of the current rate of emissions (IPCC, 1992) combined with the presently observed trend of its atmospheric concentration, so we still use the 1990 value for the lifetime. Equation (9) describes the evolution of the concentration of N_2O; parameter values are listed in Table 1.

$$dpN_2O = f_{N_2O} - \frac{1}{lft_{N_2O}} \cdot pN_2O \qquad (9)$$

CFCs: Chlorofluorocarbons and CCl_4 are modeled in the same way as N_2O. The lifetimes are taken from WMO (1992): 55 years for CFC-11, 116 years for CFC-12, 110 years for CFC-113, 220 years for CFC-114, 550 years for CFC-115 and 47 years for CCl_4. Again Equation (9) describes the evolution of the concentrations (with CFC being substituted for N_2O), while Table 1 lists the values of the parameters appearing in the equation for the different compounds.

HCFC-22: HCFC-22, methylchloroform (MCF) and methylchloride (CH_3Cl) are treated similar to CH_4. Emissions are converted into tropospheric concentrations. The stratospheric loss is proportional to the tropospheric concentration. The loss due to oxidation by OH is proportional to both the concentration of the trace gas and the concentration of OH. Transport loss rates and reaction rates are taken from Prather and Spivakovsky (1990). The changing concentrations follow from Equation (3) and the data in Table 1. The other H(C)FCs appearing in the IS92 scenarios are treated similarly. Reaction data are taken from WMO (1992) and Talukdar *et al.* (1992).

Cl: The concentration of free lower stratospheric chlorine is calculated from the tropospheric concentrations of Cl-containing compounds. This concentration represents the Cl which is actively involved in the depletion of ozone. A timelag of 4 years is included to represent the transportation time from the troposphere to the stratosphere. The Cl yield of the various compounds is taken to be 2.7 for CFC-11, 1.08 for CFC-12, 1.8 for CFC-113, 1.16 for CFC-114, 0.12 for CFC-115, 3.8 for CCl_4, 0.32 for HCFC-22, 2.94 for MCF and 0.99 for CH_3Cl (WMO, 1992).

3. Validation and uncertainty considerations

The Atmospheric Composition model was developed to simulate the tropospheric concentrations of greenhouse gases more accurately than an ordinary box model in situations not too far from the present state of the atmosphere. This was tested by comparison of our results with observations and model results from 2D and 3D troposphere models.

Model simulations for CH_4 concentrations were compared to the historical data set compiled by Khalil and Rasmussen. Model simulations gave an excellent fit (within 0.02 ppm) to the mean trend observed over the past two decades and a reasonable fit (within 10 %) in the first half of this century (Rotmans et al., 1992). However, this is not a firm validation of the modeling of atmospheric oxidant levels since the historical increase in CH_4 concentration has been dominated by the increase in emission.

Our model does not reproduce the observed slowing down of CH_4 accumulation between 1983 and 1990 (Steele et al., 1992). Since the global budget of atmospheric CH_4 is uncertain both an increase in the loss rate due to increasing levels of OH radical or a reduction in the emission rate can possibly be the cause. In case of an emission reduction, as suggested by Steele et al., our model does not simulate the slowing down because the CH_4 emissions are increasing in the 1980s in all scenarios. An increase of the global average OH concentration, for which there is no conclusive evidence (Prinn et al., 1992), might partly be the result of increase in photochemical OH production (Equation 6) as the result of ozone depletion in the stratosphere (Madronich and Granier, 1992). A better test of the global sensitivities of the atmosphere, especially the sensitivity to NO_x emissions, came from model simulations for future CH_4 concentrations resulting from the 1990 IPCC scenarios (IPCC, 1991). They were compared to simulations obtained by the models contributing to (IPCC, 1990), as gathered by Guthrie and Yarwood (1991). The range of the results for the IPCC contributing models is quite large (20 %), mainly because of difficulties in describing the structure in NO_x regimes and the weighted influence on O_3 and OH. We found the results of the Atmospheric Composition model of IMAGE 2.0 to fit into this range of results for control and accelerated policy scenarios. Within the range, the results of the present model are on the lower edge after 2050; this also holds for the simulations using the GSFC model, from which we derived the sensitivities for the Atmospheric Composition model of IMAGE 2.0.

A more detailed comparison was made to results of the 2D TNO-Isaksen model (Roemer, 1991). This model is an adaption of the Isaksen model (Isaksen and Hov, 1987) with extended NMHC chemistry that calculates the monthly average evolution for about 90 species in the troposphere on a latitude (10°), altitude (0.5-1.0 km) grid. We compared

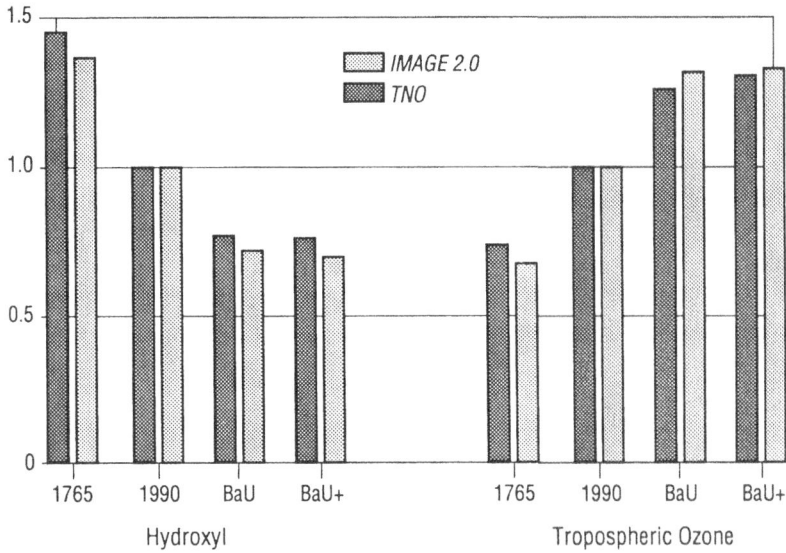

Figure 2: Comparison of the trend in OH and O_3 for the TNO-Isaksen model and the IMAGE 2.0 Atmospheric Composition model. Atmospheric concentrations of the two oxidants are scaled to their 1990 simulated levels. '1765' denotes a preindustrial situation, BAU and BAU+ denote two Business as Usual 2100 situations (BAU + with additional NMHCs).

model results for O_3 and OH for four equilibrium scenarios, see Figure 2. The scenarios represented the pre-industrial situation, the present and two versions of a 'Business as Usual' future (Roemer, 1991). For O_3 and OH we only made a comparison of the ratios of the concentrations of these gases in the different equilibria.

The TNO-Isaksen model results showed an increase in O_3 by 35 % since pre-industrial times, while our model gave a 47% increase. Comparison with historical observations are somewhat difficult because the evolution of the oxidizing capacity (OH and O_3) are predominantly determined by the tropical free troposphere (see Appendix 1) and not the boundary layer at mid-latitudes. However, the TNO-Isaksen results (Roemer, 1991) show an increase from 10-15 ppbv to 30-40 ppbv at 50° N near ground level, similar to the Montsouris data (see e.g. Thompson, 1992). Also for the future BAU scenarios we found the Atmospheric Composition model to predict a larger increase (32%) in O_3 than the 2D calculations (27%). The evolution of the global OH concentrations was calculated almost identical by the two models. Since pre-industrial times OH was found to have decreased by 31 %. This number is just above the mean of 10-30%, but not out of range with an estimate from model intercomparison (Thompson, 1992). The BAU scenario calculations gave a further OH loss of 23% (2D) and 24 % (IMAGE 2.0).

In an uncertainty analysis we quantified the uncertainties in the model results due to uncertainties in emissions, model parameters and feedbacks (Krol, 1994). The analysis consisted of a Monte Carlo experiment, in which 150 samples were taken from the uncertainty ranges of the basic parameters of the model, its input and feedback processes.

This resulted in frequency distributions of the simulated greenhouse gas concentrations. The dominant sources of uncertainty in the simulations were resolved by using statistical techniques. We found that uncertainties in model parameters are a very significant source of uncertainty. Especially over the long term, the uncertainties in the sensitivity coefficients of the atmospheric oxidants tend to dominate the total uncertainty in the atmospheric concentration of CH_4. A typical value for the uncertainties found was 20% for the concentration of CH_4 in 2100, assuming emissions from the IS92a scenario (Guthrie and Yarwood, 1991; IPCC, 1992). Some 60% of this uncertainty is due to uncertainties in the sensitivity coefficients for OH; uncertainties in emissions and uncertainties in climate sensitivity both account for 15% of the total uncertainty (Krol, 1994).

4. Results for the 1992 IPCC scenarios.

In 1992 IPCC published a set of greenhouse gas emissions scenarios (IS92a-f) as an update of the 1990 IPCC scenarios (IPCC, 1992; IPCC, 1991). The 1990 IPCC scenarios are defined in terms of the year when the CO_2 concentration would double. The 1992 IPCC scenarios are defined purely by socio-economic inputs. The IS92a scenario is the reference scenario of (IPCC, 1992). It is based on central estimates for population growth, economic growth, resource availability, existing policies and moderate (H)CFC emissions. The other five IS92 emissions scenarios differ only slightly from the IS92a scenario regarding the assumptions on control policies for reducing greenhouse gas emissions. The differences lie in the assumptions on the driving forces: population growth, economic growth and availability of resources. The five additional scenarios illustrate the sensitivity of the emissions of greenhouse gases to uncertainties in economic and population growth assumptions and to small changes in government policies related to greenhouse gas issues. In this sense, the range of emissions from the IS92 scenarios can be interpreted as an uncertainty range for the IS92a scenario, even when the subject of the probability of the different scenarios has not been addressed by IPCC (1992). There it is also noted, that the range of emissions from the IS92 scenarios is (somewhat large, but) similar to the range of independent estimates of emissions assuming central estimates for tendencies in the driving forces. Characteristics of the scenarios are given in Table 2.

Up to now little attention has been paid to the impacts resulting from the updated scenarios. Wigley and Raper used the scenarios for a climate-impact assessment (Wigley and Raper, 1992). The model we use describes the atmospheric chemistry with some more detail. For the calculations in this section some parts of the IMAGE 2.0 model are by-passed. The energy and industry emissions of greenhouse gases and ozone precursors are directly taken from the IS92 scenarios, rather than simulated by the Energy and Industry System of IMAGE 2.0. Similarly the calculation of land cover-related emissions is taken from the IS92 scenarios and not from the Terrestrial Environment System. On the other hand, the IMAGE 2.0 simulations of the terrestrial uptake of CO_2 are used, under the assumptions of the Conventional Wisdom scenario driving the land cover change. The Conventional Wisdom scenario is based on the same assumptions of population and

TABLE 2

Basic assumptions on the driving forces and emissions of halogenated species of the IS92 emissions scenarios (IPCC, 1992).

	IS92a	IS92b	IS92c	IS92d	IS92e	IS92f
population growth	medium	medium	medium/low	medium/low	medium	medium/high
economicgrowth	mid	mid	low	low/mid	high	mid
resource avail.	mid	mid	low	low	high	high
policies	existing	proposed	existing	optimistic	existing	existing
CFCs	moderate	low	moderate	low	low	moderate
HCFCs	moderate	moderate	moderate	low	low	moderate
HFCs	moderate	moderate	moderate	high	high	moderate

economic growth as IS92a, but feedback processes are taken into account more comprehensively (Alcamo et al., 1994b). The terrestrial CO_2 uptake depends on the impacts of climate and CO_2-fertilization on vegetation growth. The use of these simulated data is justified since the land cover changes found using the Conventional Wisdom scenario lead to global deforestation fluxes similar to those prescribed by the IS92 scenarios.

The emissions of the different greenhouse gases as given by the IS92 emissions scenarios are strongly coupled. In the IS92e and IS92f scenario medium to high estimates are assumed for all driving forces. As a result the emission rates are high for all greenhouse gases. In IS92c and IS92d the assumptions on the driving forces are quite moderate, leading to relatively low rates of emissions. An exception to this is the situation for the halogenated species. Here all scenarios show a fast decrease in the emissions of chlorinated carbons, assuming at least a partial compliance to the Montreal Protocol. Moreover emissions scenarios for the halogenated species are shared by some of the IS92 scenarios which allows for drawing conclusions on the cross impact of the emissions of hydrocarbons and NO_x on the ozone depletion.

The IS92b scenario is very similar to the IS92a scenario, assuming additional policies on CO_2 stabilization in the OECD countries and a global compliance of the Montreal Protocol only.

Figure 3 shows the results for CH_4 for all six scenarios. Scenarios IS92a and IS92b have identical emissions for CH_4, CO and NO_x. As a result the CH_4 concentrations are very close. The difference between the two is caused by a small difference in the emissions of VOC and by the effect of CFCs. In the IS92a scenario VOC emissions are slightly higher, which leads to higher CH_4 concentrations. In the IS92b scenarios however CFC emissions are lower, so that the ozone layer will recover somewhat quicker than for the IS92a scenario. Due to this, radiation levels will be lower in the troposphere for the IS92b scenario so the CH_4 sink will be lower as well. This difference appears to be the strongest, giving rise to slightly higher CH_4 concentrations for the IS92b scenario. The scenarios IS92c and IS92d have similar CH_4 emissions, but at a much lower level than the IS92a scenario. As a result the CH_4 concentrations are much lower too. This is especially

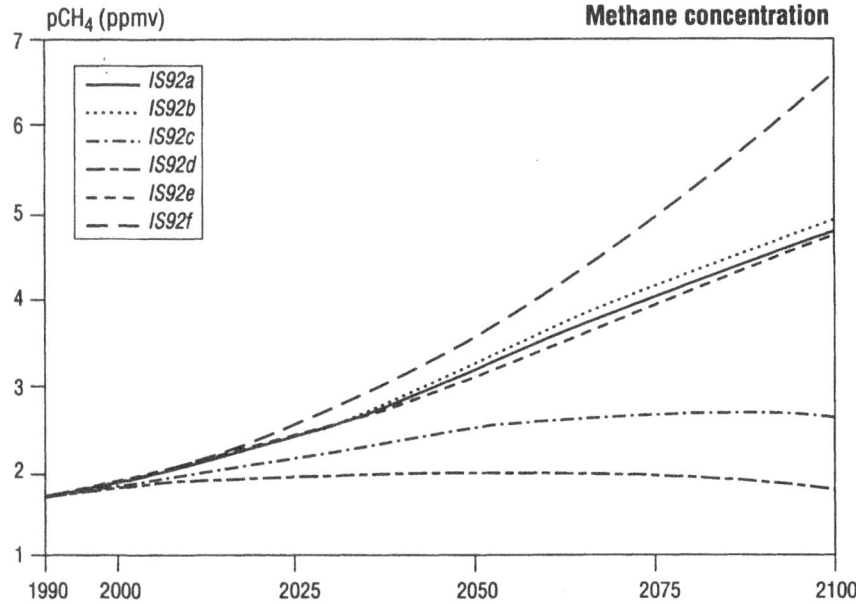

Figure 3: Simulated CH4 concentrations for the six IS92 scenarios.

true for the IS92d scenario, where the emissions for CO and VOC are smaller than or almost equal to the present day emissions over the whole period considered, while the emissions of NO_x go up by 40 %. For the IS92c scenario the CO emissions are somewhat higher, while the emissions of NO_x is reduced to present day levels in 2100. Both differences lead to lower levels of OH, compared to IS92d. The IS92e and IS92f also have similar emissions for CH_4, both at a level significantly higher than for the IS92a scenario. So the concentrations of CH_4 will be higher too, especially for the IS92f scenario, where the concentration of CH_4 keeps on rising at ever increasing rates. In that scenario the emissions of CO also increase dramatically, while the emissions of NO_x rise more moderately than for the IS92e scenario. The IS92e scenario thus shows higher concentration of OH than IS92f. This enhances the atmospheric sink of CH_4 such that the concentration of CH_4 for this scenario is even lower then for IS92a.

Figure 4 shows the results of the scenarios calculations for the mean tropospheric OH concentration and the change in tropospheric O_3 concentration, relative to the 1990 value. For all scenarios except for IS92d, the concentration of OH was simulated to decrease substantially in the next century. For O_3 the picture looks much the same as for CH_4: stabilization of concentrations was found for the IS92c and IS92d scenario only.

Figure 5 shows the simulated concentrations of stratospheric free Cl for the six scenarios. For CFCs and HCFCs three emissions scenarios were used only: moderate CFC and HCFC emissions for IS92a, IS92c and IS92f, low CFC and moderate HCFC emissions for IS92b and low CFC and HCFC emissions for IS92d and IS92e. The main sink for HCFCs is oxidation by OH in the troposphere. Therefore the concentration of free Cl is coupled to the chemistry of the troposphere. The concentrations of stratospheric free Cl in the year 2100 is higher for the IS92f scenario than for the IS92a scenario and still lower for the IS92c scenario while all three assume the same emissions for the halogenated species.

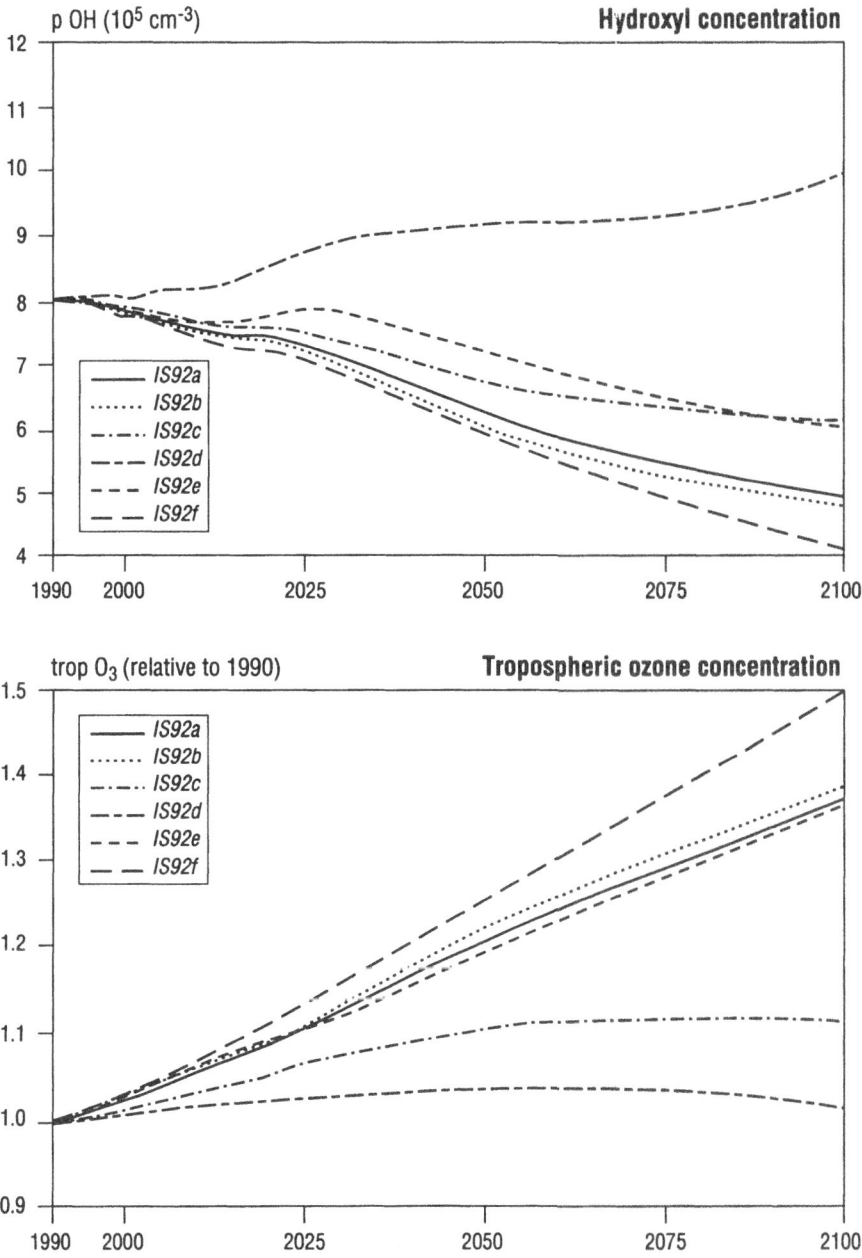

Figure 4: Simulated concentrations of OH and tropospheric O_3 for the six IS92 scenarios.

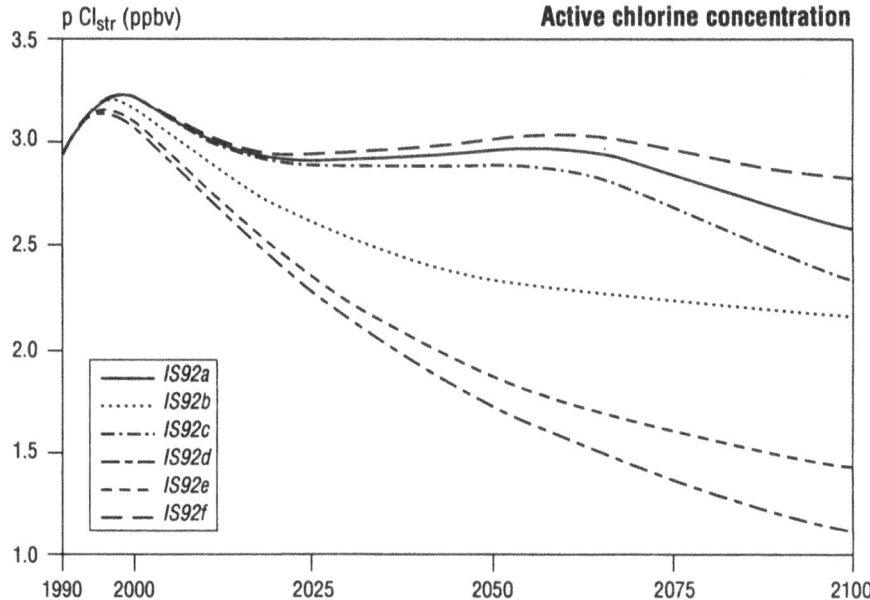

Figure 5: Simulated concentrations of active Cl in the stratosphere for the six IS92 scenarios.

5. Results for IMAGE 2.0 scenarios

The Atmospheric Composition model is part of the IMAGE 2.0 linked set of models (Alcamo *et al.*, 1994a). We now describe results for a "Conventional Wisdom" scenario and an "Ocean Realignment" scenario, in which the entire set of IMAGE models were used. These scenarios are based on the same assumptions for population and economic growth as IS92a, but emissions are calculated by IMAGE, rather than being taken from the IS92a scenario (Alcamo *et al.*, 1994b). The Ocean Realignment scenario differs from the Conventional Wisdom scenario in the assumptions on the ocean circulation only. Both emissions and trends in the emissions are similar for IS92a and the Conventional Wisdom scenario between 1990 and 2025. Between 2025 and 2100 large differences arise, especially in the emissions rates of CH_4 and CO. These differences are mainly land use-related: biomass burning sources decrease in the Conventional Wisdom scenario due to large scale land conversions in the tropical regions decreasing for instance the area of seasonally burning savanna (Alcamo *et al.*, 1994b; Kreileman and Bouwman, 1994). In the IS92a scenario this effect was not taken into account.

Figure 6 presents results for CO_2 and CH_4 for the Conventional Wisdom scenario. The level of CO_2 reaches about 775 ppm in 2100, falling between the estimates of Wigley and Raper (1992) for the IS92a and IS92f emission scenarios (740 and 850 ppmv in 2100). This is consistent with their findings since computed CO_2 emissions in the Conventional Wisdom scenario (24 Pg/a in 2100) also fall between emission estimates of the two IPCC scenarios (20 and 27 Pg/a in 2100). The IMAGE simulations of CO_2 for the IS92a and

IS92f scenarios are lower (670 and 770 ppmv in 2100). The IMAGE simulations of the CO_2 concentrations for Conventional Wisdom and IS92f are very close in 2100, even when the energy-related emissions are significantly higher for the IS92f scenario. The difference is small, mainly because of temperature increases enhance the biospheric sink of CO_2 in the Northern Hemisphere. For the IS92f scenario temperature increase is stronger after 2025, because of the higher concentrations of CH_4 and O_3. This also explains the difference between the CO_2 simulations of Wigley and Raper (1992) and IMAGE 2.0: Wigley and Raper model the terrestrial carbon sink to depend on the CO_2 concentration only.

Figure 6 also depicts the sensitivity of global CO_2 to a major change in the circulation of the world's oceans, as taken from calculations of Mikolajewicz *et al.* (1990). These investigators analyzed the sensitivity of the thermohaline oceanic circulation to future global warming, and their results were implemented in the IMAGE ocean model as an Ocean Realignment scenario (Alcamo *et al.*, 1994; de Haan *et al.*, 1994). As Figure 6 indicates, the reduced deep-water formation in this scenario significantly reduces the global uptake of atmospheric CO_2, resulting in an increase of 90 ppm in 2100 over the Conventional Wisdom scenario (Figure 6). The change in uptake is largely due to a difference in the climate feedback. In the Ocean Realignment scenario the northward transport of heat is decreased substantially. Therefore the atmospheric surface temperature is on the average 1 °C lower for the Northern Hemisphere (much more at higher latitudes), even when the CO_2 concentration is significantly higher (de Haan *et al.*, 1994). As a result, the climate feedback on carbon uptake in the Northern Hemisphere boreal forests is much lower. Differences in the oceanic uptake of carbon appear to be small.

The atmospheric concentrations of CH_4 for the IMAGE 2.0 scenarios are substantially below the values found for the IS92a emission scenario (Figure 3) after 2025. This is largely due to lower emissions of CO (and, to a lesser degree, CH_4) in the Conventional

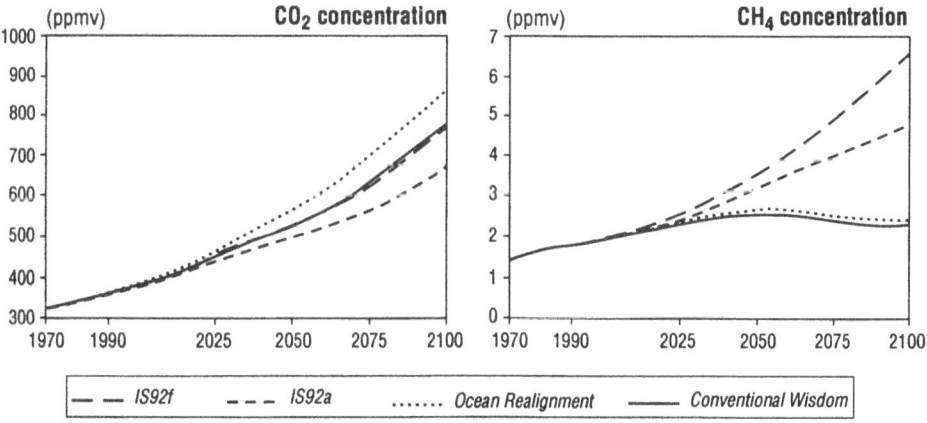

Figure 6: Simulated concentrations of CO_2 and CH_4 for the IMAGE 2.0 Conventional Wisdom and Ocean Realignment scenarios and for the IS92a and IS92f scenarios (IPCC, 1992).

Wisdom and Ocean Realignment scenarios; up to 2025 the concentrations are similar, as are the emissions.

The large differences in the emissions of CO, NO_x and VOCs between the IS92a scenario and the IMAGE scenarios are a consequence of the human-induced land cover changes. These changes will be considerable and should be taken into account in a consistent integrated assessment of climate change, even when a quantification of the changes is uncertain.

The differences in CH_4 concentrations between the Conventional Wisdom and Ocean Realignment scenario are small, only some 0.15 ppmv. The difference is mainly due to the feedback of H_2O on OH: the global troposphere warms more for the Conventional Wisdom scenario, increasing the concentration of H_2O and OH and thus increasing the sink for CH_4. Therefore the CH_4 concentration is slightly lower for the Conventional Wisdom scenario.

6. Discussion and Conclusions

Since most of the emissions scenarios considered differ from each other for more than one of their basic assumptions it is hard to draw conclusions on the effects of the individual basic inputs. The situation is even more unclear regarding stabilization targets for individual greenhouse gases. As is obvious from the chemical assumptions of the model, a strong coupling exists between the processes governing the fate of the different greenhouse gases. The most important features of the scenarios are the following:

For the IS92a scenario simulation results showed an increase of the CH_4 concentration up to 3.2 ppmv in 2050 (4.7 ppmv in 2100). The increase in CH_4 concentration in the coming 20 years is 1% a^{-1}, in the upper range of the observed present trend (0.5-1% a^{-1}).

Results for the IS92b scenario are different due to the effects on the concentration of OH of emitting 5 Tg VOC less, and of a global compliance to the Montreal Protocol rather than partial compliance. The emissions of CH_4, CO and NO_x are identical for IS92a and IS92b. The two effects are opposite: less VOC increases OH, while less CFCs lead to a more rapid recovery of the ozone layer, leading to lower UV radiation and OH concentrations. The effect of the CFCs appears to be slightly stronger, illustrating the interrelation between ozone layer policies and climate policies.

Results for the IS92c and IS92d scenario show that stabilization of CH_4 (and tropospheric O_3) at 2050 could occur with little additional climate policies, provided that both population growth and economic growth are low and that resources are relatively sparse. For the IS92d scenario, CH_4 is stabilized at the turn of the century at a level of 2.0 ppmv, slightly decreasing afterwards. For the IS92c scenario CH_4 concentrations stabilize at 2.6 ppmv.

For the IS92a scenario the increase in CH_4 concentration is not only due to emissions but also due to a significant increase in its atmospheric lifetime (25% in 2050 and even 50% at 2100). This is indicated in Figure 4, which shows the corresponding decrease of the OH concentration. The global trend is an OH loss of 17 % in the next 50 years. O_3, the other important tropospheric oxidant, was found to increase by 20 % up to 2050 (35% in 2100).

The trend in global tropospheric O_3 equals approximately 16 % in 50 years. Thompson *et al.* (1990) calculated the global oxidant concentrations for the period 1985-2035 with a BAU-type scenario. Their calculations, based on a model approach similar to ours, showed an identical behavior as our IS92a results; a global increase in O_3 by 10-15% together with a 10-15% decrease of OH.

CFC-11 concentrations reached their maximum of 330 pptv in 2005 and decreased to 140 pptv in 2100. For CFC-12 concentrations rose up to 600 pptv in 2015 and only decreased to 455 pptv in 2100. HCFC-22 concentrations rose up to 535 pptv in 2065 before decreasing to 500 pptv in 2100. Some of the Cl-free replacements increased to substantial levels (850 pptv for HFC-134 in 2100), enough to make them important for radiative forcing. The concentration of the ozone-depleting stratospheric free Cl reached a maximum of 3.2 ppbv in 2000 but decreased very slowly; it reached its 1990 level only in 2065.

The scenarios IS92a, IS92c and IS92f have identical emissions data for the halogenated species. Therefore the differences in the simulated stratospheric free Cl loading for these scenarios (Figure 5) results from the differences in the oxidizing capacity of the troposphere, affecting the lifetimes of HCFCs. The difference are in the order of magnitude of hundreds of pptv by the year 2100. The year at which the free Cl level reaches its present day level again lie decades apart for the three scenarios (2064 for IS92a, 2015 for IS92c and 2077 for IS92f). Comparing these results to the results for the IS92b scenario we observe that, for decreasing the free Cl concentrations, the effect of low emissions of hydrocarbons is about half as large as the effect of global compliance of the Montreal Protocol rather than partial compliance. This again illustrates a connection between ozone layer policies and climate policies.

The IMAGE 2.0 Conventional Wisdom scenario clearly illustrates the enormous potential effect of land cover changes. Built on assumptions similar to IS92a both greenhouse gas emissions and concentrations are close for these scenarios before the year 2025. Afterwards differences appear in the emissions of hydrocarbons and NO_x, due to expanded human land demands. This leads to a changed trend, slowing down the atmospheric accumulation of CH_4 and tropospheric O_3. Even while the uncertainties in the land cover changes are large, the considerable effects on atmospheric composition cannot be neglected.

Acknowledgements

The analysis of the CO yield of NMHCs and the comparison of the Atmospheric Composition model with the 2-D TNO-Isaksen model has been done in cooperation with dr. M.G.M. Roemer, TNO Institute of Environmental Sciences, Delft, the Netherlands. The work in this paper was supported by the Dutch National Research Program on Global Air Pollution and Climate Change (NOP grant 851042) and by the Dutch Ministry of Housing, Physical Planning and Environment (MAP grant 482507).

Appendix 1

THE GLOBAL NO_x - OH - O_3 RELATIONS

In analyzing the global dependence of OH on NO_x we observe that almost all data have large local differences: local values for NO_x-emission levels, trends in the NO_x-emissions, and OH concentrations. Simply taking an area-weighted value for all these data would lead to an overestimation of the impact of additional NO_x emissions on concentrations of O_3 and OH by 25% or more, as can be seen below.

In determining the global dependence of OH on NO_x, we started with considering the distribution of NO_x emissions over the various emission sources as given for the base year 1990 by IPCC in their emission scenarios (IPCC, 1992), see Table A.1. Not included in these estimates is the small source of odd nitrogen emitted by airplane traffic, potentially important for O_3 concentrations around the tropopause (Beck et al., 1992)

Parameterizations of the trends were taken from Brühl and Crutzen (1988) for each of the individual sources. Natural emissions are assumed to be constant, while energy-related NO_x emissions are scaled linearly to the CO_2 emissions from fossil fuels. NO_x emissions from biomass burning are scaled linearly to the population in the developing countries. The allocation of these global emissions to the chemically coherent regions was based on the latitudinal distributions of the emissions from the individual sources (Hough and Derwent, 1990). The distributions are given on a 7.5° latitudinal grid for 10 categories of emissions, including the categories appearing in the IPCC scenario.

Combining these data we obtain a relation between the global trend in the NO_x emissions and the regional trends of the NO_x emissions in each of the chemically coherent regions; see Table A.2. From this table it can be seen that a global increase in NO_x emission is dominated by an increase in urban and continental regions. Note that this distribution of NO_x trends for the five chemically coherent regions is used to derive the best estimate of global sensitivity coefficients for the NO_x - OH - O_3 relations and is not used explicitly in the IMAGE 2.0 reference run or in the climate scenarios.

TABLE A.1

Distribution of NO_x emissions over the various sources of emission, according to IPCC (1992).

Emission source	NO_x emission (TgN a^{-1})
Commercial Energy	25
Biomass Burning	9
Natural Land	12
Lightning	9
Total	55

TABLE A.2

Data for chemically coherent regions, from Thompson et al. (1989). The regions are Urban (U), Clean continental (C) and Marine (M) in the 30°-60°N latitude belt, Low latitude belt (L) (30°N-30°S), Southern hemisphere mid latitude belt (S) (30°-60°S) and the global troposphere. Given are the NO_x emissions in these regions, the area fraction, the local trends in the NOx emissions, the OH and tropospheric O_3 contents (in percents of the total between 60°N and 60°S) and the sensitivities of OH and tropospheric O_3 to changes in the NO_x emissions.

	U	C	M	L	S	global
em. NO_x (TgN)	10.2	13.0	0.3	27.5	2.4	53.4
area fraction (%)	3.1	7.8	10.2	57.8	21.1	100
NO_x trend (% a^{-1})	1.39	1.27	0.0	0.67	0.95	0.97
OH contents (%)	2.8	6.5	4.8	72.0	14.0	100
O_3 contents (%)	6.8	10.0	8.8	57.0	17.0	100
$dln(OH)/dln(NO_x)$	-0.3	0.4	0.3	0.55	0.75	0.4
$dln(O_3)/dln(NO_x)$	0.11	0.1	0.11	0.07	0.09	0.064

TABLE A.3

Present day trends in OH and tropospheric O_3 due to the changing NO_x emissions, combining data from Thompson et al. (1989), IPCC (1991), Brühl and Crutzen (1988) and Hough and Derwent (1990).

	U	C	M	L	S	Total
OH-trend % a^{-1}	-0.42	0.51	0.00	0.39	0.71	0.39
O_3-trend % a^{-1}	0.15	0.13	0.00	0.05	0.09	0.06
$dln(OH)/dln (NO_x)$		local data given in Table A.2				0.40
$dln(O_3)/dln (NO_x)$						0.064

For present day conditions the sensitivity coefficients are given in Table A.2. Some values follow from extrapolations from the data in Thompson et al. (1989). The NO_x emission levels given there are quite low: the global NO_x emission following from the base figures in Thompson et al. (1989) is only 15.3 TgN a^{-1} (Guthrie and Yarwood, 1991), roughly one third of the IPCC estimate. These data are used to convert the local trends in NO_x emission into the resulting local trends in the concentrations of OH and trop O_3. Knowing the OH and tropospheric O_3 contents of the coherent regions this gives us the global trends in OH and trop O_3, see Tables A.2 and A.3.

The resulting estimates of the global sensitivity coefficient imply that a 1% increase in NO_x emission corresponds to an increase of OH by 0.4% and for tropospheric O_3 by 0.064%. Both numbers follow from the observation that OH and O_3 are found dominantly at low latitudes (L-region), while the changes in NO_x emission in that region is relatively small at present. Both the IS92 and the IMAGE 2.0 Conventional Wisdom scenarios project a larger part of the NO_x emissions to take place in the low latitude region in the second half of the next century. This would imply an important shift in the distribution of the emissions. The impact on the global sensitivities to the emissions may however be modest, since the low latitude sensitivities (Thompson et al., 1989) are relatively close to the global sensitivities, even for increased levels of NO_x emissions.

Sensitivities of tropospheric oxidants to a doubling of NO_x emissions are calculated by various modeling groups (WMO, 1992). The results found lead to sensitivity coefficients ranging from 0.14 to 0.18 for OH and 0.05 to 0.19 for O_3. For small increases of the NO_x emissions however, the sensitivity is found to be much stronger: O_3 formation is 2.5 time as effective for small changes compared to large changes. So for small changes in the NO_x emissions the sensitivities used in this paper compare to sensitivities found in literature. For large increases in the emissions the present sensitivities are probably too large. As a result our simulations of the concentrations of CH_4 and H(C)FCs will be a bit low. However, the sensitivity coefficient we use are appropriate for the historical period, from pre-industrial to present.

References

Alcamo, J., G.J.J. Kreileman, M.S. Krol and G. Zuidema: 1994a, Modeling the global society-biosphere-climate system, Part 1: Model description and testing. *Wat. Air Soil Pollut.*, **76** (this volume).

Alcamo, J., G.J. van den Born, A.F. Bouwman, B. de Haan, K. Klein Goldewijk, O. Klepper, J. Krabec, R. Leemans, J.G.J. Olivier, A.M.C. Toet, H.J.M. Vries, H.J.M. van der Woerd: 1994b, Modeling the global societybiosphere-climate system, Part 2: Computed scenarios, *Wat. Air Soil Pollut.*, **76** (this volume).

Atkinson, R.: 1985, Kinetics and mechanisms of the gas-phase reactions of the hydroxyl radical with organic compounds under atmospheric conditions, *Chem. Rev.*, **85**: 69-201.

Beck, J.P., C.E. Reeves, F.A.A.M. De Leeuw and A. Penkett: 1992, The effect of aircraft emissions on tropospheric ozone in the northern hemisphere, *Atm. Env.*, **26A**(1): 17-29.

Brühl, C. and P. J. Crutzen: 1988, Scenarios of possible changes in atmospheric temperatures and ozone concentrations due to man's activities, estimated with a one-dimensional coupled photochemical climate model, *Clim. Dyn.*, **2**: 173-203.

Guthrie, P. D. and G. Yarwood: 1991, *Analysis of the Intergovernmental Panel of Climate Change (IPCC) Future Methane Emissions*, Report SYS-APP-91/114, Systems Applications International.

Haan, B.J. de, Jonas, M., Klepper, O., Krabec, J., Krol, M.S. and Olendrzyński, K.: 1994, An atmosphere-ocean model for integrated assessment of global change, *Wat. Air Soil Pollut.*, **76** (this volume).

Hough, A. M.: 1991, Development of a two-dimensional global tropospheric model: model chemistry. *J. Geophys. Res.*, **96**(D4) : 7325-7362.

Hough, A. M. and R. G. Derwent: 1990, Changes in the global concentration of tropospheric ozone due to human activities, *Nature*, **344**: 645-648.

IPCC: 1990, J.T. Houghton, G.J. Jenkins and J.J. Ephraums (eds), *Climate Change. The IPCC Scientific Assessment*, Cambridge Univ. Press.

IPCC: 1991, *Climate Change: The IPCC Response Strategies*, Island Press.

IPCC: 1992, J.T. Houghton, B.A. Callander and S.K. Varney (eds), *Climate Change 1992. The Supplementary Report to the IPCC Scientific Assessment*, Cambridge Univ. Press.

Isaksen, I.S.A. and Ø. Hov: 1987, Calculation of trends in the tropospheric concentration of O3, OH, CO, CH4 and NOx, *Tellus*, **39B**: 271-285.

Klein Goldewijk, K., J.G. van Minnen, G.J.J. Kreileman, M. Vloedbeld and R. Leemans: 1994, Simulating the carbon flux between terrestrial environment and the atmosphere, *Wat. Air Soil Pollut.*, **76** (this volume).

Kreileman, G.J.J. and A.F. Bouwman: 1994, Computing land use emissions of greenhouse gases, *Wat. Air Soil Pollut.*, **75** (this volume).

Krol, M.S.: 1994, Uncertainty analysis for the computation of greenhouse gas concentrations in IMAGE. In: J. Grasman and G. van Straten (eds), *Predictability and Nonlinear Modelling in Natural Sciences and Economics*, Kluwer.

Liu, S.C., M. Trainer, F.C. Fehsenfeld, D.D. Parrish, E.J. Williams, D.W. Fahey, G. Hubler and P.C. Murphy: 1987, Ozone production in the rural troposphere and the implications for regional and global ozone distributions, *J. Geoph. Res.*, **92**(D4): 4191-4207.

Logan, J.A., M.J. Prather, S.C. Wofsy and M.S. McElroy: 1981, Tropospheric chemistry: a global perspective. *J. Geophys. Res.*, **86**: 7210-7254.

Madronich, S. and C. Granier: 1992, Impact of recent total ozone changes on tropospheric photodissociation, hydroxyl radicals, and methane trends, *Geophys. Res. Lett.*, **19**: 465-467.

Mikolajevich, U., B.D. Santer and E. Maier-Reimer: 1990, Ocean response to greenhouse warming, *Nature* **345**: 589-593.

Prather, M. and C.M. Spivakovsky: 1990, Tropospheric OH and the lifetimes of hydrochlorofluorocarbons, *J. Geophys. Res.*, **95**: 18723-18729.

Prather, M. J.: 1989, *An Assessment Model for Atmospheric Composition*, NASA Conf. Publ. 3023, NASA, New York.

Prinn, R., D. Cunnold, P. Simmonds, F. Alyea, R. Boldi, A. Crawford, P. Fraser, D. Gutzler, D. Hartley, R. Rosen and R. Rasmussen: 1992, Global average concentration and trend for hydroxyl radicals deduced from ALE/GAGE trichloroethane (methyl chloroform) data for 1978-1990, *J. Geophys. Res.*, **97**(D2): 2445-2461.

Roemer, M.G.M.: 1991, *Ozone and the Greenhouse Effect*, Report No. R 91/227, IMW/TNO, Delft, the Netherlands.

Rotmans, J., M.G.J. den Elzen, M.S. Krol, R.J. Swart and H. van der Woerd: 1992, Stabilizing atmospheric concentrations: towards international methane control, *Ambio*, **21**(6) : 404-413.

Steele, L.P., E.J. Dlugokencky, P.M. Lang, P.P. Tans, R.C. Martin and K.A. Masarie: 1992, Slowing down of the global accumulation of atmospheric methane during the 1980s, *Nature*, **358**: 313-316.

Talukdar, R.K., A. Mellouki, A.-M. Schmoltner, T. Watson, S. Montzka and A.R. Ravishankara: 1992, Kinetics of the OH reaction with methyl chloroform and its atmospheric implications, *Science*, **257**: 227.

Thompson, A.M., R.W. Stewart, M.A. Owens and J.A. Herwehe: 1989, Sensitivity of tropospheric oxidants to global chemical and climate change, *Atm. Env.*, **23**(3): 519-532.

Thompson, A.M., M.A. Huntley and R.W. Stewart: 1990, Perturbations to tropospheric oxidants, 1985-2035 1. calculations of ozone and OH in chemically coherent regions, *J. Geophys. Res.*, **95**(D7): 9829-9844.

Thompson, A.M.: 1992, The oxidizing capacity of the earth's atmosphere: probable past and future changes, *Science*, **256**: 1157-1165.

Vaghjiani, G.L. and A.R. Ravishankara: 1991, New measurement of the rate coefficient for the reaction of OH with methane, *Nature*, **350**: 406-409.

Vries, H.J.M. de, R.A. van den Wijngaart, G.J.J. Kreileman, J.G.J. Olivier and A.M.C. Toet: 1994, A model for calculating regional energy use and emissions for evaluating global climate scenarios, *Wat. Air Soil Pollut.*, **76** (this volume).

Wigley, T.M.L. and S.C.B. Raper: 1992, Implications for climate and sea level of revised IPCC emissions scenarios, *Nature*, **357**: 293-300.

WMO: 1992, *Scientific Assessment of Ozone Depletion- 1991*, Global Ozone Research and Monitoring Project, Rep. No. 25., WMO.

AN ATMOSPHERE-OCEAN MODEL
FOR INTEGRATED ASSESSMENT OF GLOBAL CHANGE

B.J. DE HAAN, M. JONAS*, O. KLEPPER,
J. KRABEC*, M.S. KROL, K. OLENDRZYŃSKI*

*National Institute of Public Health and Environmental Protection, (RIVM)
P.O. Box 1, 3720 BA Bilthoven, the Netherlands*

*International Institute for Applied Systems Analysis,
A-2361 Laxenburg, Austria*

Abstract. This paper describes the atmosphere-ocean system of the integrated model IMAGE 2.0. The system consists of four linked models, for atmospheric composition, atmospheric climate, ocean climate and for ocean biosphere and chemistry. The first model is globally averaged, the latter are zonally averaged with additional resolution in the vertical. The models reflect a compromise between describing the physical, chemical and biological processes and moderate computational requirements. The system is validated with direct observations for current conditions (climate, chemistry) and is consistent with results from General Circulation Model experiments. The system is used in the integrated setting of the IMAGE 2.0 model to give transient climate projections. Global surface temperature is simulated to increase by 2.5 K over the next century for socio-economic scenarios with continuing economic and population growth. In a scenario study with reduced ocean circulation, the climate system and the global C cycle are found to be appreciably sensitive to such changes.

Keywords: climate model, atmospheric chemistry, oceanic C cycle, scenario evaluation, integrated model.

1. Introduction

Integrated global modeling of global change involves describing a broad variety of processes like changing energy demands and changing land use, having effects like changing anthropogenic concentrations of greenhouse gases and a change in surface albedo. These changes affect the climate, which in turn may provide feedbacks (positive or negative) to the system (for example, further changes in vegetation, effects on sources and sinks of greenhouse gasses). In this series of processes the characterization of the atmosphere and ocean (consisting of atmospheric chemistry, the atmospheric climate, the ocean climate and oceanic C cycle) can be grouped together in a natural way. This paper describes the atmosphere-ocean system of IMAGE 2.0 (Alcamo *et al.*, 1994a). The atmosphere-ocean system of models consists of four submodels: atmospheric composition, atmospheric climate, ocean climate and ocean biosphere and chemistry. After an introduction to the entire atmosphere-ocean system, each of the component models is described in turn, together with their validation. The linkage of these models is then reported and results from the four linked models are compared to equilibrium General Circulation Model (GCM) results. Finally, results from transient runs are reported.

The emphasis on the atmosphere-ocean calculations in an integrated model depends strongly on the goals of such a model. In a policy oriented model one may be tempted to describe the atmosphere-ocean system quite coarsely, since the management possibilities

Water, Air, and Soil Pollution **76**: 283–318, 1994.
© 1994 *Kluwer Academic Publishers.*

for this part of the system are obviously quite limited. However, to assess the effect of policy measures on "manageable" parts of the system (e.g., fossil fuel emissions, land use changes) one does need to predict the response of other parts. In this respect it is important to note that this response may change over time. For example, a change in the amount of sea-ice or snow cover changes the heat balance of the polar zones, and constitutes one of the many feedbacks in the climate system.

This implies that the atmosphere-ocean system would require a predictive rather than just an empirical ("black box") model. At present, a predictive model based on first principles of physics, chemistry and biology is not possible, both because we don't sufficiently understand these basic principles and because such a "molecular" approach would require too much computer resources. Nevertheless, we can achieve predictive "grey box" models by including a sufficient amount of physical and chemical realism and by checking the model against a variety of observations.

The level of detail depends on the purposes of the individual models, and existing models range from elaborate three-dimensional coupled ocean-atmosphere models to simple zero-dimensional (globally averaged) energy balance models. An example of the latter is the IMAGE 1.0 model (Rotmans, 1990), which is zero-dimensional in its atmosphere and one-dimensional in the ocean. Model experiments with coupled GCM's demand months of computation time on the fastest computers available, while with the simple models experiments can be carried out on a personal computer within a few minutes. In the middle of this range two-dimensional zonally and yearly averaged energy balance models (Peng *et al.*, 1982; Harvey and Schneider, 1985; Peng *et al.*, 1987) play their role with typical running times of one day per hundred years of simulation on a workstation. For our present purpose a zonally averaged 2-dimensional (i.e. latitude and height/depth) model for the atmosphere and ocean has been developed in order to include a number of latitudinally and vertically heterogeneous processes with sufficient realism, for example radiative forcing, precipitation, ocean circulation, etc. A further increase in complexity (e.g., 3-dimensional resolution) was not thought to be necessary as existing 2-dimensional models are generally able to represent the processes relevant for our purpose in sufficient detail.

In fact, apart from its ease of use and analysis, a 2-dimensional model may have a distinct advantage over 3-dimensional models in terms of flexibility of operation. In three-dimensional GCM the linkage between atmosphere and ocean models is a major modeling problem, the solution of which is still in its early stages (Neelin *et al.*, 1992). The atmosphere and ocean models are individually stable when forced by present climate observations. However, when linked together via the surface fluxes at their interface, they often exhibit a climate drift, which quickly deteriorates the simulation. As a remedy, an artificial flux correction has to be introduced to obtain an acceptable simulation of the present climate. In a more aggregated model like the present Atmosphere Ocean System, no flux correction is required and calibration of the linked model can be done more easily.

The system parts have not been selected to be the best available on its terrain. They are to form, when linked, a harmony of models, which enable transient climate projections in the context of policy analysis with IMAGE 2.0. In the following sections of this paper the

TABLE 1

Main features of the Atmospheric Composition Model.

spatial resolution	tropospheric average
temporal resolution	annual average
governing equations	mass balance equations for greenhouse gases
prescribed characteristics	sensitivities of atmospheric oxidants
main processes	tropospheric oxidation, photodissociation, O_3, OH formation and O_3 depletion
explicit feedbacks related to temperature	oxidation rates, impact humidity on OH
compounds	CO_2, CH_4, N_2O, O_3, CFCs, HCFCs, HFCs, OH, NO_x, Cl, NO_y
running time	less than 1 second on a SUN SPARC 2 workstation for one model year

system parts are described. Consecutively, examples of the application of the system are given. The first example compares the "Conventional Wisdom" scenario with the IS92a scenario of IPCC. In the second example ocean circulation is prescribed to change according to model results of Mikolajewicz *et al.* (1990). In this sensitivity study the impact of ocean circulation change on surface temperature and C cycle is demonstrated.

2. Description of the Atmospheric Composition Model

The atmospheric composition model connects the emission models of IMAGE 2.0 (energy, industrial and land-use emissions) with the radiative forcing module in the zonally-averaged atmospheric climate model (ZACM). In the model the accumulation of the most important greenhouse gases (CO_2, CH_4, N_2O, O_3 and the halogenated carbons) is accounted for; the fate of H_2O is modeled in the Zonal Atmospheric Climate model. The model simulates the interaction between greenhouse gases, and the exchange of these gases with the ocean and biosphere. To simulate the effect of increasing sulphur emissions a simple S budget equation is also included.

An extensive description of the model is given elsewhere (Krol and van der Woerd, 1994). Here, a summary of the model in connection with the other models of the atmosphere-ocean system will be presented; the main features of the model are given in Table 1.

2.1. MODEL DESCRIPTION

The atmospheric composition model of IMAGE is a box model for the troposphere similar to the AMAC model (Prather, 1989). The goal of the model is to simulate the changes in the concentrations of the main greenhouse gases that are important for climate change. As most important greenhouse gases are distributed nearly homogeneously over the atmosphere (except for O_3) and have long atmospheric residence times, the concentrations can be represented by both tropospheric and annual averages. As a consequence, no transport of chemical species is described except for the fluxes between the atmosphere and the oceans.

This highly aggregated description is satisfactory for the greenhouse gases with long atmospheric residence times, but has to be modified for O_3 and its precursors and some of the compounds involved in the oxidation of CH_4. In describing changes in the concentrations of these short-living, heterogeneously distributed compounds, parameterizations are used to connect global averages to local descriptions of chemically coherent regions (Thompson et al., 1989). The parameterization takes into account the distributions of the short-living compounds and temperature over the troposphere (Krol and van der Woerd, 1994). In this way the composition of the atmosphere and current changes in the composition of the atmosphere are represented satisfactory. It may be noted that the averaging procedure may not be valid for conditions deviating strongly from present-day conditions.

CO_2 is a chemically inert species: consequently, the model performs a mass balance of emissions and the fluxes between atmosphere, biosphere and ocean. Atmospheric oxidation of hydrocarbons is an additional small source for CO_2. Details on the exchange process with the ocean will be given later.

CH_4 is emitted by such heterogeneously distributed sources as bogs, marshes, fossil fuel and biomass burning, cattle and rice paddies. It has an atmospheric life time of about ten years, which allows the gas to spread homogeneously over the atmosphere. Three sinks are modeled: oxidation in the troposphere, absorption by the biosphere, and loss to the stratosphere. The oxidation of CH_4 by OH is by far the largest sink. OH is produced indirectly from either the photo-dissociation of O_3 and H_2O or from processes involving NO_x. Most of the OH is consumed by the oxidation of CO. CO is predominantly emitted by natural sources such as biomass burning, but it is also inadvertently emitted by human activities. In the chemical cycle of our model it is also an intermediate product of the atmospheric oxidation of CH_4 and NMHCs.

For N_2O and CFCs a fixed atmospheric lifetime (WMO, 1992) is used in the calculations. The model takes into account the main sink of HCFCs and HFCs through their oxidation by OH and a minor sink through transport to the stratosphere.

The processes affecting stratospheric O_3 are entirely different from the processes in the troposphere. In the troposphere hydrocarbons and NO_x enhance O_3 concentrations, while in the stratosphere Cl and odd-N deplete O_3. Stratospheric O_3 is important for its role in the absorption of UV radiation. An increase of UV light has the potential to influence both tropospheric chemical reaction rates and the welfare of living creatures at the Earth's surface. To compute stratospheric O_3, and to evaluate its impact, two boxes were added to the original zero-dimensional structure. The basic O_3 chemistry in the lower and upper stratosphere is parameterized, mainly as a function of the concentration of active Cl radicals.

SO_4^- aerosols reflect short wave radiation and stimulate the formation of clouds. Directly and indirectly the aerosols seem to counteract the potential global warming induced by the aforementioned greenhouse gases by increasing the Earth's albedo

(Charlson *et al.*, 1991). The Zonal Atmospheric Climate Model simulates the direct radiative effect of SO_4^- column amounts with a zonal resolution. The SO_4^- column amounts are derived from the result of studies with the 3-D atmospheric chemistry model MOGUNTIA (Langner *et al.*, 1992).

2.2. VALIDATION

A comparison was made between the results of the Atmospheric Composition Model and observed CH_4 concentrations (Rasmussen and Khalil, 1981) as well as simulations of the concentrations of CH_4, OH and O_3 with models contributing to the IPCC Assessment (IPCC, 1990; Guthrie and Yarwood, 1991) and a comparison with the 2-D TNO-Isaksen model (Roemer, 1991) has been given elsewhere in this issue (Krol and Van der Woerd, 1994). It revealed that the concentration of CH_4 and the trends in the concentrations of OH and O_3 are simulated well by the model.

In an uncertainty analysis of the model (Krol, 1994) it was found that after a century of simulation with the so-called IS92a emission scenario (IPCC, 1992), the coefficient of variation for the concentration of CH_4 in 2100 is some 20%. The uncertainty in the increase in CH_4 between 1990 and 2100 is thus as high as 40%. This uncertainty range is of the same order of magnitude as the range of model results compiled for the IPCC 1990 scenarios (IPCC, 1990; Guthrie and Yarwood, 1991). For other greenhouse gases the uncertainties are of similar size. The uncertainty is partly due to uncertainties in the emissions and uncertainties in the CH_4 climate feedback, but uncertainties in the parameters of the chemical description are equally important, especially over long integration periods.

3. Description of the Zonal Atmosphere Climate model

With respect to the greenhouse problem the atmosphere and its underlying surface play two dominant roles in the global climate: at first by the way the atmosphere transmits and remits radiation (of the sun) and subsequently by the way it redistributes heat between land and sea masses and between tropical, temperate and polar regions. These roles have been translated into an atmosphere energy balance model that is described here.

The physical processes in the atmosphere are realized by a two-dimensional Zonally averaged Atmosphere Climate Model (ZACM) which follows the Multilayer Energy Balance Model approach taken by Peng and co-workers (Chou *et al.*, 1982; Peng *et al.*, 1982; Peng *et al.*, 1987). The ZACM resolves 18 latitude belts of 10° width each and up to 18 layers vertically for radiation transfer processes. The latter are realized by means of a recent 1-D Radiative Convective Model (RCM) (MacKay and Khalil, 1991) which was integrated into the ZACM. In this version the 2-D ZACM describes zonally and annually averaged thermodynamic conditions (Jonas *et al.*, 1992). The main features of the model are summarized in Table 2. A schematic illustration of one latitude belt of the model is shown in Figure 1.

2-D ZCM

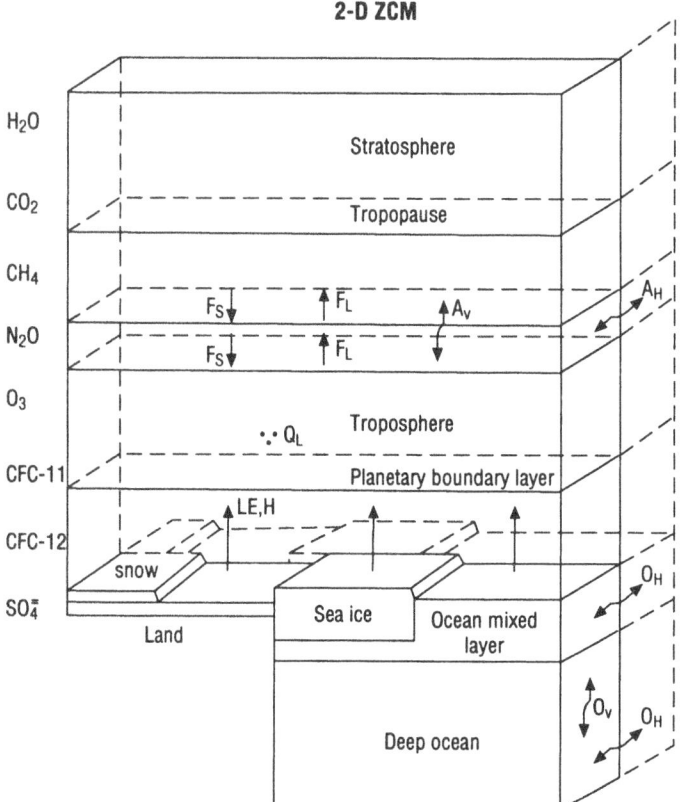

Figure 1: Schematic illustration of one latitude belt of the 2-D Zonal Atmosphere Climate model. F_S, F_L denote short-and long-wave radiation. A_V and A_H vertical and horizontal heat transport; Q_L latent heat release; LE and H latent and sensible heat fluxes for land (with and without snow), sea ice and ocean; O_V and O_H vertical and horizontal heat transport in the ocean.

3.1.   MODEL DESCRIPTION

In this section an outline of the radiative transfer and heat transport submodels of the Zonally averaged Atmosphere Climate Model are presented. The section concludes with a discussion of the most important tunable parameters.

3.1.1. Radiative transfer

The model is forced by the concentrations of a number of greenhouse gases. With respect to the absorption and emission of terrestrial radiation, the model accounts for H_2O, CO_2, CH_4, N_2O, O_3 and CFC-11 and CFC-12. With respect to the solar spectrum, the principal absorbers are H_2O in the troposphere and O_3 in the stratosphere. In addition, CO_2 and O_2 are also accounted for as minor absorbers. Both the long-wave and the shortwave parameterizations employed by Mackay and Khalil are all taken from literature; they

TABLE 2

Main features of the ZACM.

spatial resolution	18 latitudinal belts of 10° width, 8 vertical layers in the atmosphere (18 layers for radiative transfer calculations)
temporal resolution	annual average (seasonality envisaged)
governing equations	energy balance equations for atmospheric, land and sea ice surface temperatures
prescribed characteristics	relative humidity, clouds (height, cover, optical depth), surface albedos of land and free ocean, water availability
main processes	radiation transfer, diffusive horizontal and vertical heat and moisture transport in the atmosphere, convection by means of convective adjustment, latent heat release
explicit feedbacks related to temperature	H_2O, snow-albedo, sea ice-albedo, horizontal and vertical transport of heat and moisture, latent heat release
surface types	four with respect to surface albedo: free land, land covered by snow, free ocean and sea ice
orography	none (envisaged)
greenhouse gases	LW calculations: H_2O, CO_2, CH_4, N_2O, O_3, CFC-11, CFC-12 SW calculations: H_2O, O_3, CO_2, O_2
aerosols	direct radiative effect of SO_4^-
running time	ca. 5 minutes on a SUN SPARC 2 workstation for one model year

represent the state-of-the-art. Their model has been extensively tested against other 1-D RCMs in regard to CO_2 increase, variation of the solar constant and aerosol emissions from volcanic eruptions.

The cooling effect of tropospheric SO_4^- aerosols is included in the model by accounting for the backscattering of solar radiation under cloud-free conditions, i.e. their direct radiative influence. We follow the methodology described by Coakley *et al.* (1983) and Charlson *et al.* (1991). The concentration data for SO_4^- are taken from Langner *et al.* (1992). The optical depth of the aerosol layer and the resulting change of the reflected solar flux at the top of the atmosphere (TOA) is a function of latitude. The global mean of the change (0.83 W/m^2) compares well with the value given in Charlson *et al.* (1991) i.e. 1.02 W/m^2, who used slightly higher emissions in their calculations.

Radiative transfer calculations require a fine vertical resolution. Therefore the atmosphere is divided into 18 layers including 6 layers above 250 hPa. However, for non-radiative processes a coarser resolution of 8 vertical layers is used. For this study the four lowest layers of MacKay and Khalil's RCM were combined into one extending up to 875 hPa which represents the planetary boundary layer in the model. The remaining 14 layers of their model were grouped into pairs thus forming seven thicker layers in the free atmosphere. Three of the seven layers are below 500 hPa, two between 500 and 100 hPa and the last two represent the stratosphere. Surface pressure is set to 1000 hPa.

3.1.2. Non-radiative processes
The governing equations of the model are heat balance equations for atmosphere, land and sea ice. Energy conservation is also assured at the atmosphere-ocean interface and is described below. The heat balance equations are used to determine the temperature in each grid element of the 2-D space. Temperature is the only prognostic variable of the model, i.e. all heat transport processes, the release of latent heat due to condensation, or the extent of snow and ice cover including their associated albedo feedbacks are parameterized in terms of a temperature field which depends on latitude and altitude.

The surface heat balance equation for land is given by

$$C_L D \frac{\partial T_L}{\partial t} = S - I - H - LE \qquad [W/m^2] \tag{1}$$

where C_L, D and T_L are the heat capacity of land, the effective depth of the surface and the temperature of land, S is the solar radiation absorbed by the surface, and I, H and LE denote upward fluxes of long-wave radiation, sensible and latent heat, respectively. In case of sea ice, an additional term is added on the right hand side, which describes the vertical heat flux through sea ice. The latter equation has not been employed in this study and sea ice temperature is set equal to land surface temperature.

The atmospheric heat balance equation can be written as

$$\frac{\partial T}{\partial t} = Q_s + Q_I + Q_L + A \qquad [K/day] \tag{2}$$

where T is the atmospheric temperature, and Q_S, Q_I and Q_L are heating rates due to solar and long-wave radiation, latent heat release, respectively. A is a heating rate due to the dynamical redistribution of heat. This process is a rather complex atmospheric process. A full consideration of it requires no less than a solution of the full set of equations of fluid motion as in a GCM. In the context of this model in which velocity is not a variable, the dynamic transport is treated in a simplified manner. The horizontal heat and moisture transport due to large-scale eddies is parameterized by means of meridional diffusion with constant diffusion coefficients, $0.15 \cdot 10^7$ m²/s and $0.5 \cdot 10^7$ m²/s, respectively. In case of vertical heat and moisture transport, only the contributions from small and mesoscale eddies are accounted for in the model. The eddy diffusion coefficient decreases linearly with height from 1 m²/s at the top of the boundary layer to zero at the tropopause.

The precipitation rate is computed as the residual of the surface evaporation rate and the divergence of H_2O in a vertical column since both must balance in a steady-state atmosphere. In the tropics the total amount of precipitation is calculated for the entire region (i.e. for all tropical latitude belts) and then redistributed in a weighted fashion according to observed values (Jaeger, 1976). Once the precipitation rate is known, it is used in the computation of the heating rate due to atmospheric latent heat release Q_L.

With respect to surface albedo the model accounts for four surface types: free land, land covered by snow, open ocean and sea ice. The parameterizations of snow cover on land and sea ice extent are empirically related to surface air temperature and oceanic mixed layer temperature respectively. In the case of sea ice extent the relationship is derived separately for each hemisphere.

At the end of each integration step a convective adjustment scheme with a prescribed critical lapse rate is applied which prevents the model from reaching unrealistic vertical temperature profiles.

3.1.3. Prescribed parameters

Clouds in the radiative transfer scheme are represented by a single effective layer. Its maximum height is in the tropics and decreases towards the poles. The height of clouds and their optical thickness are the main tunable parameters of the model. Cloud cover is prescribed according to Harvey (1988). Annual relative humidity by latitude and height is derived from climatological data obtained from the European Centre for Medium Range Weather Forecasts. The albedo of open ocean and free land are taken from Curran et al. (1978) and Sellers (1965) respectively.

3.2. VALIDATION AGAINST PRESENT CLIMATE

The atmospheric part of the climate system is validated using prescribed oceanic mixed layer temperatures taken from Levitus (1982). Following the regional structure of the ocean model, the mixed layer temperature is specified separately for Atlantic and Pacific in each latitude belt. The model atmosphere is then forced to adjust to this boundary condition.

In Figure 2, the difference between model simulated temperatures and observed values (Oort, 1983) is shown. Only the values of the regions from 75°S to 75°N are considered because of missing observations for latitude belts centred at 85°S and 85°N. The current temperature distribution is simulated with good approximation in the troposphere, where the globally averaged absolute difference is less than 1.8 K. However, the situation is worse in the stratosphere and around the tropopause level with pronounced warming in the tropics and cooling at high latitudes. This model result is similar to GCM results in which the lower stratosphere and upper troposphere of high latitudes are systematically cooler than in reality, the difference exceeding 20 K in some models (e.g. Boer et al., 1991).

Surface evaporation (Figure 3) and precipitation rates (Figure 4) are compared with observed mean values (Sellers, 1965; Baumgartner and Reichel, 1975). The evaporation rates are somewhat overestimated in the tropics and, at the same time, underestimated in mid latitudes of both hemispheres and at high northern latitudes. The local maxima of the precipitation rates at mid latitudes are shifted towards the equator. In both hemispheres precipitation is thus underestimated in mid latitudes and, at the same time, overestimated in high latitudes. The globally averaged value is 2.73 mm/day, which corresponds to other estimates (Jaeger, 1976). The hemispherical distribution of both precipitation and evaporation is in agreement with present climate. Precipitation of the northern hemisphere exceeds the precipitation of the southern hemisphere and the opposite is true for the evaporation.

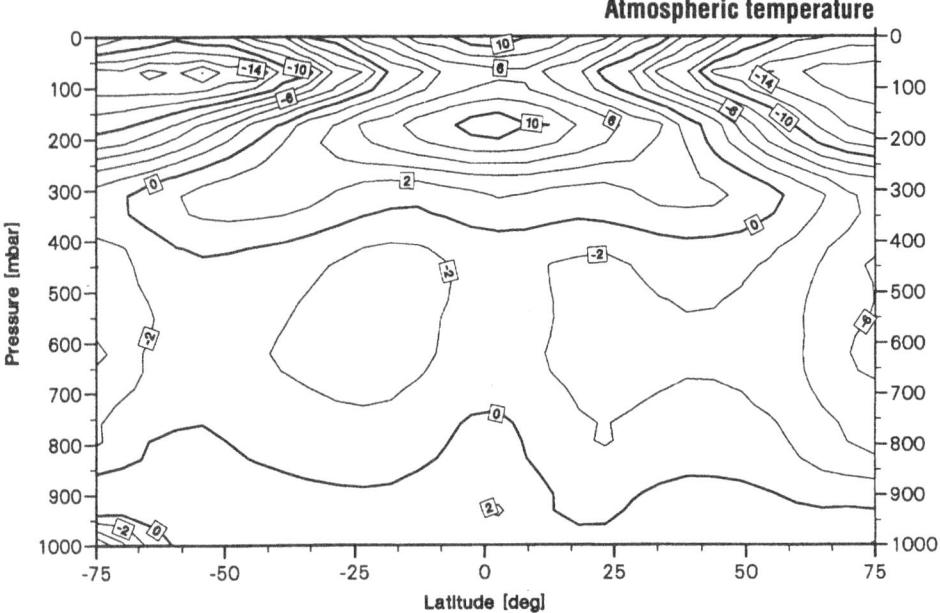

Figure 2: Temperature difference (K) between model simulation and observations for present climate.

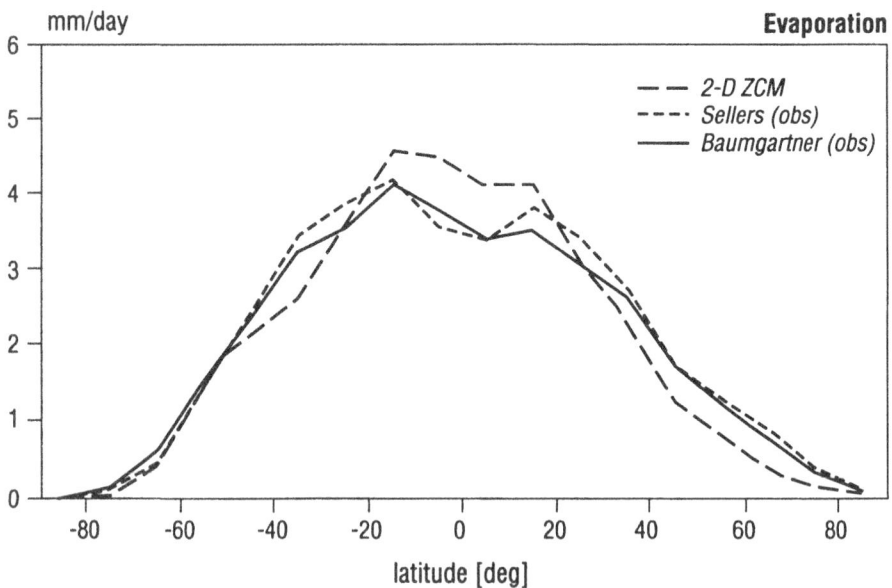

Figure 3: Evaporation rate (mm/day) for present climate (Sellers, 1965; Baumgartner and Reichel, 1975).

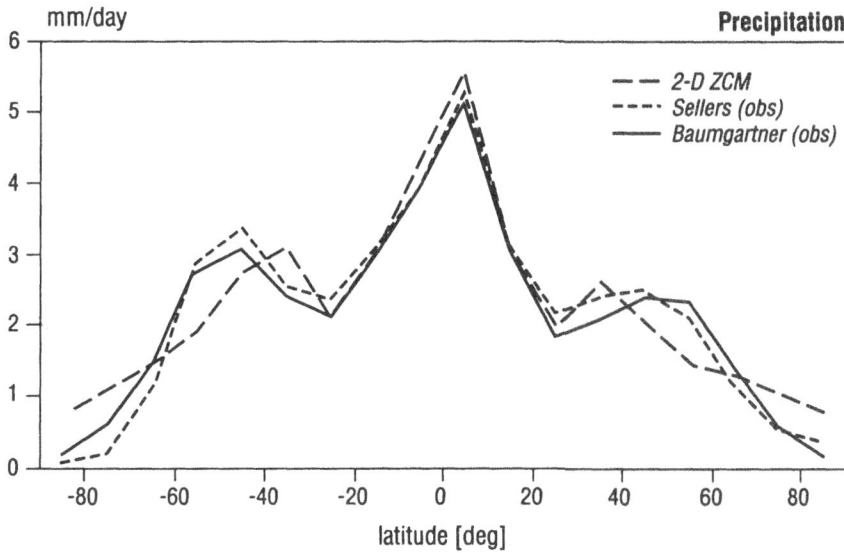

Figure 4: Precipitation rate (mm/day) for present climate (Sellers, 1965; Baumgartner and Reichel, 1975).

The computed meridional heat transport in the atmosphere (positive northward) is shown in Figure 5 in comparison with global estimates (Carissimo *et al.*, 1985) and hemispheric estimates (Oort and Vonder Haar, 1976). The location of the computed local maxima agrees well with the estimates. According to Carissimo, the uncertainty of the estimates is about 0.5 PW in mid latitudes, which seems to suggest that the model results are a small underestimate. However, the shape of the curve and its values in low and high latitudes agree well with the estimates.

The outgoing long-wave radiation at the top of the atmosphere (flux positive upward) is presented in Figure 6. The distribution of this variable, which reflects the atmospheric thermal structure, is in good agreement with observations (Ellis and Vonder Haar, 1976; Smith and Smith, 1987) for the northern hemisphere. In the case of the southern hemisphere, the observed maximum in the tropics is missing in the model and, also, the model values in mid latitudes are underestimated by some 10-15%.

4. Description of the Ocean Climate Model and Ocean Biosphere and Chemistry Model

The ocean plays two major roles in the global climate: a direct role by storing and transporting heat, and an indirect role as the largest pool of C in the C cycle, potentially available on the time scale considered. This chapter describes both the Ocean Climate Model and the Ocean Biosphere and Chemistry Model as parts of the IMAGE 2.0 model. These two models are treated together as they are closely linked: the

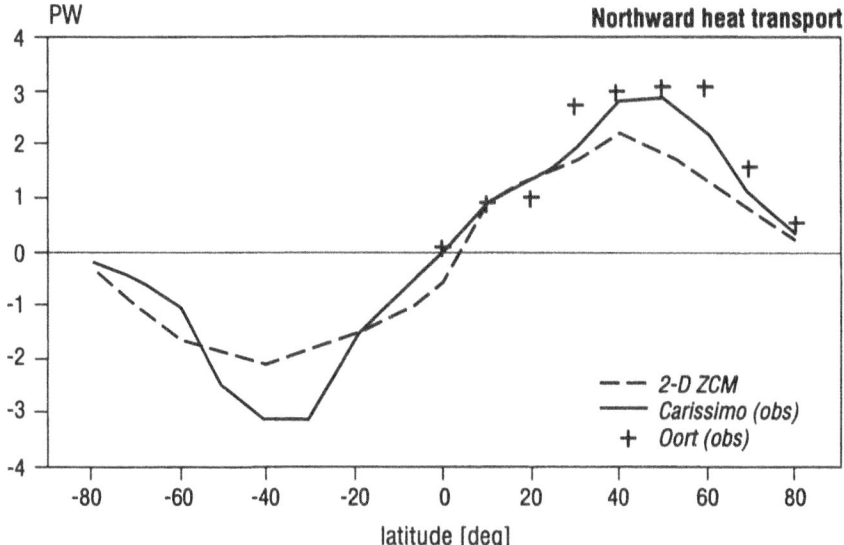

Figure 5: Meridional heat transport in the atmosphere (PW) for present climate (Oort and Vonder Haar, 1976; Carissimo *et al.*, 1985).

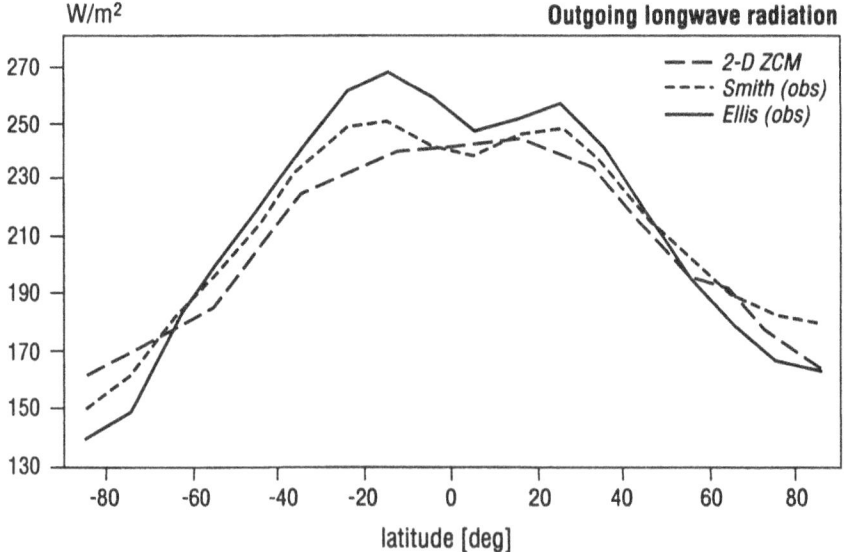

Figure 6: Outgoing longwave radiation at the top of the atmosphere (W/m²) for present climate (Ellis and Vonder Haar, 1976; Smith and Smith, 1987).

TABLE 3

Main features of the Ocean Climate and the Ocean Biosphere and Chemistry Models.

spatial resolution	two branches representing the Atlantic and Indo-Pacific Ocean respectively. 18 latitudinal belts of 10° width, 8 vertical layers.
temporal resolution	annual average
governing equations	water mass balance equation, advection-diffusion equation for temperature steady state assumptions for "fast" processes governed by "slow" transport and transformation equations
prescribed characteristics	circulation pattern: 'conveyor belt' consisting of North Atlantic and Antarctic Ocean deep water production rates and upwelling velocity, horizontal and vertical diffusion rates
main processes	surface exchange of heat and CO_2, transport of heat and CO_2, chemical equilibria and biological transformation
explicit feedbacks related to temperature	chemical equilibria constants, piston velocity
constituents	temperature C, N, O_2, DOC, Ca, ^{14}C (natural and bomb-produced), ^{14}DOC, ^{3}H
running time	ca. 0.75 seconds on a SUN SPARC 2 workstation for one model year

models share spatial resolution as well as the description ofoceanic transport. The main features of the two models are listed in Table 3.

The present model is a 2.5-dimensional model, i.e. 2-dimensional models of both Atlantic and Indo-Pacific ocean joined by an Antarctic circumpolar ocean. Such a two-basin model is needed in representing the ocean "conveyor belt" (Broecker and Peng, 1982) transporting large quantities of water southward in the Indo-Pacific surface waters and northward in the Atlantic surface waters, and returning through the deeper waters.

Although there is some evidence that the thermohaline circulation is variable on the time scales of our interest (Lenderink and Haarsma, 1993), the circulation pattern used in the model is not a prognostic variable but prescribed as input. It is either kept constant over time (equal to present-day conditions, based on the distribution of tracers) or prescribed by a scenario.

The Ocean Biosphere and Chemistry model describes the oceanic carbon pool, which plays a key role in the global C budget. Its fate is determined mainly by chemical buffering and transport, with phytoplankton primary production as an important additional process in the C budget, influencing both pool size and equilibrium time scale. The model includes the elements C (as inorganic, particulate and dissolved organic), N, O, Ca and ^{14}C, and models the fluxes between the various compartments. Although some of these elements are not directly involved in CO_2 uptake, they have been included to allow for a better validation of the model. Processes that are described are exchange with the atmosphere, physical transport, primary production, calcification, decomposition and dissolution. The ocean model draws from the work of Keir (1988) and Goudriaan (1990).

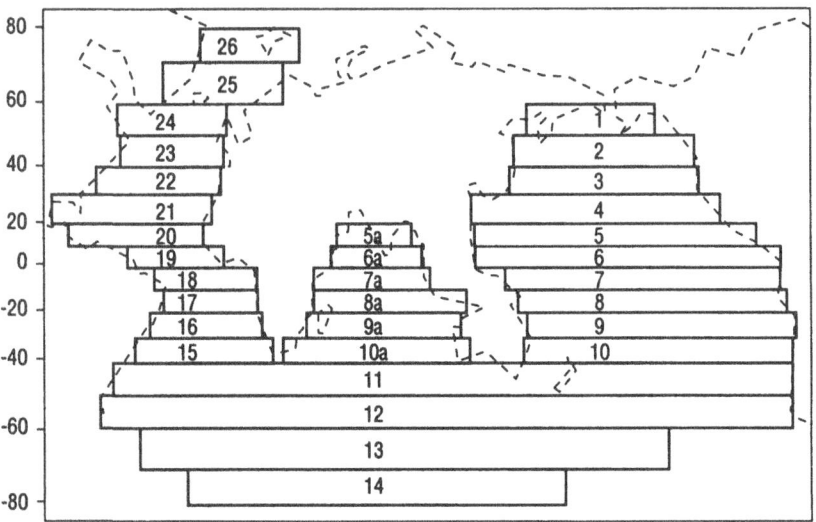

Figure 7: Layout and numbering of ocean compartments. Indian and Pacific oceans are lumped. Note distortion in surface areas due to map projection.

4.1. MODEL DESCRIPTION

In this section an outline of the ocean transport modeling, the Ocean Climate model and the Ocean Biosphere and Chemistry model is given. It also contains details on the external data sets and parameter values.

4.1.1. Ocean transports

The modeling of the ocean transport (circulation and diffusion) is used in both the Ocean Climate Model and the Ocean Biosphere and Chemistry Model. It can be viewed as part of either of the two models. Ocean circulation is described on a 10° latitudinal grid in two branches: an Indo-Pacific and an Atlantic branch, linked in the Antarctic Ocean south of 40°S (Figure 7). Vertical resolution consists of a surface layer of variable depth between 25 and 100 m depth based on observed temperature profiles (Levitus, 1982), a second layer making up the difference to 400 m and a further 9 layers of 400 m each.

The schematization of the advective flows in the model is based on the work of Keir (1988), illustrated in Figure 8. Four types of advective flows can be distinguished: horizontal lows in the upper water layers and in the deeper water layers, concentrated deep water formations and global upwelling. A Southward flow is found in the surface layer and thermocline of the Indo-Pacific basin, a northward flow in the Atlantic surface waters. The deep water horizontal flows are globally in the opposite direction. In the schematization the deep water formation is concentrated in two spots: the North Atlantic and to the South of the Antarctic Circumpolar joining Atlantic and Indo-Pacific. The upwelling of the deep water is diffuse.

In addition to advective flows there is horizontal and vertical mixing, represented by

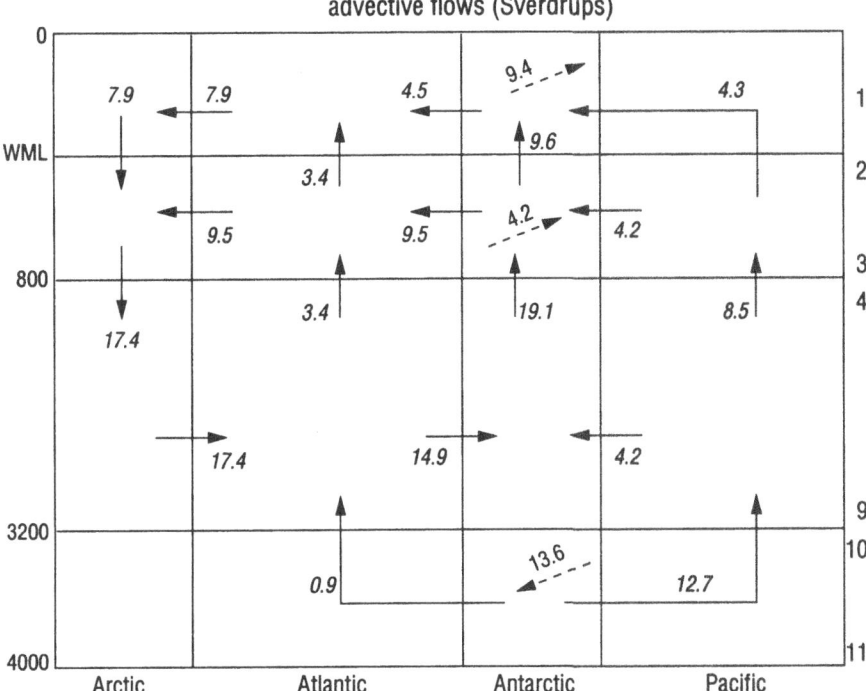

Figure 8: Schematisation of the advective flows adapted from (Keir, 1988).

dispersion (diffusion) coefficients. These model the (advective) mixing processes that cannot be represented in the 2-D grid like gyre-circulation and sub-grid mixing. Horizontal dispersion has a high value at the surface, a lower value below 400 m and an intermediate value in the transition layer. Vertical dispersion is assumed to be the same at all depths, but the polar (>60° latitude) columns have a higher value, consistent with the higher velocities of the convection in these columns.

4.1.2. The Ocean Climate Model

The temperature profile in the ocean changes due to two processes: heat exchange with the atmosphere and redistribution of heat by transport. Heat exchanges with sea-ice are not considered in the present model.

The surface heat fluxes are a result of temperature gradients between sea surface and the lower atmospheric layers. This temperature gradient results in fluxes of sensible and latent heat as well as heat exchange due to long-wave radiation. These heat fluxes are simulated by the ZACM for each 10° latitudinal belt, for Atlantic and Indo-Pacific ocean basins separate. The fluxes range from 30 W/m² from the atmosphere to the ocean in the tropical zones to 50 W/m² in the opposite direction close to the poles.

Heat in ocean is redistributed according to the transport pattern described in the previous section. Cold polar waters are convected to the ocean bottom in the North

Atlantic and Southern Circumpolar. Heat is efficiently transported from the tropics to the poles both by the 'conveyor belt'.

The change in the oceanic temperature profile is thus described by an advection-diffusion equation, forced by the surface fluxes as boundary conditions. This equation was discretized using a finite volume method in the spatial domain and implicit in time. This resulted in a matrix equation for the temperature.

4.1.3. The Ocean Biosphere and Chemistry Model

The elements considered in the Ocean Biosphere and Chemistry model are affected by the same transport process prescribed by the circulation model; they differ because of their exchange with the atmosphere and possible vertical transport by so-called "biological pumps".

These biological pumps cause a transport of substances downward as a result of uptake by phytoplankton and the subsequent settling of faeces from the various grazers. The settling of particles does not occur with a constant rate, but is highly variable, especially at the temperate and polar latitudes. This is a result of the fluctuating environmental conditions (light, mixed layer depth, temperature) and the population dynamics of the various species involved. For the present model, we are interested in yearly-averaged rather than in short-term fluxes only. This makes it possible to simplify the calculations considerably. It has been shown (Klepper, 1994) that the yearly averaged downward particle flux calculated by a dynamical model of the ocean foodweb is almost identical to the yearly averaged quasi steady state (i.e. the steady state that corresponds to the environmental conditions at a particular time of the year) particle flux. This result implies that we can speed up the calculations considerably but also that we can use a highly simplified model of the ocean food web (Klepper *et al.*, 1994): more complicated models (several species of phytoplankton and grazers, explicit descriptions of life histories) may show a different dynamical behaviour, but yield almost identical steady state results.

The simplified ocean food web needs to contain only two species: phytoplankton and detritus. The phytoplankton balance equation is then:

$$\frac{dB}{dt} = B \left(P_{max} f(I) \frac{N}{N+k} - r - m \right) \tag{3}$$

where B denotes the phytoplankton biomass (mole.kg^{-1}), N the inorganic N (Nitrate+ Ammonium; mole.kg^{-1}), P_{max} the maximum production rate (in day^{-1}, with a value of 1.5 at 10° C), f(I) the light-limitation function (a dimensionless, saturation curve with an initial slope of 0.07 m^2/W), k the half-saturation for N (with a value of 14 μmole.kg^{-1}), m the mortality rate (with a value of 0.09 day^{-1} at 10° C) and r the respiration rate (0.01 day^{-1} at 10° C). The temperature-dependent rates (P_{max}, m, r) increase with temperature with a Q10 of 2.

The mortality term in effect includes all transfer of organic matter to the rest of the food chain like excretion, senescence and grazing. The detritus is subject to (local) mineralization and removal:

$$\frac{dD}{dt} = m\ B - s\ D \tag{4}$$

where D denotes the detritus concentration (mole.kg^{-1}) and s the removal rate (day^{-1}). The removal rate is a composite of settling to deeper layers (4 m day^{-1}) and transfer of particulate matter to DOC (0.015 day^{-1} and 0.001 day^{-1} for two DOC fractions).

For a particular set of external conditions (surface light intensity, depth of well-mixed layer, temperature) one can directly determine the steady-state inorganic N concentration from Equation (3). Equation (4) then prescribes the ratio phytoplankton: detritus. In a situation of (near) steady state, removal of organic nutrients should (almost) equal its supply in inorganic form: on a short time scale the sum of inorganic N, phytoplankton-N and detritus-N can be considered constant. Using these assumptions we can calculate the three N fractions and thus the transfer of N to the deep sea and the two DOC fractions.

After determining the particulate organic matter in phytoplankton and detritus we can determine inorganic C, again assuming a constant total of the three forms and steady state between them. The alkalinity is calculated from inorganic N and Ca concentrations, and this determines the partitioning H_2CO_3:HCO_3^-:CO_3^-, which in combination with the piston velocity determines the gas exchange rate. The different seasons give different steady states for the N and C partitioning. Yearly averaged fluxes are calculated by averaging separate seasonal steady states.

From these basic calculations we can determine a number of fluxes between the substances in the model. The settling detritus is distributed over the various depths below it according to an empirical double-exponential model:

$$F(z) = F(z_0)\,(a\,e^{c_1 z} + (1-a)\,e^{c_2 z}) \tag{5}$$

where $F(z)$ denotes the flux of organic matter at depth z (mole.m^{-1}.day^{-1}), z_0 the flux at the bottom of surface mixed layer, a and (1-a) the fractions (0.79 and 0.21) in two classes of particles and c_1, c_2 the decay rates (with values $6.5 \cdot 10^{-3}$ and $2.3 \cdot 10^{-4}$ m^{-1}).

This model is -for the range of depth over which there are measurements- nearly indistinguishable from an empirical model (Martin et al., 1987) but predicts a lower settling rate at the bottom. Apart from a better fit to the deep-sea part of the nutrient and O_2 profiles the main advantage of Equation (5) is that it is possible to interpret the decay rate of the two fractions in terms of actual properties of the detritus. The decay rate per meter is the quotient of the microbial decay (day^{-1}) and settling rate (m.day^{-1}); the power function of Martin et al. lacks such a clear interpretation.

Using Equation (5) and standard stoichiometry of C:N:O (the so-called Redfield ratios of 7:1:8 in moles) we can calculate the production of inorganic C and N and the consumption of O_2 associated with the decrease in particulate matter flux. For the conversion of N into labile DOC a higher C:N ratio (10:1) and for refractory DOC an even higher C:N (25:1) has been used. The decay of DOC has been modeled as a temperature-dependent first order decay. The O_2 calculations differ from the other compounds because we don't have a closed cycle but assume atmospheric equilibrium. This is reasonable in view of the surface gas exchange rate which is in the order of 10 days for O_2.

The cycle of Ca is governed by the so-called "hard-tissue pump": some of the organisms in the food web produce shells or skeletons of $CaCO_3$ (calcite or aragonite). This settles down and most of it dissolves again below several kilometres depth. In the model it has been assumed that $CaCO_3$ production is proportional to gross primary production. This assumption has the effect that the ratio of the production of calcareous to organic material is higher in the tropics than near the poles: in the tropics the productivity is mainly driven by recycled nutrients, while at high latitudes the supply of "new" (i.e. deep-sea) nutrients is most important. Because the calcareous material is not recycled at the surface this means a higher tropical flux.

The cycles of the different fractions of ^{14}C (TIC, particulate matter, refractory and labile dissolved organic C) are simply grafted on those of their respective "carriers".

4.1.4. Input data and parameter values

The parameters determining oceanic circulation were based on those given by Keir (1988), adjusting them to obtain a better agreement with the observed distribution of ^{14}C and the northward heat flux in the ocean (Klepper et al., 1993). The main oceanic parameters influencing these variables were found to be the horizontal and vertical eddy diffusivities, the volumes of the deep water formations, some of the fractions in the oceanic flow pattern and the coefficient for gas exchange between sea and air. The advective flows are given in Figure 8. For horizontal dispersion coefficients values of $6.0 \cdot 10^3$ m^2.s^{-1} (surface) and $7.1 \cdot 10^2$ m^2.s^{-1} (deep ocean) were used; for vertical dispersion values of $6.5 \cdot 10^{-5}$ m^2.s^{-1} (temperate and tropical) and $1.2 \cdot 10^{-3}$ m^2.s^{-1} (polar) were adopted.

Input data for the Biosphere and Chemistry model are surface irradiance (from the ZACM) and the input of several substances from rivers and atmosphere. For the riverine and atmospheric fluxes of N and DOC it has been assumed that these are proportional to runoff per 10° latitude band (Baumgartner and Reichel, 1975). Pre-industrial total land-sea fluxes have been adopted from Wollast (1981), which is generally in close agreement with other summaries of literature data (Sarmiento and Sundquist, 1992; Wollast et al., 1993). For the input of DOC it has been assumed that this is entirely refractory (Mantoura and Woodward, 1983). Both the biological fixation of N_2 and denitrification fluxes are highly uncertain (Wollast, 1981). Assuming them to be equal is well within the range of literature values; in the model both fluxes are omitted.

For the radioactive tracers it has been assumed that the atmospheric concentration of ^{14}C is homogeneously distributed (with time-dependent concentrations according to Broecker and Peng, 1982), and the atmospheric deposition of ^3H has been modeled by scaling the zonal distribution by the time course at a single station in Ireland (Broecker and Peng, 1982).

Equilibrium constants and solubilities for the carbonate-borate-calcium chemistry have been adopted from Gieskes (1974). The parameters P_{max} (maximal production rate) and k (half-saturation for inorganic N) in the biological model, Equation (3), are taken from literature (Eppley, 1972; Parsons et al., 1984). The remaining parameters (light limited productivity, mortality, excretion to labile and refractory DOC) are calibrated to fit the model to observations of primary productivity, ratio of new to total production (f-ratio) and horizontal and vertical distributions of N and DOC (both amount and ^{14}C content). The Ca cycle is linked to the organic C cycle by a fixed ratio of total (gross) production to calcification. By adopting a constant fraction of $CaCO_3$ that dissolves at the bottom compartment (83%, Wollast (1981)) and assuming steady state (riverine input equals sedimentation) we can determine the proportionality constant.

4.3. VALIDATION

The Ocean Climate Model and the Ocean Biosphere and Chemistry Model were validated against a variety of data sets (Klepper *et al.*, 1993). The validation included temperature, alkalinity, nitrate concentration, primary production, dissolved organic C, inorganic C, natural ^{14}C (both inorganic and in dissolved organic C), ^3H and bomb-produced ^{14}C. The individual tracers are generally most useful for a particular process or time scale; a coupled model such as the present makes it possible to combine these sources of information.

The agreement between calculated and measured (Levitus, 1982) ocean temperatures is on the average good, but shows disagreement in some regions, see Figure 9. The ocean surface temperatures are within 2.5 K from the Levitus data; in the deep ocean the deviations are in the order of magnitude of 2 K. The largest deviation is 8 K in the

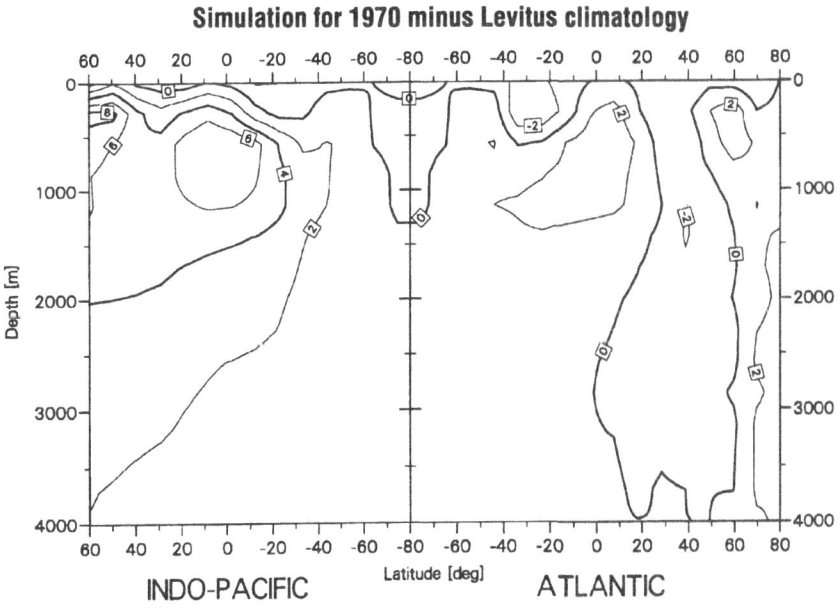

Figure 9: Difference between calculated and measured ocean temperature.

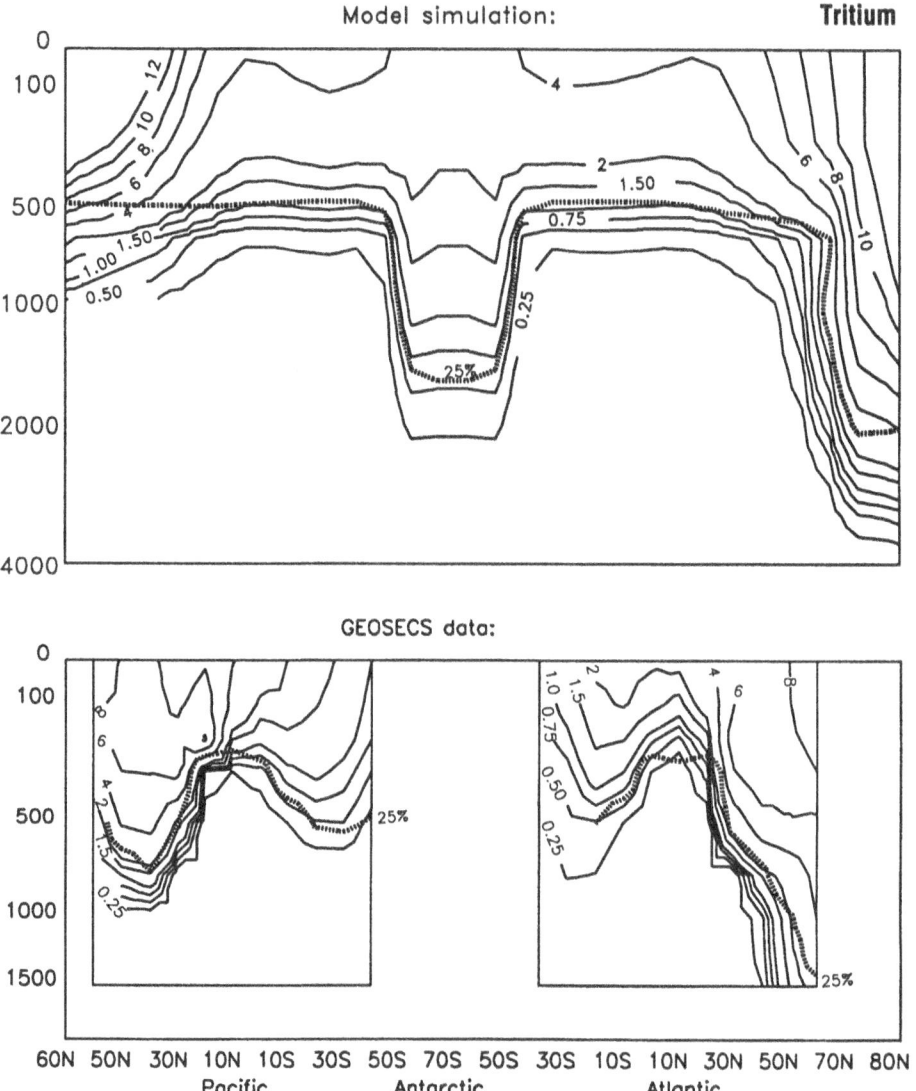

Figure 10: Calculated distribution of ^{3}H (in T.U.) in 1973 and GEOSECS measurements (Broecker and Peng, 1982). The heavy lines in both graphs denote the depth at which the ^{3}H concentration reaches one-quarter of that in the overlying surface water.

Northern Pacific, but this concerns only a single cell (data point). A more serious problem is the deviation above 6 K in a large part of the middle deep North Pacific; this is probably due to the absence of wind-driven vertical water motions in the upper ocean layers (Ekman pumping); to the fact that no difference was made between upwelling velocities in the tropical, temperate and polar regions; and to the constant value of the vertical diffusion coefficient. An extension of the model, to include circulation data that depend on latitude and/or depth could improve the temperature simulation considerably.

The atmospheric nuclear tests in the 1950s produced several radioactive tracers that provide information on model circulation and carbon uptake. This allows for tests of both the circulation data, shared by the two ocean models, and of the modeling of carbon uptake. ^3H, which enters the ocean as ^3H_2O in rainfall and runoff can be used to check on water movement; the distribution of excess ^{14}C (i.e., relative to pre-industrial steady state) provides additional information on atmospheric CO_2 uptake by the ocean.

The horizontal and vertical distribution of ^3H is depicted in Figure 10. There is in general a good agreement between measurements (Broecker and Peng, 1982) and model results; in particular the concentrations and penetration depth (>1.5 km) in the North Atlantic are reproduced well. The major disagreement can be noted by comparing the 25% surface concentration penetration depths: both are at approximately 400 m, but the measurements show a considerable difference between the equatorial regions (250 m) and the central gyres (600 m), which is not reproduced by the model. Again, this reflects the absence of Ekman pumping in the model. The model predicts a considerable penetration depth (albeit at low concentrations) in the Antarctic region below 50°S; although this cannot be fully checked because of a lack of data in this region, this high penetration depth seems an overestimate.

The results of a comparison between measured and simulated uptake of bomb-^{14}C leads to similar conclusions as for ^3H: the average value is reproduced well, the simulated value is $7.6 \cdot 10^9$ atoms cm^{-2} in 1974 as compared to $8.3 \cdot 10^9$ atoms cm^{-2} estimated by Peng (1986); the average penetration depth is 300 m, 14% below the GEOSECS estimate of 350 m (Broecker et al., 1985) and North Atlantic uptake matches with observations, but the model lacks spatial detail in the tropics and appears to overestimate Antarctic uptake.

Another point that requires some attention is the low age (simulated $\Delta^{14}C$ -220 ‰ as compared to a measured -240 ‰) in the deep Pacific (Figure 11): apparently, the overall circulation appears some 10% too fast. On the average, results seem acceptable however, and it may be anticipated that the results obtained with the circulation model are reliable, apart from some spatial details in the tropics and a small overestimate of global circulation.

It appears that both the chemical and biological models give a satisfactory depiction of the uptake and transformation of carbon in the ocean. The chemical equilibrium model can be mainly judged from horizontal and vertical distribution of total inorganic C (Figure 12) and the biological model from nitrate (Figure 13), O_2, alkalinity, primary production and DOC. It appears that both the concentrations (from measurements) and turnover rates (from productivity measurements and ^{14}C data) are close to observations.

The main output of the Ocean Biosphere and Chemistry Model, important for the feedbacks in IMAGE, is the total uptake of anthropogenic CO_2 by the oceans. This CO_2 uptake was calculated to be 1.6 Gt C.a^{-1} for 1980-1989. Direct estimates from observations have been made by Tans et al. (1990); these have been corrected by Siegenthaler and Sarmiento (1993) to give a total uptake of anthropogenic of 1.8 Gt C.a^{-1}. Model results (corrected to agree with bomb ^{14}C) are also reviewed by Siegenthaler and Sarmiento (1993), giving uptake fluxes of anthropogenic CO_2 between 1.9 and 2.3 GtC.a^{-1}, with the

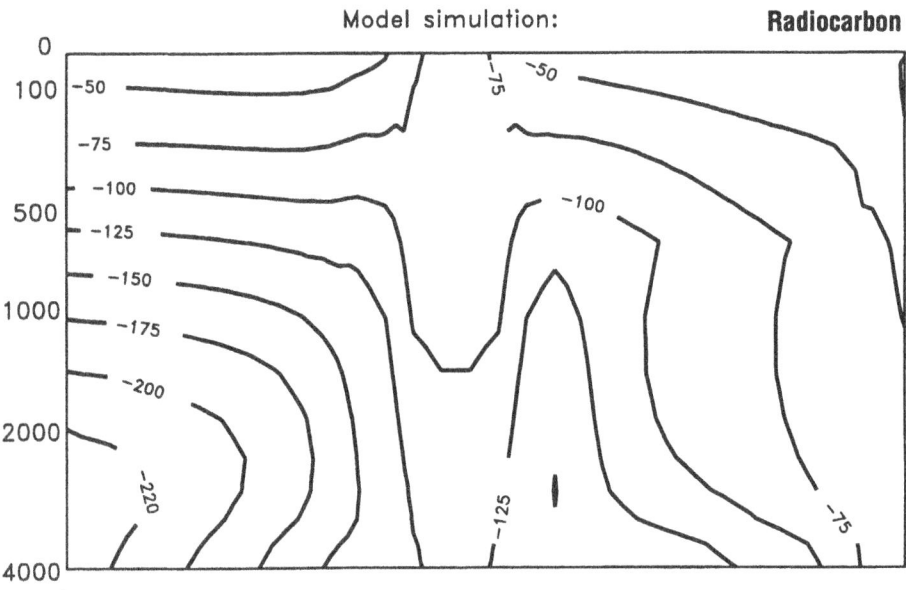

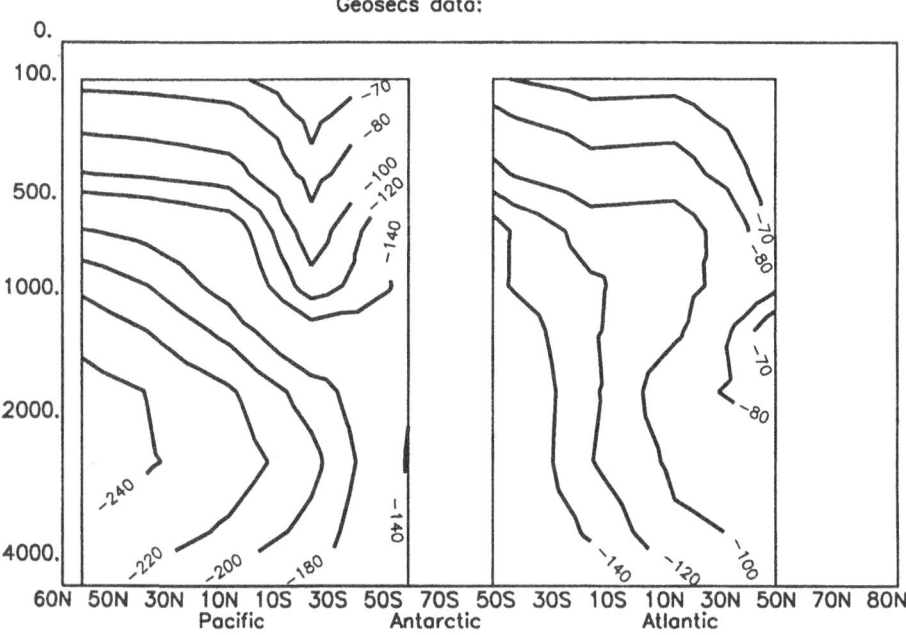

Figure 11: $\Delta^{14}C$ (in ‰, relative to preindustrial atmospheric value): calculated values and measurements of the GEOSECS program (Broecker *et al.*, 1985).

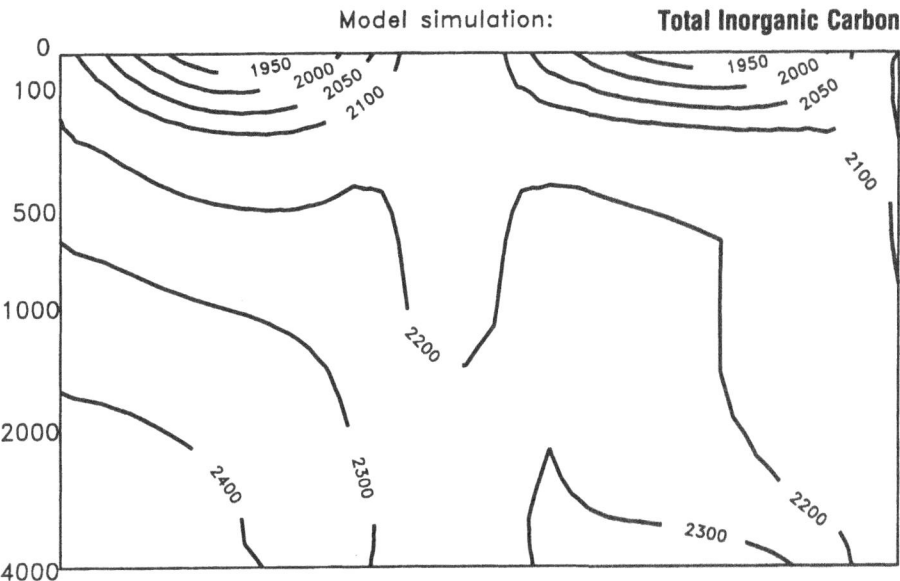

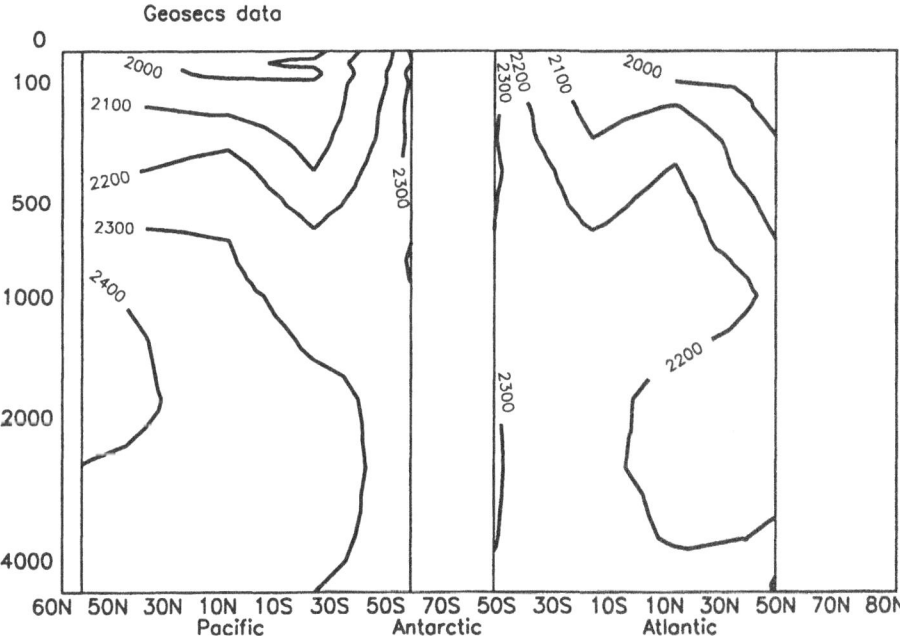

Figure 12: Total inorganic carbon concentration (μMole.kg^{-1}): calculated values and measurements of the GEOSECS program (Takahashi *et al.*, 1981).

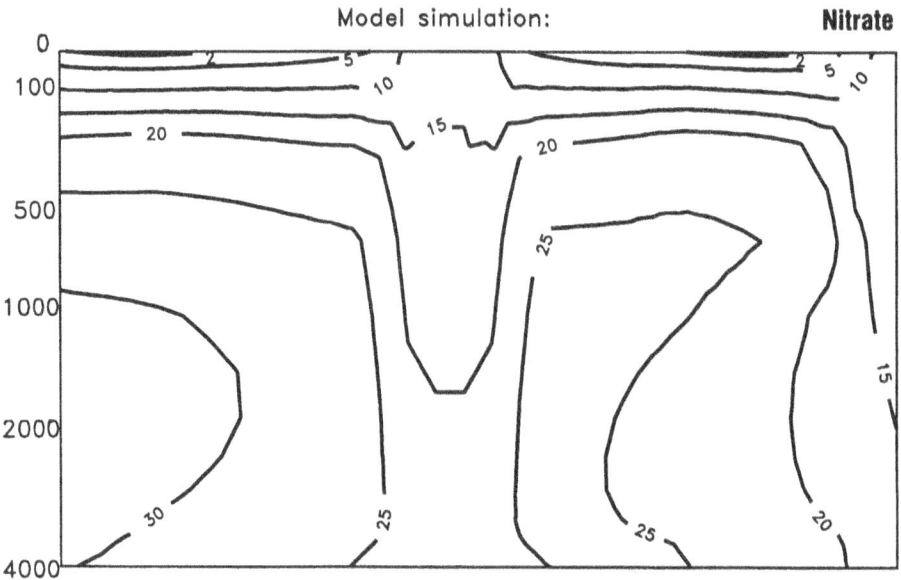

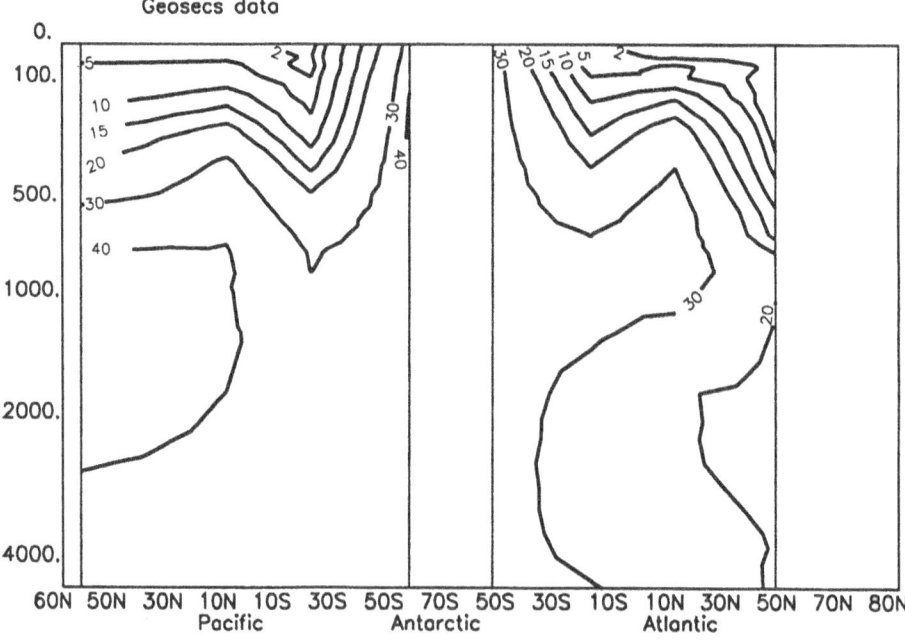

Figure 13: Nitrate concentration (μMole.kg^{-1}): calculated values and measurements of the GEOSECS program (Takahashi *et al.*, 1981).

exception of the "Hamburg" GCM (Maier-Reimer and Hasselmann, 1987), which gives the same figure as the present model. A factor that may partly explain the difference between our results and most other models is that we did not correct for our 8% underestimate of bomb ^{14}C.

Although there are considerable differences between existing physics-based circulation models in total carbon uptake, these differences are even larger if we consider details of circulation patterns. Another aspect of the use of physics-based circulation models for the calculation of carbon uptake is that they can not be (reliably) used for a prognosis of ocean circulation. For this reasons, the present choice for an empirical circulation pattern (inferred from tracer distributions) can be well justified: the physics-based models are more detailed but do not appear to be more reliable, and our approach to treat circulation as an input (which can be subjected to a sensitivity study, discussed below) can at present not be replaced by an actual prediction.

5. Linkage of submodels

The Atmosphere Ocean System consists of the four autonomous models described above. Each of these models can be used as a stand-alone model. In the Atmosphere-Ocean System of IMAGE 2.0, however, the four models are closely linked. Typically the time step taken within the zonal atmospheric climate model is a few days, smaller than in the other models, but information on the atmosphere-ocean heat flux is only exchanged at a somewhat greater time step of once every month of simulation. The three other models each use a time step of one month, partly due to numerical stability, partly to considerations of consistency. In this section we will describe the nature of the information exchanges and a validation against current climate. Next a comparison with the results of other climate models forced with a doubled CO_2 concentration will be given. Finally results will be presented for two full model IMAGE scenarios: the "Conventional Wisdom" scenario and the "Ocean Realignment" scenario, compared to results for the 1992 IPCC reference scenario, IS92a (IPCC, 1992). The Conventional Wisdom scenario is built on central estimates for the driving forces (population, economic growth etc.) and assumes that no specific greenhouse policies will influence trends. The Ocean Realignment scenario investigates the consequences of a prescribed slowing down of the global ocean circulation.

5.1. LINKING PROCEDURES

This section highlights the main linkages between the four atmosphere and ocean models; some details of the linkage between the models were already given in previous sections.

The main connections between the Atmosphere Ocean System and the remaining parts of IMAGE concern emission data and climatic data. The emissions of greenhouse gases and related compounds are simulated by the Energy and Industry System and the Terrestrial Environmental System of IMAGE (de Vries *et al.*, 1994; Kreileman and

Bouwman, 1994). As an output to the Terrestrial Environmental System zonally averaged temperatures and precipitation are provided on yearly basis. These data are used to simulate changes in crop productivity and in natural vegetation due to climate change. GCM output and a climate database have been used for downscaling climate simulations with the present model to a small geographic resolution and a yearly cycle (Alcamo *et al.*, 1994a). Two other major feedbacks require linkages between the Atmosphere Ocean System and the remaining parts of IMAGE: the atmospheric CO_2 concentration is used in the simulation of the CO_2 fertilization effect in the terrestrial C cycle calculations (Klein Goldewijk *et al.*, 1994) while simulated land cover changes (Zuidema *et al.*, 1994) affect the surface albedo used in the climate simulation.

5.1.1. Linkage of Atmospheric Composition Model and ZACM
The Zonal Atmospheric Climate Model builds its climate simulations on the heat trapped in the atmosphere due to the radiative properties of the greenhouse gases. Therefore it requires information on the concentrations of greenhouse gases.

CO_2, CH_4, N_2O, CFCs and their replacements (HCFCs, HFCs) are distributed (nearly) homogeneously over the atmosphere, and the (globally averaged) results of the Atmospheric Composition Model can be used directly.

The effects of halogenated species other than CFC-11 and CFC-12 are not explicitly included in the radiative scheme of ZACM, but implicitly included by using an 'equivalent CFC-11' concentration. These CFC-equivalents account for some 40% of the additional radiative forcing in the period from 1970 to 1990 of CFC-11 and CFC-12 together (IPCC, 1990). The conversion to equivalent CFC-11 concentration was based on a linear relation between concentration and radiative forcing (IPCC, 1990).

In linking the models, O_3 must be handled with care. O_3 is distributed heterogeneously over the troposphere, and its distribution in the stratosphere strongly depends on latitude and height. Furthermore O_3 is radiatively active in both the visible and the infrared windows of the spectrum. In order to account for the effect of O_3 on climate, the ZACM therefore uses a 2-D O_3 field. The Atmospheric Composition Model represents the O_3 distribution by only three quantities: tropospheric, lower stratospheric and upper stratospheric O_3. These three quantities are used in parameterizing 2-D fields of O_3 concentrations. For the troposphere we use the present day O_3 field from the TNO-Isaksen model (Roemer, 1991) and the changes in this field occurring between pre-industrial and present and between present and a 'Business as Usual' scenario as basic profiles. For the stratosphere we use the present day profiles and the changes between 1980 and 1990 in these profiles (WMO, 1992). The profiles of O_3 concentration changes are used as sensitivities of the local O_3 changes to changes in the global O_3 characteristics simulated by the model.

The SO_4^- calculations performed by the present model are rather preliminary. In the ZACM only the direct radiative effect is included. In the Atmospheric Composition Model a linear relation between natural and anthropogenic S emissions and SO_4^- column amounts is implemented only. The downscaling of natural and anthropogenic S emissions to zonal SO_4^- columns relies on three dimensional atmospheric chemical computations

(Langner *et al.*, 1992).

Atmospheric chemical processes are affected by climate. The absolute humidity is important for the production of OH radicals, while reaction rates are temperature dependent. These climate feedbacks on atmospheric chemistry are included in the model.

5.1.2. Linkage of Ocean Climate Model and ZACM

The ocean climate and the atmospheric climate are linked by the fluxes of heat and water between the two. The ocean gains heat from the atmosphere in the tropics and transports it polewards; the atmosphere redistributes H_2O from the tropics to the poles where it again precipitates.

Although the latitudinal spacing in the different models of the atmosphere ocean system is the same, a resolution problem had to be solved as the ocean model distinguishes between the Atlantic and Indo-Pacific basins, while the atmosphere model does not. The model however does include parameterizations for surface air temperatures and heat fluxes for different underlying surfaces within each zonal band. Parallel computations distinguish between snow covered and snow free land and ice covered and ice free sections of the Atlantic and Indo-Pacific ocean. The resulting heat fluxes are used directly in the ocean model and an area-weighted average is used in the ZACM.

5.1.3. Linkages involving the Ocean Biosphere and Chemistry Model.

The main output of the ocean biosphere and chemistry model to the other models is the net uptake of CO_2 by the ocean. The temperature of the ocean water is simulated by the Ocean Climate Model and used in the Ocean Biosphere and Chemistry Model in determining the equilibrium concentrations of dissolved inorganic carbon in the surface waters and in a number of biological process rates.

The atmospheric CO_2 concentration determines the equilibrium concentration of CO_2 in the surface waters. The value of ΔpCO_2 varies with latitude, causing an atmospheric flux from (C emitting) tropical waters to (C absorbing) polar waters. This is one of the causes of the atmospheric pCO_2 gradient, which is not taken into account in the atmospheric composition model. As the gradients in the atmospheric concentration of CO_2 are small, this reflects a minor inaccuracy only. The simulated net C flux into the ocean is used in the atmospheric carbon mass balance of the Atmospheric Composition Model.

A direct link between ZACM and the Ocean Biosphere and Chemistry Model is provided by the simulation of the fractional sea ice coverage: it is assumed that gas exchange and biological uptake are restricted to the open sea only.

5.2   VALIDATION

The linked climate model has been tested extensively, with the values of different parameters varied within their uncertainty ranges. In the calibration process equilibrium runs were performed using the constant 1900 concentrations of greenhouse gases in the radiation calculations, followed by a transient simulation over the period 1900 to 1970 with increasing greenhouse gas concentrations. The resulting temperature profiles were compared to the Oort and Rasmussen atmospheric temperature data and to the Levitus ocean climatology. In order to improve agreement with measurements, cloud height and cloud optical depth were varied within the ranges given by MacKay and Khalil (1991) and were taken the same for Northern and Southern hemisphere. The ocean temperatures of this coupled run have been discussed in previous sections, and generally show a good agreement with measurements. The simulated air temperatures agree adequately with measurements (Oort, 1983), similar to the results of section 3.

The linked climate model was validated against the CO_2 doubling experiments performed using GCMs. The model was initialized with an equilibrium climate simulation for present day greenhouse gas concentrations; the CO_2 concentration was doubled and the linked model was ran into equilibrium again. The equilibria were obtained using an asynchronous integration, taking smaller time steps for the atmosphere than for the ocean and required some 2000 oceanic years of integration (150 atmospheric years). The global climate sensitivity of the model was found to be 1.9 K. The climate sensitivity is well within the range of 1.5 to 4.5 K given by IPCC (IPCC, 1992). The simulated patterns of climate change were similar to those found by GCMs, both for atmosphere and ocean (IPCC, 1992). The atmospheric surface temperature increase ranges from 2 K in the tropics to over 5 K in the polar regions, with a similar cooling at the top of the atmosphere. In the ocean surface waters, temperature increases were smaller, ranging from 0.7 K in the tropics to 1 K at high latitudes.

6. Results for IMAGE 2.0 scenarios

The Atmosphere-Ocean model was used within the framework of IMAGE 2.0 for the evaluation of scenarios. Here a scenario consists of a set of assumptions on the driving forces of greenhouse gas emissions. The scenarios specify basic data like the growth of population and economy in 13 world regions as well as more detailed information like market shares and prices of energy carriers; other data relate to consumption of and by animals or net imports of crops. The scenarios are not merely a specification of emissions of greenhouse gases, nor a set of parameters that directly determine these emissions. Within the IMAGE framework land cover changes are simulated. These changes depend on both basic driving factors (determining the demand for crop products) and climate (determining potential crop yields). In this way the land use related emissions are simulated to be consistent with both the increasing land requirements and the changing climate.

6.1   REFERENCE SCENARIO

The reference scenario for IMAGE 2.0 is the "Conventional Wisdom" scenario (Alcamo *et al.*, 1994b). This scenario assumes central estimates for the basic driving forces, similar to the assumptions of the IS92a reference scenario of IPCC (1992). As a result, the energy and industry-related emissions simulated by IMAGE are generally consistent with the

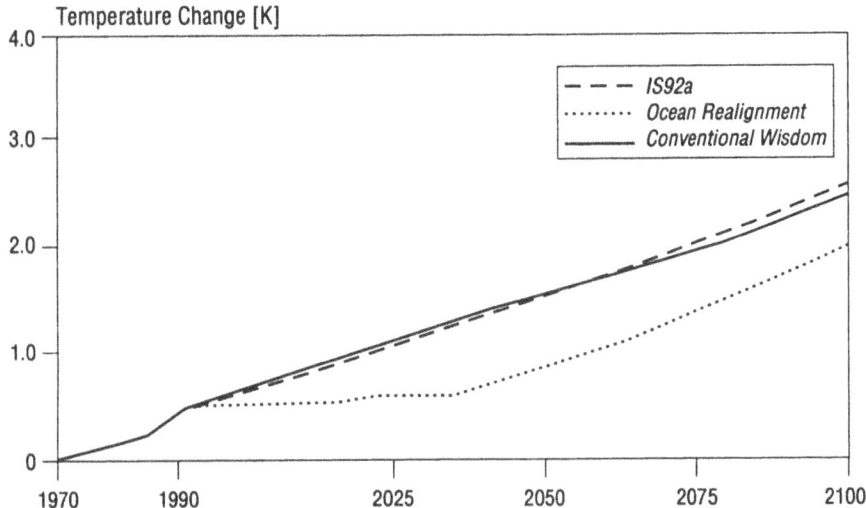

Figure 14. Increase in surface air temperature since 1970 for the Conventional Wisdom scenario, the Ocean Realignment scenario and the IS92a scenario.

emissions in the IS92a scenario (de Vries *et al.*, 1994); CO_2 emissions are 10 to 20% higher in the Conventional Wisdom scenario. The land use related emissions for the Conventional Wisdom scenario are similar to the IS92a emissions up to 2025. After 2025 land use-related emissions are significantly lower for the Conventional Wisdom scenario, especially the emissions of CO, NO_x and VOCs. The difference is largely due to the changes in land cover, for instance the decrease of the area of seasonally burning savanna.

The atmospheric concentrations of greenhouse gases are also similar for the Conventional Wisdom scenario and the IS92a scenario up to 2025 (Krol and van der Woerd, 1994). After 2025 the concentrations of CH_4 and tropospheric O_3 stabilize for the Conventional Wisdom scenario, while a continuation of the increasing trend is found for the IS92a scenario. The concentration of CO_2 is higher in the Conventional Wisdom scenario.

The stabilization of CH_4 and tropospheric O_3 in the Conventional Wisdom scenario affects the trend in radiative forcing of climate. This causes a difference between the Conventional Wisdom scenario and the IS92a scenario, which is matched closely by the difference in CO_2 concentration. Figure 14 shows the globally averaged increase of surface air temperature as a function of time; warming is simulated to be very similar for the IS92a scenario (2.6 K in 2100, relative to 1970) and the Conventional Wisdom scenario (2.5 K in 2100). Warming between 1970 and 1990 is overestimated, caused by a technical implementation problem in shifting from historical data to precribed emissions scenarios. The rate of change in 2100 is higher for IS92a (0.23 K/decade) than for Conventional Wisdom (0.18 K/decade). For both scenarios the rate of temperature increase is close to 0.2 K/decade throughout the 21st century, ie. double the value of 0.1 K/decade that is roughly assumed to be the upper limit for ecosystems to adapt to. The

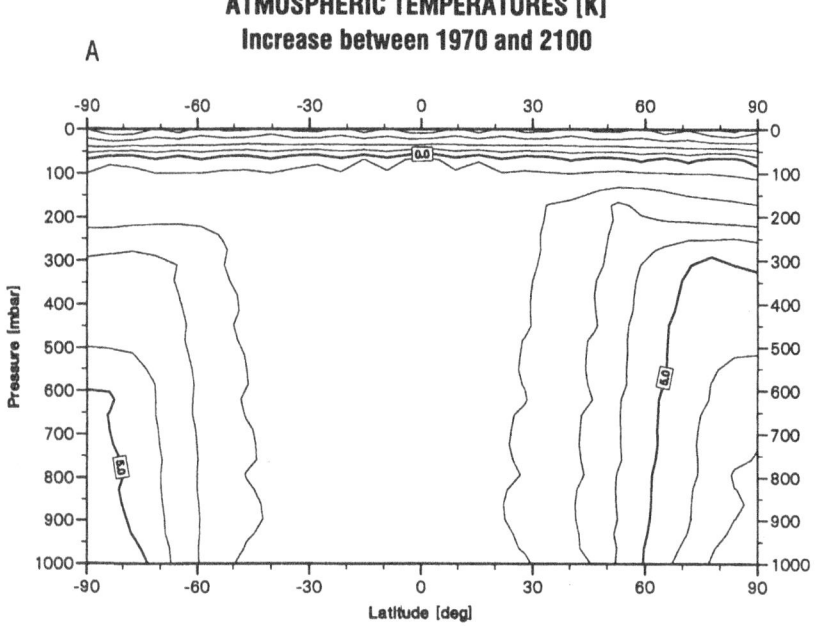

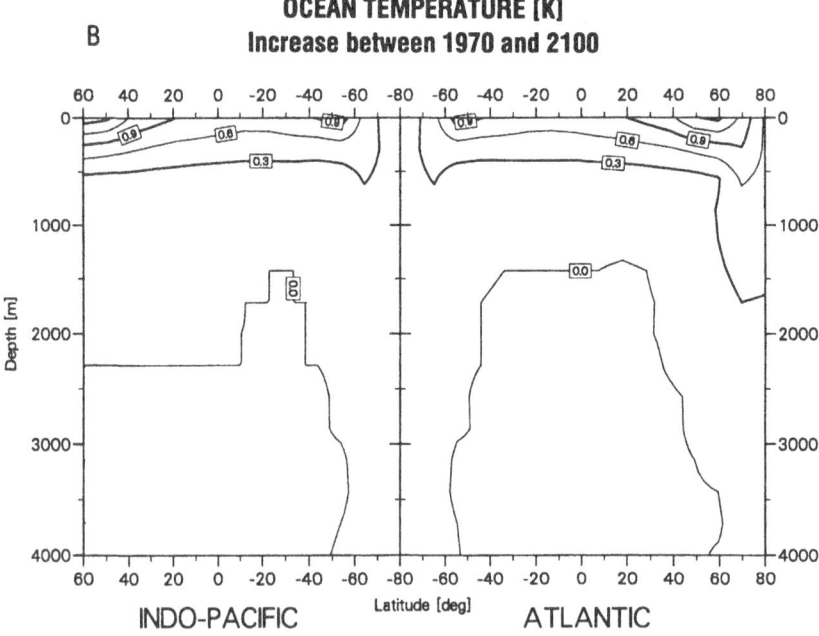

Figure 15: Increase over the simulation period in the temperatures of (a) atmosphere and (b) ocean for the Conventional Wisdom scenario.

simulated pattern of temperature increase between 1970 and 2100 is shown in Figure 15[a].

Ocean surface temperatures are simulated to increase by some 1 K. Warming is somewhat stronger in the Indo-Pacific and smallest in the regions of deep water formation, in the North Atlantic and close to Antarctica, see Figure 15[b]. The uptake of CO_2 by the ocean increases steadily from 1.6 Gt C a^{-1} in 1990 to 4.2 Gt C a^{-1} in 2100.

6.2 SENSITIVITY TO THE OCEAN CIRCULATION

Within the ocean model of IMAGE 2.0, the ocean circulation is not simulated, but prescribed. An "Ocean Realignment" scenario was used to investigate the models sensitivity to the ocean circulation. In this scenario the rates of deep water formation were changed according to the results of Mikolajewicz et al. (1990). These investigators studied the sensitivity of ocean circulation to changes in its forcing. The ocean was forced by prescribed steadily increasing atmospheric temperatures, distributed according to atmospheric GCM output, ending up with a 2.9 K temperature increase after 50 years; feedbacks between ocean temperatures and atmospheric temperatures were not considered, while such feedbacks tend to stabilize the ocean circulation significantly (IPCC, 1992). So the experiment should only be interpreted as a sensitivity study. The present use of the rates of deep water formation in the IMAGE ocean model should be viewed in the same way. The rates of deep water formation are used as prescribed parameters of the ocean circulation and no feedback of the resulting climate simulation is considered. The prescribed circulation and the resulting climate are not guaranteed to be consistent with the atmospheric forcing assumed by Mikolajewicz et al. (1990). In the simulations of Mikolajewicz et al. the North Atlantic deep water formation gradually decreases to one third of its original volume in the 50 year of the simulation period, the Southern ocean deep water formation decreases by one third of its volume in 35 years but afterwards it recovers half of this loss. In the Ocean Realignment scenario the deep water formation is assumed to change in the period of 1990 to 2040 and to remain constant afterwards.

The slowing down of the ocean circulation has a very significant effect on climate. The northward transport of heat in the Atlantic is strongly reduced. The surface air temperature in the Northern Hemisphere mid and high latitudes remain almost the same and even go down up to 2050. Climate thus behaves different than was assumed by Mikolajewicz et al., so results are only indicative for the sensitivity of the simulations to changes in the ocean circulation. In the year 2100 the temperature has only increased by 1.5 K, throughout the Northern Hemisphere. For the Conventional Wisdom scenario the increase was some 3.5 K at mid latitudes and 6.5 K at high latitudes. The globally averaged surface air temperature is stable up to the year 2035, see Figure 14. At that moment the average temperature is 0.7 K below the temperature for the Conventional Wisdom at that time. In the second half of the 21st century temperature increases at a rate of 0.22 K/decade for the Ocean Realignment scenario, faster than in the Conventional Wisdom scenario.

The ocean surface temperatures show similar behavior, but more persistent; the sea surface temperatures in the regions of deep water formation in the North Atlantic stay bellow their initial values throughout the simulation. This does not have a very large effect on the oceanic uptake of CO_2. The uptake of CO_2 in the Ocean Realignment case is enhanced by the lower temperatures of the ocean surface, compared to the Conventional Wisdom scenario. Furthermore, the oceanic carbon uptake for the Ocean Realignment scenario is stimulated by relatively high atmospheric CO_2 concentrations, due to a small terrestrial carbon sink, compared to the Conventional Wisdom scenario. On the other hand the advective transport of ocean surface carbon to the deeper ocean is reduced, so the surface waters will saturate faster. The net effect of these three processes is a difference of at most 0.15 Gt C a^{-1} in the C uptake in the two scenarios, the sink being somewhat larger for the Ocean Realignment scenario up to 2075 and somewhat smaller afterwards. The cumulative uptake of CO_2 over the entire simulations period is 1.4 Gt C a^{-1} larger for the Ocean Realignment scenario. This is roughly equivalent to 0.7 ppmv CO_2 in the atmosphere.

Feedbacks of climate on vegetation and chemistry have two interesting effects on atmospheric composition (Krol and van der Woerd, 1994). The strong increase of temperature in the Northern Hemispheric regions covered with boreal forests enhances the carbon uptake by vegetation. This sink is thus much stronger for the Conventional Wisdom scenario than for the Ocean Realignment scenario. The difference reaches a maximum of 3 Gt C a^{-1} in the year 2030 and slowly decreases to 1.5 Gt C a^{-1} in 2100. Integrated over the simulation period, the terrestrial sink accounts for 650 Gt C in the Conventional Wisdom case, and for 185 Gt C less in the Ocean Realignment case. The sum of sources and sinks leads to a difference of 90 ppmv in the atmospheric CO_2 concentration in 2100 for the scenarios. A second climate feedback on atmospheric composition is caused by increasing atmospheric H_2O. This increase enhances the production of OH and thus the sink for CH_4 and other hydrocarbons. This feedback is strongest for the Conventional Wisdom scenario, where the temperature increase is largest. Consequently, the atmospheric concentration of CH_4 is lower for the Conventional Wisdom scenario. The difference is 0.1 ppmv in 2100.

The radiative forcing of climate is stronger for the Ocean Realignment scenario, due to the difference in the CO_2 concentration. Nevertheless the average temperature increase is smaller for this scenario. This apparent contradiction can be explained by the reduced oceanic transport of heat from the tropical zone to the higher latitudes. In the tropical zone a small increase in temperature has a much stronger effect on the absolute humidity than at higher latitudes, assuming that relative humidity remains constant. The sensitivity of climate is lower when more heat is trapped in the tropical zone; this is what happens in the Ocean Realignment scenario.

The differences in climate change thus found have an important effect on the land cover simulations in other parts of the IMAGE 2.0 framework. The suitability (or lack of suitability) of land for crop growth are among the factors determining where agricultural land will be expanded or where agricultural land will be abandoned for reforestation. Here distinct differences appear between the Conventional Wisdom scenario and the Ocean Realignment scenario: the reforested area is simulated to be significantly smaller in the Ocean Realignment scenario for all regions at the high latitudes of the Northern

Hemisphere. The difference in forest area in 2100 is $6 \cdot 10^5$ km^2, ie. the total area of France or Madagascar (Alcamo *et al.*, 1994b).

7. Conclusions

The present coupled atmosphere-ocean model is able to simulate the main features of the climate system on a latitudinally averaged basis. It combines the possibilities to incorporate the main mechanisms and feedbacks of the ocean-atmosphere climate system with moderate computational requirements. As such, it is a convenient tool to perform transient simulations of climate change in an integrated model of global change. Furthermore, the model is well-suited to perform stand-alone studies such as sensitivity analyses. The climate sensitivity of the model for CO_2 doubling is 1.9 K, inside the plausible range of 1.5 K to 4.5 K (IPCC, 1990); the Ocean Biology and Chemistry model simulates a global oceanic C uptake of 1.6 Gt C a^{-1} in 1990, inside the range found using state of the art models of 2.0 ± 0.5 Gt C a^{-1} (Orr, 1993).

The reference scenario for IMAGE 2.0, the Conventional Wisdom scenario, closely resembles the IS92a scenario from the climate perspective. While the atmospheric concentrations of greenhouse gases are quite different, the net effect on the radiative forcing of these differences is small: CO_2 concentrations are higher for Conventional Wisdom, CH_4 and tropospheric O_3 concentrations are higher for the IS92a scenario. The oceanic uptake of carbon grows steadily from 1.6 Gt C a^{-1} in 1990 to 4.2 Gt C a^{-1} in 2100. For both scenarios a globally averaged increase of surface air temperature of around 2.5 K is found between 1970 and 2100. This is slightly lower than the 2.5 K that Wigley and Raper (1992) find for the IS92a scenario for the period 1990 to 2100 (Wigley and Raper, 1992). The difference is consistent with the difference in the climate sensitivities of the models used (1.9 K for IMAGE vs. 2.5 K for Wigley and Raper). The pattern of temperature increase found is consistent with the patterns found by GCMs. The relatively strong increase of temperature in the mid and high latitudes of the Northern Hemisphere directly affect the carbon uptake by vegetation, seriously enhancing the terrestrial C sink.

The Ocean Realignment scenario illustrates the sensitivity of climate to changes in the ocean circulation. Furthermore it illustrates the effects this has on the C cycle and land use. Decreased deep water formation can lead to differences in temperature of 0.5 K globally averaged, or 2 K to 5 K at mid or high latitudes Northern Hemisphere. Differences in the oceanic carbon uptake are found to be small only. The increase of the terrestrial C sink can be 3 Gt C a^{-1} smaller than for a climate with fixed ocean circulation, adding up to a difference of almost 200 Gt C up to 2100. The effects of climate change on crop productivity can influence the location of expansion of agricultural land or reforestation as well as the total area required for crop growth.

Acknowledgements

The 2-D fields of tropospheric O_3 concentrations and trends in this concentrations used in the radiative calculations were obtained from dr. M.G.M. Roemer, TNO Institute of Environmental Sciences, Delft, the Netherlands. The SO_4^- columns were derived from 3-D seasonal SO_4^- fields, provided by drs. F. Dentener and dr. J.Lelieveld, MPI Mainz, Germany. The work in this paper was supported by grants number 851040, 851042 and 851045 of the Dutch National Research Program on Global Air Pollution and Climate Change (NOP) and by grants from the Dutch Ministry of Housing, Physical Planning and Environment (MAP 482506, 482507 and 482509).

References

Alcamo, J., G.J.J. Kreileman, M.S. Krol and G. Zuidema: 1994a, Modeling the global society-biosphere-climate system, Part 1. Model description and testing, *Wat. Air Soil Pollut.*, **76** (this volume).

Alcamo, J., G.J. van den Born, A.F. Bouwman, B.J. de Haan, K. Klein Goldewijk, O. Klepper, J. Krabec, J.G.J. Olivier, A.M.C. Toet, H.J.M. de Vries and H.J. van der Woerd: 1994b, Modeling the global society-biosphere-climate system, Part 2. Computed scenarios, *Wat. Air Soil Pollut.*, **76** (this volume).

Baumgartner, A. and E. Reichel: 1975, *Die Weltwasserbilanz*, Oldenbourg.

Boer, G.J., K. Arpe, M. Blackburn, M. Déqué, W.L. Gates, T.L. Hart, H. le Treut, E. Roeckner, D.A. sheinin, I. Simmonds, R.N.B. Smith, T. Tokioba, R.T. Wetherald and D. Williamson: 1991, *An Intercomparison of the Climates Simulated by 14 Atmospheric General Circulation Models*, Report No. 15, WMO/ISU World Climate Research Programme.

Broecker, W.S. and T.H. Peng: 1982, *Tracers in the Sea*, Eldigo Press.

Broecker, W.S., T.H. Peng, G. Ostlund and M. Stuiver: 1985, The distribution of bomb radiocarbon in the ocean, *J. Geophys. Res.*, **90**(C4): 6953-6970.

Carissimo, B. C., A.H. Oort and T.H. Vonder Haar: 1985, Estimating the meridional energy transport in the atmosphere and ocean, *J. Phys. Ocean.*, **15**: 82-91.

Charlson, R.J., J. Langner, H. Rodhe, C.B. Leovy and S.G. Warren: 1991, Perturbation of Northern Hemisphere radiative balance by backscattering from anthropogenic sulfate aerosols, *Tellus*, **43AB**: 152-163.

Chou, M.-D., L. Peng and A. Arking: 1982, Climate studies with a multi-layer energy balance model. Part II: the role of feedback mechanisms in the CO_2 problem, *J. Atm. Sci.*, **39**: 2657-2666.

Coakley, J.A. jr, R.D. Cess and F.B. Yurevitch: 1983, The effect of tropospheric aerosols on the Earths radiation budget: a parameterization for climate models, *J. Atm. Sci.*, **40**: 116-138.

Curran, R.J., R. Wexler and M.L. Nack: 1978, *Albedo Climatology Analysis and the Determination of Fractional Cloud Cover*, Technical Memorandum 79576, NASA, 52 pp.

Ellis, J.S. and T.H. Vonder Haar: 1976, *Zonal Average Earth Radiation Budget Measurements from Satelites for Climate Studies*, Paper 240, Colorado State Univ.

Eppley, R.W.: 1972, Temperature and phytoplankton growth in the sea, *Fish. Bull.*, **70**: 1063-1085.

Gieskes, J.M.: 1974, The carbonate chemistry of the ocean, in: E.D. Goldberg (ed), *The Sea*, Wiley & Sons, pp. 125-131.

Goudriaan, J.: 1990, Atmospheric CO_2, global carbon fluxes and the biosphere, in: R. Rabbinge (ed), *Theoretical Production Ecology: Reflections and Prospects*, Wageningen, pp. 17-37.

Guthrie, P.D. and G. Yarwood: 1991, *Analysis of the Intergovernmental Panel of Climate Change (IPCC) Future Methane Emissions*, Report SYS-APP-91/114, Systems Applications International.

Harvey, L.D.D.: 1988, A semianalytic energy balance climate model with explicit sea ice and snow physics, *J. Clim.*, **1**: 1065-1085.

Harvey, L.D.D. and S.H. Schneider: 1985, Transient climate response to external forcing on 10^4-10^4 year time scales, Part 1: Experiments with globally averaged, coupled, atmosphere and ocean energy balance models, *J. Geophys. Res.*, **90**(D1): 2191-2205.

IPCC: 1990, J.T. Houghton, G.J. Jenkins and J.J. Ephraums (eds), *Climate Change. The IPCC Scientific Assessment*, Cambridge Univ. Press.

IPCC: 1992, J.T. Houghton, B.A. Callendar and S.K. Varney (eds), *Climate Change 1992. The Supplementary Report to the IPCC Scientific Assessment*, Cambridge Univ. Press.

Jaeger, L.: 1976, Monatskarten des Niederslags für die ganze Erde, *Ber. d. Dt. Wetterdienstes*, **139**(18).

Jonas, M., K. Olendrzyński, J. Krabec and R. Shaw: 1992, *IIASA's Work on Climate Change: Assessing Environmental Impacts*, Report SR-92-9, IIASA, Laxenburg, Austria.

Keir, R.S.: 1988, On the late Pleistocene ocean geochemistry and circulation, *Paleoceanography*, **3**(4): 413-445.

Klein Goldewijk, K., J.G. van Minnen, G.J.J. Kreileman, M. Vloedbeld and R Leemans: 1994, Simulating the carbon flux between terrestrial environment and the atmosphere, *Wat. Air Soil Pollut.*, **76** (this volume).

Klepper, O.: 1994, Modelling the oceanic food web using a quasi steady state approach, *Ecol. Modelling*, (in press).

Klepper, O., B.J. de Haan and H. van Huet: 1994, Biochemical feedbacks in the oceanic carbon cycle, *Ecol. Modelling*, (in press).

Klepper, O., B.J. de Haan, P. Saager and M.S. Krol: 1993, *Oceanic uptake of anthropogenic CO_2*, Report 481507004, RIVM, Bilthoven, the Netherlands.

Kreileman, G.J.J. and A.F. Bouwman: 1994, Computing land use emissions of greenhouse gases, *Wat. Air Soil Pollut.*, **76** (this volume).

Krol, M.S.: 1994, Uncertainty analysis for the computation of greenhouse gas concentrations in IMAGE, in: J. Grasman and G. van Straten (eds), *Predictability and nonlinear modelling in natural sciences and economics*, Kluwer.

Krol, M.S. and H.J. van der Woerd: 1994, Atmospheric computations for the evaluation of climate scenarios, *Wat. Air Soil Pollut.*, **76** (this volume).

Langner, J., H. Rodhe, P.J. Crutzen and P. Zimmermann: 1992, Anthropogenic influence on the distribution of tropospheric sulphate aerosol, *Nature*, **359**: 712-716.

Lenderink, G. and R. J. Haarsma: 1993, Variability and multiple equilibria of the thermohaline circulation, associated with deep water formation, *J. Phys. Ocean.*, submitted.

Levitus, S.: 1982, *Climatological Atlas of the World Ocean*, Prof. Paper No.13, NOAA, US Government Printing Office, Washington.

MacKay, R.M. and M.A.K. Khalil: 1991, Theory and development of a one dimensional time dependent radiative convective climate model,*Chemosphere*, **22**(3-4): 383-417.

Maier-Reimer, E. and K. Hasselmann: 1987, Transport and storage of CO_2 in the ocean - an inorganic ocean-circulation carbon cycle model, *Clim. Dyn.*, **2**: 63-90.

Mantoura, R.F.C. and E.M.S. Woodward: 1983, Conservative behaviour of riverine dissolved organic carbon in the Severn Estuary: chemical and geochemical implications, *Geochimica et Cosmochimica Acta*, **47**: 1293-1309.

Martin, J.H., G.A. Knauer, D.M. Karl and W.W. Broeknow: 1987, VERTEX: carbon cycling in the northeast Pacific, *Deep-Sea Res.*, **34**(2): 267-285.

Mikolajewicz, U., B.D. Santer and E. Maier-Reimer: 1990, Ocean response to greenhouse warming, *Nature*, **345**: 589-593.

Neelin, J.D., M.A.F. Latif, M.A. Allart, M.A. Cane, U. Cubash, W.L. Gates, P.R. Gent, M. Ghil, C. Gordon, N.-C. Lau, C.R. Mechoso, G.A. Meehl, J.M. Oberhuber, S.G.H. Philander, P.S. Schopf, K.R. Sperber, A. Sterl, T. Tokioba, J. Tribbia, S.E. Zebiak: 1992, Tropical air-sea interaction in general circulation models, *Clim. Dyn.*, **7**: 73-104.

Oort, A.H.: 1983, *Global Atmospheric Circulation Statistics*, Prof. Paper No. 14, NOAA, U.S. Dept. Comm., Rockville, Md. U.S.A.

Oort, A.H. and T.H. Vonder Haar: 1976, On the observed annual cycle in the ocean-atmosphere heat balance over the northern hemisphere, *J. Phys. Ocean.*, **6**(6): 781-800.

Orr, J.C.: 1993, Accord between ocean models predicting uptake of anthropogenic CO_2, *Water, Air, Soil Pollut.*, **70**: 465-481.

Parsons, T.R., M. Takahashi and B. Hargrave: 1984, *Biological oceanographic processes*, Pergamon Press.

Peng, L., M.-D. Chou and A. Arking: 1982, Climate studies with a multi-layer energy balance model. Part I: model description and sensitivity to the solar constant, *J. Atm. Sci.*, **39**(12): 2639-2656.

Peng, L., M.-D. Chou and A. Arking: 1987, Climate warming due to increasing atmospheric CO_2: simulations with a multilayer coupled atmosphere-ocean seasonal energy balance model, *J. Geophys. Res.*, **92**(D5):

5505-5521.

Peng, T.S.: 1986, Uptake of anthropogenic CO_2 by lateral transport models of the ocean based on the distribution of bomb-produced ^{14}C, *Radiocarbon*, **28(2A)**: 363-375.

Prather, M.J.: 1989, *An Assessment Model for Atmospheric Composition*, Conf. Publ. 3023, NASA, New York.

Rasmussen, R.A. and M.A.K. Khalil: 1981, Atmospheric methane (CH_4): trends and seasonal cycles, *J. Geophys. Res.*, **86**: 9826-9832.

Roemer, M.G.M.: 1991, *Ozone and the Greenhouse Effect*, Report R91/227, IMW-TNO, Delft, the Netherlands.

Rotmans, J.: 1990, *IMAGE: an Integrated Model to Assess the Greenhouse Effect*, Kluwer.

Sarmiento, J.L. and E.T. Sundquist: 1992, Revised budget for the oceanic uptake of anthropogenic carbon dioxide, *Nature*, **356**: 589-593.

Siegenthaler, U. and J.L. Sarmiento: 1993, Atmospheric carbon dioxide and the ocean, *Nature*, **365**: 119-125.

Sellers, W.D.: 1965, *Physical Climatology*, Univ. Chicago Press.

Smith, E.A. and M.R. Smith: 1987, *J. Atmos. Sci.*, **44**: 3210-3224.

Takahashi, T., W.S. Broecker and A.E. Bainbridge: 1981, Supplement to the alkalinity and total carbon dioxide concentration in the world oceans, in: *Carbon Cycle Modeling*, Wiley & Sons, pp. 159-199.

Tans, P.P., I.Y. Fung and T. Takahashi: 1990, Observational constraints on the global atmospheric CO_2 budget, *Science*, **247**: 1431-1438.

Thompson, A.M., R.W. Stewart, M.A. Owens and J.A. Herwehe: 1989, Sensitivity of tropospheric oxidants to global chemical and climate change, *Atm. Env.*, **23**(3): 519-532.

de Vries, H.J.M., R.A. van den Wijngaart, G.J.J. Kreileman, J.G.J. Olivier and A.M.C. Toet: 1994, A model for calculating regional energy use and emissions for evaluating global climate scenarios, *Wat. Air Soil Pollut.*, **76** (this volume).

Wigley, T.M.L. and S.C.B. Raper: 1992, Implications for climate and sea level of revised IPCC emissions scenarios, *Nature*, **357**: 293-300.

WMO: 1992, *Scientific Assessment of Ozone Depletion- 1991*, Global Ozone Research and Monitoring Project, report No. 25, WMO.

Wollast, R.: 1981, Interactions between major biochemical cycles in marine ecosystems, in: E. Likens (ed), *Some Perspectives of Major Biochemical Cycles*, Wiley & Sons, pp. 125-142.

Wollast, R., F.T. Mackenzie and L. Chou (eds): 1993, *Interactions of C, N, P and S Biogeochemical Cycles and Global Change*, NATO ASI Series, Springer-Verlag.

Zuidema, G., G.J. van den Born, J. Alcamo and G.J.J. Kreileman: 1994, Simulating changes in global land cover as affected by economic and climatic factors, *Wat. Air Soil Pollut.*, **76** (this volume).

AUTHOR INDEX

Alcamo, J. 1, 37, 163
van den Born, G.J. 37, 133, 163
Bouwman, A.F. 31, 231
de Haan, B.J. 37, 283
Jonas, M. 283
Klein Goldewijk, K. 37, 199
Klepper, O. 37, 283
Krabec, J. 37, 283
Kreileman, G.J.J. 1, 79, 163, 199, 231
Krol, M.S. 1, 259, 283

Leemans, R. 37, 133, 199
Minnen, J.G. 199
Olivier, J.G.J. 37, 79
Toet, A.M.C. 37, 79
Vloedbeld, M. 163
de Vries, H.J.M. 37, 79
van den Wijgaart, R.A. 79
van den Woerd, H.J. 37, 259
Zuideman, G. 163
Zuldema, G. 1

SUBJECT INDEX

The manufacturer's authorised representative in the EU is Springer
Nature Customer Service Centre GmbH, Europaplatz 3, 69115 Heidelberg,
Germany. If you have any concerns regarding our products, please
contact ProductSafety@springernature.com

Printed and bound by CPI Group (UK) Ltd, Croydon, CR0 4YY

04/05/2026

02102432-0001